DRINKING WATER DISTRIBUTION SYSTEMS

ASSESSING AND REDUCING RISKS

Committee on Public Water Supply Distribution Systems:
Assessing and Reducing Risks

Water Science and Technology Board

Division on Earth and Life Studies

NATIONAL RESEARCH COUNCIL
OF THE NATIONAL ACADEMIES

THE NATIONAL ACADEMIES PRESS
Washington, D.C.
www.nap.edu

THE NATIONAL ACADEMIES PRESS 500 Fifth Street, N.W. Washington, DC 20001

Support for this project was provided by EPA Contract No. 68-C-03-081. Any opinions, findings, conclusions, or recommendations expressed in this publication are those of the author(s) and do not necessarily reflect the views of the organizations or agencies that provided support for the project.

International Standard Book Number-10 0-309-10306-1 (Book)
International Standard Book Number-13 978-0-309-10306-0 (Book)
International Standard Book Number-10 0-309-66432-2 (PDF)
International Standard Book Number-13 978-0-309-66432-5 (PDF)

Drinking Water Distribution Systems: Assessing and Reducing Risks is available from the National Academies Press, 500 Fifth Street, N.W., Lockbox 285, Washington, DC 20055; (800) 624-6242 or (202) 334-3313 (in the Washington metropolitan area); Internet, http://www.nap.edu.

Illustration on the book cover is designed by Alight Tsai, MWH Soft.

Printed in the United States of America

THE NATIONAL ACADEMIES

Advisers to the Nation on Science, Engineering, and Medicine

The **National Academy of Sciences** is a private, nonprofit, self-perpetuating society of distinguished scholars engaged in scientific and engineering research, dedicated to the furtherance of science and technology and to their use for the general welfare. Upon the authority of the charter granted to it by the Congress in 1863, the Academy has a mandate that requires it to advise the federal government on scientific and technical matters. Dr. Ralph J. Cicerone is president of the National Academy of Sciences.

The **National Academy of Engineering** was established in 1964, under the charter of the National Academy of Sciences, as a parallel organization of outstanding engineers. It is autonomous in its administration and in the selection of its members, sharing with the National Academy of Sciences the responsibility for advising the federal government. The National Academy of Engineering also sponsors engineering programs aimed at meeting national needs, encourages education and research, and recognizes the superior achievement of engineers. Dr. Wm. A. Wulf is president of the National Academy of Engineering.

The **Institute of Medicine** was established in 1970 by the National Academy of Sciences to secure the services of eminent members of appropriate professions in the examination of policy matters pertaining to the health of the public. The Institute acts under the responsibility given to the National Academy of Sciences by its congressional charter to be an adviser to the federal government and, upon its own initiative, to identify issues of medical care, research, and education. Dr. Harvey V. Fineberg is president of the Institute of Medicine.

The **National Research Council** was organized by the National Academy of Sciences in 1916 to associate the broad community of science and technology with the Academy's purposes of furthering knowledge and advising the federal government. Functioning in accordance with general policies determined by the Academy, the Council has become the principal operating agency of both the National Academy of Sciences and the National Academy of Engineering in providing services to the government, the public, and the scientific and engineering communities. The Council is administered jointly by both Academies and the Institute of Medicine. Dr. Ralph J. Cicerone and Dr. Wm. A. Wulf are chair and vice-chair, respectively, of the National Research Council.

www.national-academies.org

COMMITTEE ON PUBLIC WATER SUPPLY DISTRIBUTION SYSTEMS: ASSESSING AND REDUCING RISKS

VERNON L. SNOEYINK, *Chair*, University of Illinois, Urbana-Champaign
CHARLES N. HAAS, *Vice-Chair,* Drexel University, Philadelphia, Pennsylvania
PAUL F. BOULOS, MWH Soft, Broomfield, Colorado
GARY A. BURLINGAME, Philadelphia Water Department, Philadelphia, Pennsylvania
ANNE K. CAMPER, Montana State University, Bozeman
ROBERT N. CLARK, Environmental Engineering and Public Health Consultant, Cincinnati, Ohio
MARC A. EDWARDS, Virginia Polytechnic and State University, Blacksburg
MARK W. LECHEVALLIER, American Water, Voorhees, New Jersey
L. D. MCMULLEN, Des Moines Water Works, Des Moines, Iowa
CHRISTINE L. MOE, Emory University, Atlanta, Georgia
EVA C. NIEMINSKI, Utah Department of Environmental Quality, Salt Lake City
CHARLOTTE D. SMITH, Charlotte Smith and Associates, Inc., Orinda, California
DAVID P. SPATH, California Department of Health Services (Retired), Sacramento
RICHARD L. VALENTINE, University of Iowa, Iowa City

National Research Council Staff

LAURA J. EHLERS, Study Director
ELLEN A. DE GUZMAN, Research Associate

Preface

The distribution system is a critical component of every drinking water utility. Its primary function is to provide the required water quantity and quality at a suitable pressure, and failure to do so is a serious system deficiency. Water quality may degrade during distribution because of the way water is treated or not treated before it is distributed, chemical and biological reactions that take place in the water during distribution, reactions between the water and distribution system materials, and contamination from external sources that occurs because of main breaks, leaks coupled with hydraulic transients, and improperly maintained storage facilities, among other things. Furthermore, special problems are posed by the utility's need to maintain suitable water quality at the consumers tap, and the quality changes that occur in consumers' plumbing, which is not owned or controlled by the utility.

The primary driving force for managing and regulating distribution systems is protecting the health of the consumer, which becomes more difficult as our nation's distribution systems age and become more vulnerable to main breaks and leaks. Certainly factors that cause water of poor aesthetic quality to be delivered to the tap, or that increase the cost of delivering water, are also important. Possibly because they are underground and out of sight, it is easy to delay investments in distribution systems when budgets are considered. Rather than wait for further deterioration, however, there is an urgent need for new science that will enable cost-effective treatment for distribution, and design, construction, and management of the distribution system for protection of public health and minimization of water quality degradation.

This report was undertaken at the request of the U.S. Environmental Protection Agency (EPA) and was prepared by the Water Science and Technology Board (WSTB) of the National Research Council (NRC). The committee formed by the WSTB conducted a study of water quality issues associated with public water supply distribution systems and their potential risks to consumers. Although the report focused on public systems that serve at least 25 people, much that is said in the report is also applicable to private, individual distribution systems. The study considered regulations and non-regulatory approaches to controlling quality; the health effects of distribution system contamination; physical, hydraulic, and water quality integrity; and premise plumbing issues. Important events that constitute health risks, such as cross connections and backflow, pressure transients, nitrification and microbial growth, permeation and leaching, repair and replacement of water mains, aging infrastructure, corrosion control, and contamination in premise plumbing, were examined. The activities of the Committee included the following tasks:

1—As background and based on available information, identification of trends relevant to the deterioration of drinking water in water supply distribution systems.

2—Identification and prioritization of issues of greatest concern for distribution systems based on review of published material.

3—Focusing on the highest priority issues as revealed by task #2, (a) evaluation of different approaches to characterization of public health risks posed by water-quality deteriorating events or conditions that may occur in public water supply distribution systems; and (b) identification and evaluation of the effectiveness of relevant existing codes and regulations and identification of general actions, strategies, performance measures, and policies that could be considered by water utilities and other stakeholders to reduce the risks posed by water-quality deteriorating events or conditions. Case studies were identified and recommendations were presented in their context.

4—Identification of advances in detection, monitoring and modeling, analytical methods, information needs and technologies, research and development opportunities, and communication strategies that will enable the water supply industry and other stakeholders to further reduce risks associated with public water supply distribution systems.

The Committee prepared an interim report entitled "Public Water Supply Distribution Systems: Assessing and Reducing Risks, First Report" in March 2005 that dealt with the first two tasks listed above; the interim report has been incorporated into this report in order to make this report a complete compilation of Committee's activities. The third and fourth tasks constitute the subject matter of the present report; an explanation of where individual issues are discussed in the report can be found at the end of Chapter 1.

The EPA is in the process of considering changes to the Total Coliform Rule (TCR), which is one of the existing rules governing water quality in distributions systems. This report does not include a comprehensive evaluation of the science behind the TCR, a critique of that science, or specific suggestions on how to change the Rule. However, the Committee believes that this report should be considered when developing changes to the Rule, in order to determine whether the revised Rule could better encompass distribution system integrity.

When preparing the report the committee made a series of assumptions that affected the outcome of the report. First, it was assumed that both treated and distributed water has to meet U.S. water quality standards. Second, water distribution will almost certainly be accomplished with the existing infrastructure in which the nation has invested billions of dollars and which is continuously being expanded. Thus, the report focuses on how to best use traditionally designed distribution systems in which potable water is distributed for all uses. These assumptions led the Committee to devote only a small section of the report to non-traditional distribution system design (such as dual distribution systems), investigation of which was not in the Committee's charge. The Committee believes that alternative methods of distributing water, including dual distribution systems, point-of-use and point-of-entry treatment systems, and community-based treatment systems need more research and evaluation to determine their effectiveness and applicability, both in the United States and elsewhere in the world. The Committee did not consider lead and copper corrosion because this subject is part of the Lead and Copper Rule and for this reason was intentionally excluded from the committee's charge by the study sponsor. Corrosion in distribution systems, in general, has very important impacts on water quality in distribution systems, and the committee believes that state-of-the-art internal and external corrosion control procedures should be made available to the industry, perhaps in the form of a manual of practice. Finally, at the request of EPA, the committee did not consider issues surrounding the security of the nation's distribution systems, including potential threats and monitoring needed for security purposes.

In developing this report, the Committee benefited greatly from the advice and input of EPA representatives, including Ephraim King, Yu-Ting Guilaran, Elin Betanzo, and Kenneth Rotert and from presentations by Russ Chaney, IAPMO; Barry Fields, CDC; Johnnie Johannesen, Matt Velardes, and Chris Kinner, Irvine Ranch Water District; Laura Jacobsen, Las Vegas Valley Water District; Dan Kroll, HACH HST; Kathy Martel, Economic and Engineering Services; Pankaj Parehk, LA Department of Water and Power; Paul Schwartz, USC Foundation for Cross-Connection Control and Hydraulic Research; and Walter J. Weber, Jr., University of Michigan. We also thank all those who took time to share with us their perspectives and wisdom about the various issues affecting the water resources research enterprise.

The Committee was ably served by the staff of the Water Science and Technology Board and its director, Stephen Parker. Study director Laura Ehlers kept the Committee on task and on time, provided her own valuable insights which have improved the report immeasurably, and did a superb job of organizing and editing the report. Ellen de Guzman provided the Committee with all manner of support in a timely and cheerful way. This report would not have been possible without the help of these people.

This report has been reviewed in draft form by individuals chosen for their diverse perspectives and technical expertise, in accordance with procedures approved by the NRC's Report Review Committee. The purpose of this independent review is to provide candid and critical comments that will assist the institution in making its published report as sound as possible and to ensure that the report meets institutional standards for objectivity, evidence, and responsiveness to the study charge. The review comments and draft manuscript remain confidential to protect the integrity of the deliberative process. We wish to thank the following individuals for their review of this report:

Gunther F. Craun, Gunther F. Craun and Associates;
Stephen Estes-Smargiassi, Massachusetts Water Resources Authority;
Timothy Ford, Montana State University;
Jerome B. Gilbert, J. Gilbert, Inc.;
Gregory J. Kirmeyer, HDR;
Michael J. McGuire, McGuire Environmental Consultants, Inc.;
Danny D. Reible, University of Texas;
Philip C. Singer, University of North Carolina; and
James Uber, University of Cincinnati.

Although the reviewers listed above have provided many constructive comments and suggestions, they were not asked to endorse the conclusions and recommendations nor did they see the final draft of the report before its release. The review of this report was overseen by Edward Bouwer, Johns Hopkins University. Appointed by the National Research Council, he was responsible for making certain that an independent examination of this report was carried out in accordance with institutional procedures and that all review comments were carefully considered. Responsibility for the final content of this report rests entirely with the authoring committee and institution.

Vernon Snoeyink,
Committee Chair

Contents

Summary

Water distribution systems carry drinking water from a centralized treatment plant or well supplies to consumers' taps. These systems consist of pipes, pumps, valves, storage tanks, reservoirs, meters, fittings, and other hydraulic appurtenances. Spanning almost 1 million miles in the United States, distribution systems represent the vast majority of physical infrastructure for water supplies, and thus constitute the primary management challenge from both an operational and public health standpoint. Public water supplies and their distribution systems range in size from those that can serve as few as 25 people to those that serve several million.

The issues and concerns surrounding the nation's public water supply distribution systems are many. Of the 34 billion gallons of water produced daily by public water systems in the United States, approximately 63 percent is used by residential customers. More than 80 percent of the water supplied to residences is used for activities other than human consumption such as sanitary service and landscape irrigation. Nonetheless, distribution systems are designed and operated to provide water of a quality acceptable for human consumption. Another important factor is that in addition to providing drinking water, a major function of most distribution systems is to provide adequate standby fire-flow. In order to satisfy this need, most distribution systems use standpipes, elevated tanks, storage reservoirs, and larger sized pipes. The effect of designing and operating a distribution system to maintain adequate fire flow and redundant capacity is that there are longer transit times between the treatment plant and the consumer than would otherwise be needed.

The type and age of the pipes that make up water distribution systems range from cast iron pipes installed during the late 19th century to ductile iron pipe and finally to plastic pipes introduced in the 1970s and beyond. Most water systems and distribution pipes will be reaching the end of their expected life spans in the next 30 years (although actual life spans may be longer depending on utility practices and local conditions). Thus, the water industry is entering an era where it will have to make substantial investments in pipe assessment, repair, and replacement.

Most regulatory mandates regarding drinking water focus on enforcing water quality standards at the treatment plant and not within the distribution system. Ideally, there should be no change in the quality of treated water from the time it leaves the treatment plant until the time it is consumed. However, in reality substantial changes can occur to finished water as a result of complex physical, chemical, and biological reactions. Indeed, data on waterborne disease outbreaks, both microbial and chemical, suggest that distribution systems remain a source of contamination that has yet to be fully addressed. As a consequence, the U.S. Environmental Protection Agency (EPA) has renewed its interest in water quality degradation occurring during distribution, with the goal of defining the extent of the problem and considering how it can be addressed during rule revisions or via non-regulatory channels. To assist in this process, EPA requested that the National Academies' Water Science and Technology Board conduct a study of water quality issues associated with public water supply distribution systems and their potential risks to consumers. The following statement of task guided the expert committee formed to conduct the study:

1) Identify trends relevant to the deterioration of drinking water in water supply distribution systems, as background and based on available information.

2) Identify and prioritize issues of greatest concern for distribution systems based on a review of published material.

3) Focusing on the highest priority issues as revealed by task #2, (a) evaluate different approaches for characterization of public health risks posed by water quality deteriorating events or conditions that may occur in public water supply distribution systems; and (b) identify and evaluate the effectiveness of relevant existing codes and regulations and identify general actions, strategies, performance measures, and policies that could be considered by water utilities and other stakeholders to reduce the risks posed by water-quality deteriorating events or conditions. Case studies, either at the state or utility level, where distribution system control programs (e.g., Hazard Analysis and Critical Control Point System, cross-connection control, etc.) have been successfully designed and implemented will be identified and recommendations will be presented in their context.

4) Identify advances in detection, monitoring and modeling, analytical methods, information needs and technologies, research and development opportunities, and communication strategies that will enable the water supply industry and other stakeholders to further reduce risks associated with public water supply distribution systems.

The committee addressed tasks one and two in its first report, which is included as Appendix A to this report. The distribution system issues given highest priority were those that have a recognized health risk based on clear epidemiological and surveillance data, including cross connections and backflow; contamination during installation, rehabilitation, and repair activities; improperly maintained and operated storage facilities; and control of water quality in premise plumbing. This report focuses on the committee's third and fourth tasks and makes recommendations to EPA regarding new directions and priorities to consider.

This report considers service lines and premise plumbing to be part of the distribution system. Premise plumbing and service lines have longer residence times, more stagnation, lower flow conditions, and elevated temperatures compared to the main distribution system, and consequently can have a profound effect on the quality of water reaching the consumer. Also, the report focuses on traditional distribution system design, in which water originates from a centralized treatment plant or well and is then distributed through one pipe network to consumers. Non-conventional distribution system designs including decentralized treatment and dual distribution systems are only briefly considered. Such designs, which would be potentially much more complicated than traditional systems, require considerably more study regarding their economic feasibility, their maintenance and monitoring requirements, and how to transition from an existing conventional system to a non-conventional system. Nonetheless, many of the report recommendations are relevant even if an alternative distribution system design is used.

REGULATORY CONSIDERATIONS

The federal regulatory framework that targets degradation of distribution system water quality is comprised of several rules under the Safe Drinking Water Act, including the Lead and Copper Rule (LCR), the Surface Water Treatment Rule (SWTR), the Total Coliform Rule (TCR), and the Disinfectants/Disinfection By-Products Rule (D/DBPR). The LCR establishes monitoring requirements for lead and copper within tap water samples, given concern over their leaching from premise plumbing and fixtures. The SWTR establishes the minimum required detectable disinfectant residual and the maximum allowed heterotrophic bacterial plate count, both measured within the distribution system. The TCR calls for distribution system monitoring of total coliforms, fecal coliforms, and/or *E. coli*. Finally, the D/DBPR addresses the maximum disinfectant residual and concentration of disinfection byproducts like total trihalomethanes and haloacetic acids allowed in distribution systems. A plethora of state regulations and plumbing codes also affect distribution system water quality, from requirements for design, construction, operation, and maintenance of distribution systems to cross-connection control programs.

Despite the existence of these rules, programs, and codes, current regulatory programs have not removed the potential for outbreaks attributable to distribution system-related factors. Part of this can be attributed to the fact that existing federal regulations are intended to address only certain aspects of distribution system water quality and not the integrity of the distribution system in its totality. Most contaminants that have the potential to degrade distribution system water quality are not monitored for compliance purposes, or the sampling requirements are too sparse and infrequent to detect contamination events. For example, TCR monitoring encompasses only microbiological indicators and not in real time. With the exception of monitoring for disinfectant residuals and DBPs within the distribution system and lead and copper at the customer's tap, existing federal regulations do not address other chemical contaminants.

Although it is hoped that state regulations and local ordinances would contribute to public safety from drinking water contamination in areas where federal regulations are weak, the considerable variation in relevant state programs makes this impossible to conclude on a general basis. For cross-connection control programs, for the design, construction, operation, and maintenance of distribution systems, and for plumbing code components, state programs range from an absolute requirement to simply encouraging a practice to no provision whatsoever. Voluntary programs do exist to fill gaps in the federal and state regulatory requirements for distribution system operation and maintenance, most notably the G200 standard of the American Water Works Association. These programs, if adopted, can help a utility organize its many activities by unifying all of the piecemeal requirements of the federal, state, and local regulations. The following select conclusions and recommendations regarding the effectiveness of existing regulations and codes and the potential for their improvement are made, with additional detail found in Chapter 2.

EPA should work closely with representatives from states, water systems, and local jurisdictions to establish the elements that constitute an acceptable cross-connection control program. State requirements for cross-connection control programs are highly inconsistent, and state oversight of such programs varies and is subject to availability of resources. If states expect to maintain primacy over their drinking water programs, they should adopt a cross-connection control program that includes a process for hazard assessment, the selection of appropriate backflow devices, certification and training of backflow device installers, and certification and training of backflow device inspectors.

Existing plumbing codes should be consolidated into one uniform national code. The two principal plumbing codes that are used nationally have different contents and permit different materials and devices. In addition to integrating the codes, efforts should be made to ensure more uniform implementation of the plumbing codes, which can vary significantly between jurisdictions and have major impacts on the degree of public health protection afforded.

For utilities that desire to operate beyond regulatory requirements, adoption of G200 or an equivalent program is recommended to help utilities develop distribution system management plans. G200 has advantages over other voluntary programs, such as HACCP, in that it is more easily adapted to the dynamic nature of drinking water distribution systems.

PUBLIC HEALTH RISK OF DISTRIBUTION SYSTEM CONTAMINATION

Three primary approaches are available to better understand the human health risks that derive from contamination of the distribution system: risk assessment methods that utilize pathogen occurrence data, waterborne disease outbreak surveillance, and epidemiology studies. Chapter 3 extensively reviews the available information in each of these categories and its implications for determining public health risk. In the case of pathogen occurrence measurements, our understanding of the microbial ecology of distribution systems is at an early stage. Microbial monitoring methods are expensive, time consuming, require optimization for specific conditions, and currently are appropriate only for the research laboratory. Methods do not exist for routine detection and quantification of most of the microbes on the EPA's Contaminant Candidate List. Until better methods, dose-response relationships, and risk assessment data are available, pathogen occurrence measurements are best used in conjunction with other supporting data on health outcomes, such as data on enhanced or syndromic surveillance in communities, or from microbial or chemical indicators of potential contamination.

Outbreak surveillance data currently provide more information on the public health impact of contaminated distribution systems. In fact, investigations conducted in the last five years suggest that a substantial proportion of waterborne disease outbreaks, both microbial and chemical, is attributable to problems within distribution systems. The reason for these observations is not clear; outbreaks associated with distribution system deficiencies have been reported since the surveillance system was started. However, there may be more attention focused on the distribution system now that there are fewer reported outbreaks associated with inadequate treatment of surface water. Also, better outbreak investigations and reporting systems in some states may result in increased recognition and reporting of all the risk factors contributing to the outbreak, including problems with the distribution system that may have been overlooked in the past. Contamination from cross-connections and backsiphonage were found to cause the majority of the outbreaks associated with distribution systems, followed by contamination of water mains following breaks and contamination of storage facilities. The situation may be of even greater concern because incidents involving domestic plumbing are less recognized and unlikely to be reported. In general the identified number of waterborne disease outbreaks is considered an underestimate because not all outbreaks are recognized, investigated, or reported to health authorities.

A third approach for estimating public health risk is to conduct an epidemiology study that isolates the distribution system component. The body of evidence from four epidemiological studies does not eliminate the consumption of tap water that has been in the distribution system from causing increased risk of gastrointestinal illness. However, differences between the study designs, the study population sizes and compositions and follow-up periods, and the extent of complementary pathogen occurrence measurements make comparisons difficult. Although all four cohort studies used similar approaches for recording symptoms of gastrointestinal illness, different illness rates were observed, with some more than twice as high as others. One of the major challenges for designing an epidemiology study of health risks associated with water quality in the distribution system is separating the effect of source water quality and treatment from the effect of distribution system water quality.

Although there is a lack of definitive estimates, the available information seems to be implicating contamination of the distribution system in public health risk. This is particularly true for *Legionella pneumophila* in water systems, for which occurrence data, outbreak data, and epidemiological data are available. In fact, since *Legionella* was incorporated into the waterborne disease outbreak surveillance system in 2001, several outbreaks have been attributed to the microorganism, all of which occurred in large buildings or institutional settings. As discussed in Appendix A, the committee relied on the limited available outbreak and epidemiological data as well as its best professional judgment to prioritize distribution system contamination events into high, medium, and low priority. Better public health data could help refine distribution system risks and provide additional justification for the prioritization. The following select conclusions and recommendations regarding the public health risks of distribution systems are made, with additional detail found in Chapter 3.

The distribution system is the remaining component of public water supplies yet to be adequately addressed in national efforts to eradicate waterborne disease. This is evident from data indicating that although the number of waterborne disease outbreaks including those attributable to distribution systems is decreasing, the *proportion* of outbreaks attributable to distribution systems is increasing. Most of the reported outbreaks associated with distribution systems have involved contamination from cross-connections and backsiphonage. Furthermore, *Legionella* appears to be a continuing risk and is the single most common etiologic agent associated with outbreaks involving drinking water. Initial studies suggest that the use of chloramine as a residual disinfectant may reduce the occurrence of *Legionella*, but additional research is necessary to determine the relationship between disinfectant usage and the risks of *Legionella* and other pathogenic microorganisms.

Distribution system ecology is poorly understood, making risk assessment via pathogen occurrence measurements difficult. There is very little information available about the types, activities, and distribution of microorganisms in distribution systems, particularly premise plumbing. Limited heterotrophic plate count data are available for some systems, but these data are not routinely collected, they underestimate the numbers of organisms present, and they include many organisms that do not necessarily present a health risk.

Epidemiology studies that specifically target the distribution system component of waterborne disease are needed. Recently completed epidemiological studies have either not focused on the specific contribution of distribution system contamination to gastrointestinal illness, or they have been unable to detect any link between illness and drinking water. Epidemiological studies of the risk of endemic disease associated with drinking water distribution systems need to be performed and must be designed with sufficient power and resources to adequately address the deficiencies of previous studies.

PHYSICAL, HYDRAULIC, AND WATER QUALITY INTEGRITY

One of the options being considered during revision of the TCR is that it more adequately address distribution system integrity—defined in this report as having three components: (1) physical integrity, which refers to the maintenance of a physical barrier between the distribution system interior and the external environment, (2) hydraulic integrity, which refers to the maintenance of a desirable water flow, water pressure, and water age, taking both potable drinking water and fire flow provision into account, and (3) water quality integrity, which refers to the maintenance of finished water quality via prevention of internally derived contamination. The three types of integrity have different causes of their loss, different consequences once they are lost, different methods for detecting and preventing a loss, and different remedies for regaining integrity. Protection of public health requires that water professionals take all three integrity types into account in order to maintain the highest level of water quality.

Physical Integrity

The loss of physical integrity of the distribution system—in which the system no longer acts as a barrier that prevents external contamination from deteriorating the internal, drinking water supply—is brought about by physical and chemical deterioration of materials, the absence or improper installation of critical components, and the installation of already contaminated components. When physical integrity is compromised, the drinking water supply becomes exposed to contamination that increases the risk of negative public health outcomes. Most documented cases of waterborne disease outbreaks attributed to distribution systems have been caused by breaches in physical integrity, such as a backflow event through a cross connection or contamination occurring during repair or replacement of distribution system infrastructure. Selected conclusions and recommendations for maintaining and restoring physical integrity to a distribution system are given below. Additional detail is found in Chapter 4.

Storage facilities should be inspected on a regular basis. A disciplined storage facility management program is needed that includes developing an inventory and background profile on all facilities, developing an evaluation and rehabilitation schedule, developing a detailed facility inspection process, performing inspections, and rehabilitating and replacing storage facilities when needed. Depending on the nature of the water supply chemistry, every three to five years storage facilities need to be drained, sediments need to be removed, appropriate rust-proofing needs to be done to the metal surfaces, and repairs need to be made to structures. These inspections are in addition to daily or weekly inspections for vandalism, security, and water quality purposes (such as identifying missing vents, open hatches, and leaks).

Better sanitary practices are needed during installation, repair, replacement, and rehabilitation of distribution system infrastructure. All trades people who work with materials that are being installed or repaired and that come in contact with potable water should be trained and certified for the level of sanitary and materials quality that their work demands. Quality workmanship for infrastructure materials protection as well as sanitary protection of water and materials are critical considering the increasing costs of infrastructure failure and repair and increasingly stringent water quality standards.

External and internal corrosion should be better researched and controlled in standardized ways. There is a need for new materials and corrosion science to better understand how to more effectively control both external and internal corrosion, and to match distribution system materials with the soil environment and the quality of water with which they are in contact. At present the best defense against corrosion relies on site-specific testing of materials, soils, and water quality followed by the application of best practices, such as cathodic protection. Indeed, a manual of practice for external and internal corrosion control should be developed to aid the water industry in applying what is known. Corrosion is poorly understood and thus unpredictable in occurrence. Insufficient attention has been given to its control, especially considering its estimated annual direct cost of $5 billion in U.S. for the main distribution system, not counting premise plumbing.

Hydraulic Integrity

Maintaining the hydraulic integrity of distribution systems is vital to ensuring that water of acceptable quality is delivered in acceptable amounts. The most critical element of hydraulic integrity is adequate water pressure inside the pipes. The loss of water pressure resulting from pipe breaks, significant leakage, excessive head loss at the pipe walls, pump or valve failures, or pressure surges can impair water delivery and will increase the risk of contamination of the water supply via intrusion. Another critical hydraulic factor is the length of time water is in the distribution system. Low flows in pipes create long travel times, with a resulting loss of disinfectant residual as well as sections where sediments can collect and accumulate and microbes can grow and be protected from disinfectants. Furthermore, sediment deposition will result in rougher pipes with reduced hydraulic capacity and increased pumping costs. Long detention times can also greatly reduce corrosion control effectiveness by impacting phosphate inhibitors and pH management. A final component of hydraulic integrity is maintaining sufficient mixing and turnover rates in storage facilities, which if insufficient can lead to short circuiting and generate pockets of stagnant water with depleted disinfectant residual. Fortunately, water utilities can achieve a high degree of hydraulic integrity through a combination of proper system design, operation, and maintenance, along with monitoring and modeling. The following select conclusions and recommendations are made, with additional detail found in Chapter 5.

Water residence times in pipes, storage facilities, and premise plumbing should be minimized. Excessive residence times can lead to low disinfectant residuals and leave certain service areas with a less protected drinking water supply. In addition, long residence times can promote microbial regrowth and the formation of disinfection byproducts. From an operational viewpoint it may be challenging to reduce residence time where the existing physical infrastructure and energy considerations constrain a utility's options. Furthermore, limited understanding of the stochastic nature of water demand and water age makes it difficult to assess the water quality benefits of reduced residence time. Research is needed to investigate such questions, as well as how to achieve minimization of water residence time while maintaining other facets of hydraulic integrity (such as adequate pressure and reliability of supply).

Positive water pressure should be maintained. Low pressures in the distribution system can result not only in insufficient fire fighting capacity but can also constitute a major health concern resulting from potential intrusion of contaminants from the surrounding external environment. A minimum residual pressure of 20 psi under all operating conditions and at all locations (including at the system extremities) should be maintained.

Distribution system monitoring and modeling are critical to maintaining hydraulic integrity. Hydraulic parameters to be monitored should include inflows/outflows and water levels for all storage tanks, discharge flows and pressures for all pumps, flows and/or pressure for all regulating valves, and pressures at critical points. An analysis of these patterns can directly determine if the system hydraulic integrity is compromised. Calibrated distribution system models can calculate the spatial and temporal variations of flow, pressure, velocity, reservoir level, water age, and other hydraulic and water quality parameters throughout the distribution system. Such results can, for example, help identify areas of low or negative pressure and high water age, estimate filling and draining cycles of storage facilities, and determine the adequacy of the system to supply fire flows under a variety of conditions.

Water Quality Integrity

Breaches in physical and hydraulic integrity can lead to the influx of contaminants across pipe walls, through breaks, and via cross connections. These external contamination events can act as a source of inoculum, introduce nutrients and sediments, or decrease disinfectant concentrations within the distribution system, resulting in a degradation of water quality. Even in the absence of external contamination, however, there are situations where water quality is degraded due to transformations that take place within piping, tanks, and premise plumbing. These include biofilm growth, nitrification, leaching, internal corrosion, scale formation, and other chemical reactions associated with increasing water age.

Maintaining water quality integrity in the distribution system is challenging because of the complexity of most systems. That is, there are interactions between the type and concentration of disinfectants used, corrosion control schemes, operational practices (e.g., flow characteristics, water age, flushing practices), the materials used for pipes and plumbing, the biological stability of the water, and the efficacy of treatment. The following select conclusions and recommendations are made, with additional details found in Chapter 6.

Microbial growth and biofilm development in distribution systems should be minimized. Even though the general heterotrophs found in biofilms are not likely to be of public health concern, their activity can promote the production of tastes and odors, increase disinfectant demand, and may contribute to corrosion. Biofilms may also harbor opportunistic pathogens (those causing disease in the immunocompromised). This issue is of critical importance in premise plumbing where long residence times promote disinfectant decay and subsequent bacterial growth and release.

Residual disinfectant choices should be balanced to meet the overall goal of protecting public health. For free chlorine, the potential residual loss and DBP formation should be weighed against the problems that may be introduced by chloramination, which include nitrification, lower disinfectant efficacy against suspended organisms, and the potential for deleterious corrosion problems. Although some systems have demonstrated increased biofilm control with chloramination, this response has not been universal. This ambiguity also exists for the control of opportunistic pathogens.

Standards for materials used in distribution systems should be updated to address their impact on water quality, and research is needed to develop new materials that will have minimal impacts. Materials standards have historically been designed to address physical/strength properties including the ability to handle pressure and stress. Testing of currently available materials should be expanded to include (1) the potential for permeation of contaminants, and (2) the potential for leaching of compounds of public health concern as well as those that contribute to tastes and odors and support biofilm growth. Also, research is needed to develop new materials that minimize adverse water quality effects such as the high concentrations of undesirable metals and deposits that result from corrosion and the destruction of disinfectant owing to interactions with pipe materials.

INTEGRATING APPROACHES TO REDUCING PUBLIC HEALTH RISK FROM DISTRIBUTION SYSTEMS

Because only a few regulations govern water quality in distribution systems, public health protection from contamination arising from distribution system events will require that utilities independently choose to design and operate their systems beyond regulatory requirements. One voluntary standard in particular—the G200 standard for distribution system operation and management—directly addresses the issues highlighted by EPA and characterized as high priority by this committee (see Appendix A).

As for any voluntary program, it may be necessary to create incentives for utilities to adopt G200, for which several options exist. An extreme would be to create federal regulations that require adherence to a prescribed list of activities deemed necessary for reducing the risk of contaminated distribution systems; this list could partly or fully parallel the G200 standard. Another mechanism to capture elements of G200 within existing federal regulations would be via the sanitary surveys conducted by the state and required for some systems every three to five years. Sanitary surveys encompass a wide variety of activities, and could capture those felt to be of highest priority for reducing risk. Several other options are discussed, including (1) making some of the elements of G200 fall under existing federal regulations through the Government Accounting Standards Board, (2) state regulations that require adherence to G200 including building and plumbing codes and design and construction requirements, (3) linking qualification for a loan from the State Revolving Fund to a utility demonstrating that it is adhering to G200, and (4) implementation of G200 as a way to improve a drinking water utilities' access to capital via better bond ratings.

For small water systems that are resource limited, adherence to the G200 standard or its equivalent may present financial, administrative, and technological burdens. Thus, its adoption should occur using the following guidelines: (1) implement new activities using a step-wise approach; (2) provide technical assistance, education, and training; and (3) develop regulatory, financial, and social incentives. Training materials, scaled for small-size systems, are essential for operators and maintenance crew. Public education can result in an increased awareness and emphasis on the significance of implementing proactive voluntary efforts, which could help to justify increased actions.

Certain elements of G200 deserve more thoughtful consideration because emerging science and technology are altering whether and how these elements are implemented by a typical water utility. Much of the current scientific thrust is in the development of new monitoring techniques, models, and methods to integrate monitoring data and models to inform decision making. The following select conclusions and recommendations relate specifically to these techniques and methods, with additional detail found in Chapter 7.

Distribution system integrity is best evaluated using on-line, real-time methods to provide warning against any potential breaches in sufficient

time to effectively respond and minimize public exposure. This will require the development of new, remotely operated sensors and data collection systems for continuous public health surveillance monitoring. These types of systems should be capable of accurately (with sufficient precision) determining the nature, type, and location/origin of all potential threats to distribution system integrity. The availability, reliability, and performance of on-line monitors are improving, with tools now available for detecting pressure, turbidity, disinfectant residual, flow, pH, temperature, and certain chemical parameters. Although these devices have reached the point for greater full-scale implementation, additional research is needed to optimize the placement and number of monitors.

Research is needed to better understand how to analyze data from on-line, real-time monitors in a distribution system. A number of companies are selling (and utilities are deploying) multiparameter analyzers. These companies, as well as EPA, are assessing numerical approaches to convert such data into a specific signal (or alarm) of a contamination event—efforts which warrant further investigation. Some of the data analysis approaches are proprietary, and there has been limited testing reported in "real world" situations. Furthermore, when multiple analyzers are installed in a given distribution system, the pattern of response of these analyzers in space provides additional information on system performance, but such spatially distributed information has not been fully utilized. To the greatest degree possible, this research should be conducted openly (and not in confidential or proprietary environments).

ALTERNATIVES FOR PREMISE PLUMBING

Premise plumbing includes that portion of the distribution system associated with schools, hospitals, public and private housing, and other buildings. It is connected to the main distribution system via the service line. The quality of potable water in premise plumbing is not ensured by EPA regulations, with the exception of the Lead and Copper Rule which assesses the efficacy of corrosion control by requiring that samples be collected at the tap after the water has been allowed to remain stagnant.

Virtually every problem previously identified in the main water transmission system can also occur in premise plumbing. However, unique characteristics of premise plumbing can magnify the potential public health risk relative to the main distribution system and complicate formulation of coherent strategies to deal with problems. These characteristics include:

- *a high surface area to volume ratio*, which along with other factors can lead to more severe leaching and permeation;
- variable, often *advanced water age*, especially in buildings that are irregularly occupied;

- more *extreme temperatures* than those experienced in the main distribution system
- *low or no disinfectant residual*, because buildings are unavoidable "dead ends" in a distribution system;
- potentially *higher bacterial levels and regrowth* due to the lack of persistent disinfectant residuals, high surface area, long stagnation times, and warmer temperatures. *Legionella* in particular is known to colonize premise plumbing, especially hot water heaters;
- *exposure routes through vapor and bioaerosols* in relatively confined spaces such as home showers;
- *proximity to service lines*, which have been shown to provide the greatest number of potential entry points for pathogen intrusion;
- *higher prevalence of cross connections*, since it is relatively common for untrained and unlicensed individuals to do repair work in premise plumbing;
- *variable responsible party*, resulting in considerable confusion over who should maintain water quality in premise plumbing.

Premise plumbing is a contributor to the degradation of water quality, particularly due to microbial regrowth, backflow events, and contaminant intrusion, although additional research is needed to better understand its magnitude. In particular, more extensive sampling of water quality within premise plumbing by utilities or targeted sampling via research is required. The following detailed conclusions and recommendations are given.

Communities should squarely address the problem of *Legionella*, both via changes to the plumbing code and new technologies. Changes in the plumbing code such as those considered in Canada and Australia that involve mandated mixing valves would seem logical to prevent both scalding and microbial regrowth in premise plumbing water systems. On-demand water heating systems may have benefits worthy of consideration versus traditional large hot water storage tanks in the United States. The possible effects of chloramination and other treatments on *Legionella* control should be quantified to a higher degree of certainty.

To better assess cross connections in the premise plumbing of privately owned buildings, inspections for cross connections and other code violations at the time of property sale could be required. Such inspection of privately owned plumbing for obvious defects could be conducted during inspection upon sale of buildings, thereby alerting future occupants to existing hazards and highlighting the need for repair. These rules, if adopted by individual states, might also provide incentives to building owners to follow code and have repairs conducted by qualified personnel, because disclosure of sub-standard repair could affect subsequent transfer of the property.

EPA should create a homeowner's guide and website that highlights the nature of the health threat associated with premise plumbing and mitigation strategies that can be implemented to reduce the magnitude of the risk. As part of this guide, it should be made clear that water quality is regulated only to the property line, and beyond that point responsibility falls mainly on consumers. Whether problems in service lines are considered to be the homeowner's responsibility or the water utility's varies from system to system.

Research is needed that specifically addresses potential problems arising from premise plumbing. This includes the collection of data quantifying water quality degradation in representative premise plumbing systems in geographically diverse regions and climates. In addition, greater attention should be focused on understanding the role of plumbing materials. Furthermore, the role of nutrients in distributed water in controlling regrowth should be assessed for premises. Finally, the potential impacts of representative point-of-use and point-of-entry devices need to be quantified. An epidemiological study to assess the health risks of contaminated premise plumbing should be undertaken in high risk communities.

1
Introduction

The first municipal water utility in the United States was established in Boston in 1652 to provide domestic water and fire protection (Hanke, 1972). The Boston system emulated ancient Roman water supply systems in that it was multipurpose in nature. Many water supplies in the United States were subsequently constructed in cities primarily for the suppression of fires, but most have been adapted to serve commercial and residential properties with water. By 1860, there were 136 water systems in the United States, and most of these systems supplied water from springs low in turbidity and relatively free from pollution (Baker, 1948). However, by the end of the nineteenth century waterborne disease had become recognized as a serious problem in industrialized river valleys. This led to the more routine treatment of water prior to its distribution to consumers. Water treatment enabled a decline in the typhoid death rate in Pittsburgh, PA from 158 deaths per 100,000 in the 1880s to 5 per 100,000 in 1935 (Fujiwara et al., 1995). Similarly, both typhoid case and death rates for the City of Cincinnati declined more than tenfold during the period 1898 to 1928 due to the use of sand filtration, disinfection via chlorination, and the application of drinking water standards (Clark et al., 1984). It is without a doubt that water treatment in the United States has proven to be a major contributor to ensuring the nation's public health.

Since the late 1890s, concern over waterborne disease and uncontrolled water pollution has regularly translated into legislation at the federal level. The first water quality-related regulation was promulgated in 1912 under the Interstate Quarantine Act of 1893. At that time interstate railroads made a common cup available for train passengers to share drinking water while on board—a practice that was prohibited by the Act. Several sets of federal drinking water standards were issued prior to 1962, but they too applied only to interstate carriers (Grindler, 1967; Clark, 1978). By the 1960s, each of the states and trust territories had established their own drinking water regulations, although there were many inconsistencies among them. As a consequence, reported waterborne disease outbreaks declined from 45 per 100,000 people in 1938–40 to 15 per 100,000 people in 1966–70. Unfortunately, the annual number of waterborne disease outbreaks ceased to fall around 1951 and may have increased slightly after that time, leading, in part, to the passage of the Safe Drinking Water Act (SDWA) of 1974 (Clark, 1978).

Prior to the passage of the SDWA, most drinking water utilities concentrated on meeting drinking water standards at the treatment plant, even though it had long been recognized that water quality could deteriorate in the distribution system—the vast infrastructure downstream of the treatment plant that delivers water to consumers. After its passage, the SDWA was interpreted by the U.S. Environmental Protection Agency (EPA) as meaning that some federal water quality standards should be met at various points within the distribution system rather than at the water treatment plant discharge. This interpretation forced water utilities to include the entire distribution system when considering compliance with federal law. Consequently water quality in the distribution system became a focus of regulatory action and a major interest to drinking water utilities.

EPA has promulgated many rules and regulations as a result of the SDWA that require drinking water utilities to meet specific guidelines and numeric standards for water quality, some of which are enforceable and collectively referred to as maximum contaminant levels (MCLs). As discussed in greater detail in Chapter 2, the major rules that specifically target water quality within the distribution system are the Lead and Copper Rule (LCR), the Surface Water Treatment Rule (SWTR), the Total Coliform Rule (TCR), and the Disinfectants/Disinfection By-Products Rule (D/DBPR). The LCR established monitoring requirements for lead and copper within tap water samples, given concern over their leaching from premise plumbing and fixtures. The SWTR establishes the minimum required detectable disinfectant residual, or in its absence the maximum allowed heterotrophic bacterial plate count, both measured within the distribution system. The TCR calls for the monitoring of distribution systems for total coliforms, fecal coliforms, and/or *E. coli*. Finally, the D/DBPR addresses the maximum disinfectant residual and concentration of disinfection byproducts (DBPs) like total trihalomethanes and haloacetic acids that are allowed in distribution systems.

Despite the existence of these rules, for a variety of reasons most contaminants that have the potential to degrade distribution system water quality are not monitored for, putting into question the ability of these rules to ensure public health protection from distribution system contamination. Furthermore, some epidemiological and outbreak investigations conducted in the last five years suggest that a substantial proportion of waterborne disease outbreaks, both microbial and chemical, is attributable to problems within distribution systems (Craun and Calderon, 2001; Blackburn et al., 2004). As shown in Figure 1-1, the *proportion* of waterborne disease outbreaks associated with problems in the distribution system is increasing, although the total number of reported waterborne disease outbreaks and the number attributable to distribution systems have decreased since 1980. The decrease in the total number of waterborne disease outbreaks per year is probably attributable to improved water treatment practices and compliance with the SWTR, which reduced the risk from waterborne protozoa (Pierson et al., 2001; Blackburn et al., 2004).

There is, however, no evidence that the current regulatory program has resulted in a diminution in the proportion of outbreaks attributable to distribution system related factors. Therefore, in 2000 the Federal Advisory Committee for the Microbial/Disinfection By-products Rule recommended that EPA evaluate available data and research on aspects of distribution systems that may create

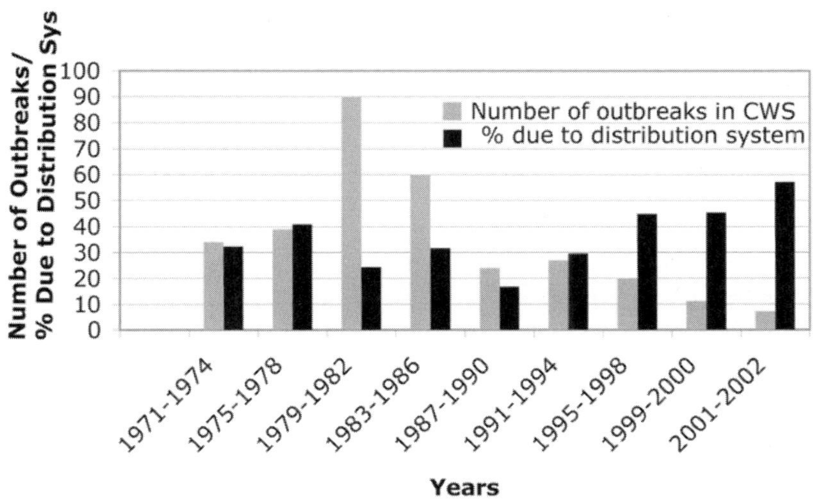

FIGURE 1-1 Waterborne disease outbreaks in community water systems (CWS) associated with distribution system deficiencies. Note that the majority of the reported outbreaks have been in small community systems and that the absolute number of outbreaks has decreased since 1982. SOURCE: Data from Craun and Calderon (2001), Lee et al., (2002), and Blackburn et al. (2004).

risks to public health. Furthermore, in 2003 EPA committed to revising the TCR—not only to consider updating the provisions about the frequency and location of monitoring, follow-up monitoring after total coliform-positive samples, and the basis of the MCL, but also to address the broader issue of whether the TCR could be revised to encompass "distribution system integrity." That is, EPA is exploring the possibility of revising the TCR to provide a comprehensive approach for addressing water quality in the distribution system environment. To aid in this process, EPA requested the input of the National Academies' Water Science and Technology Board, which was asked to conduct a study of water quality issues associated with public water supply distribution systems and their potential risks to consumers.

INTRODUCTION TO WATER DISTRIBUTION SYSTEMS

Distribution system infrastructure is generally the major asset of a water utility. The American Water Works Association (AWWA, 1974) defines the water distribution system as "including all water utility components for the distribution of finished or potable water by means of gravity storage feed or pumps though distribution pumping networks to customers or other users, including distribution equalizing storage." These systems must also be able to provide water for nonpotable uses, such as fire suppression and irrigation of landscaping.

They span almost 1 million miles in the United States (Grigg, 2005b) and include an estimated 154,000 finished water storage facilities (AWWA, 2003). As the U.S. population grows and communities expand, 13,200 miles (21,239 km) of new pipes are installed each year (Kirmeyer et al., 1994).

Because distribution systems represent the vast majority of physical infrastructure for water supplies, they constitute the primary management challenge from both an operational and public health standpoint. Furthermore, their repair and replacement represent an enormous financial liability; EPA estimates the 20-year water transmission and distribution needs of the country to be $183.6 billion, with storage facility infrastructure needs estimated at $24.8 billion (EPA, 2005a).

Infrastructure

Distribution system infrastructure is generally considered to consist of the pipes, pumps, valves, storage tanks, reservoirs, meters, fittings, and other hydraulic appurtenances that connect treatment plants or well supplies to consumers' taps. The characteristics, general maintenance requirements, and desirable features of the basic infrastructure components in a drinking water distribution system are briefly discussed below.

Pipes

The systems of pipes that transport water from the source (such as a treatment plant) to the customer are often categorized from largest to smallest as transmission or trunk mains, distribution mains, service lines, and premise plumbing. Transmission or trunk mains usually convey large amounts of water over long distances such as from a treatment facility to a storage tank within the distribution system. Distribution mains are typically smaller in diameter than the transmission mains and generally follow the city streets. Service lines carry water from the distribution main to the building or property being served. Service lines can be of any size depending on how much water is required to serve a particular customer and are sized so that the utility's design pressure is maintained at the customer's property for the desired flows. Premise plumbing refers to the piping within a building or home that distributes water to the point of use. In premise plumbing the pipe diameters are usually comparatively small, leading to a greater surface-to-volume ratio than in other distribution system pipes.

The three requirements for a pipe include its ability to deliver the quantity of water required, to resist all external and internal forces acting upon it, and to be durable and have a long life (Clark and Tippen, 1990). The materials commonly used to accomplish these goals today are ductile iron, pre-stressed concrete, polyvinyl chloride (PVC), reinforced plastic, and steel. In the past, unlined cast iron and asbestos cement pipes were frequently installed in distribution systems, and thus are important components of existing systems (see Figure 1-2). Transmission mains are frequently 24 inches (61 cm) in diameter or greater, dual-purpose mains (which are used for both transmission and distribution) are normally 16–20 inches (40.6–50.8 cm) in diameter, and distribution mains are usually 4–12 inches (10.0–30.5 cm) in diameter. Service lines and premise plumbing may be of virtually any material and are usually 1 inch (2.54 cm) in diameter or smaller (Panguluri et al., 2005).

It should be noted that this report considers service lines and premise plumbing to be part of the distribution system, and it considers the effects of service lines and premise plumbing on drinking water quality. If premise plumbing is included, the figure for total distribution system length would increase from almost 1 million miles (Grigg, 2005b) to greater than 6 million miles (Edwards et al., 2003). Premise plumbing and service lines have longer residence times, more stagnation, lower flow conditions, and elevated temperatures compared to the main distribution system (Berger et al., 2000). Inclusion of premise plumbing and service lines in the definition of a public water supply distribution system is not common because of their variable ownership, which ultimately affects who takes responsibility for their maintenance. Most drinking water utilities and regulatory bodies only take responsibility for the water delivered to the curb stop, which generally captures only a portion of the service line. The portion of the service line not under control of the utility and all of the premise plumbing are entirely the building owner's responsibility.

Pipe-Network Configurations

The two basic configurations for most water distribution systems are the branch and grid/loop (see Figure 1-3). A branch system is similar to that of a tree branch, in which smaller pipes branch off larger pipes throughout the service area, such that the water can take only one pathway from the source to the consumer. This type of system is most frequently used in rural areas. A grid/looped system, which consists of connected pipe loops throughout the area to be served, is the most widely used configuration in large municipal areas. In this type of system there are several pathways that the water can follow from the source to the consumer. Looped systems provide a high degree of reliability should a line break occur because the break can be isolated with little impact on consumers outside the immediate area (Clark and Tippen, 1990; Clark et al., 2004). Also, by keeping water moving looping reduces some of the problems associated with water stagnation, such as adverse reactions with the pipe walls, and it increases fire-fighting capability. However, loops can be dead-ends, especially in suburban areas like cul-de-sacs, and have associated water quality problems. Most systems are a combination of both looped and branched portions.

Design of water networks is very much dependent on the specific topography and the street layout in a given community. A typical design might consist

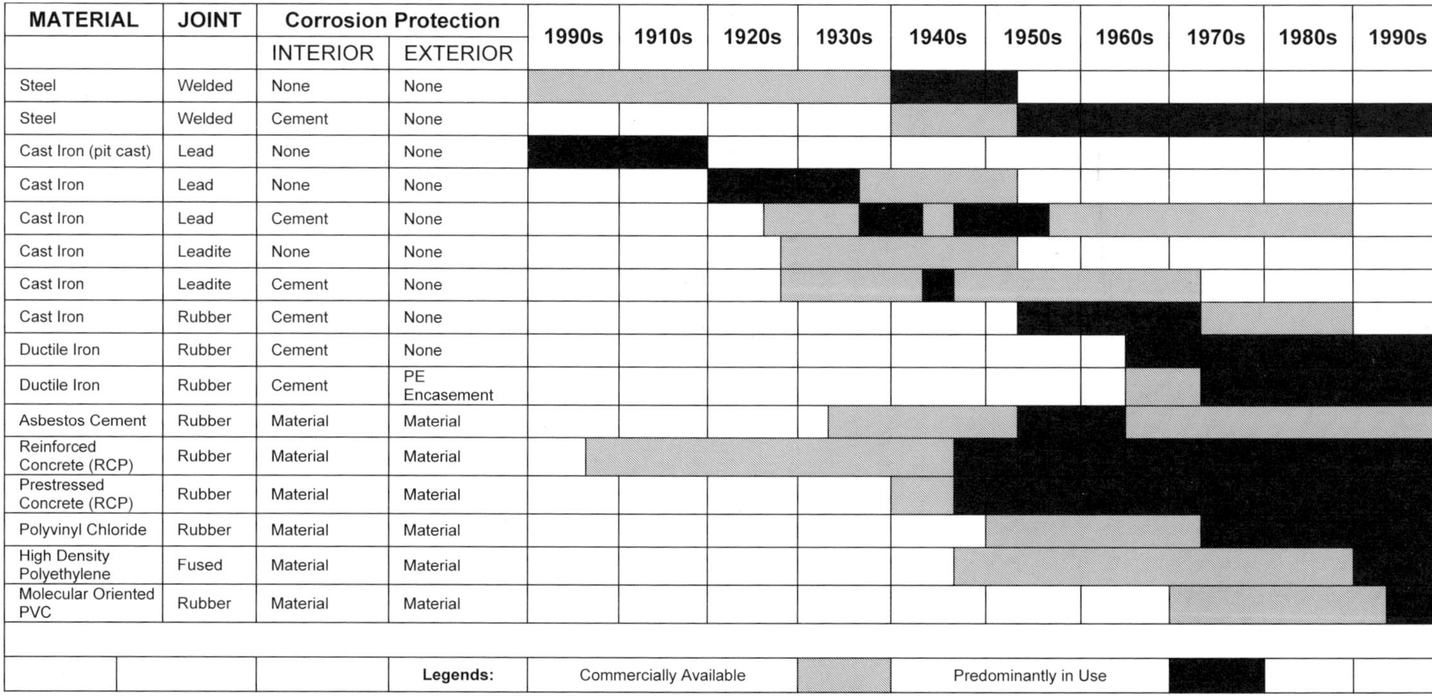

FIGURE 1-2 Timeline of pipe technology in the United States. SOURCE: Reprinted, with permission, from AWWSC (2002). © 2002 American Water.

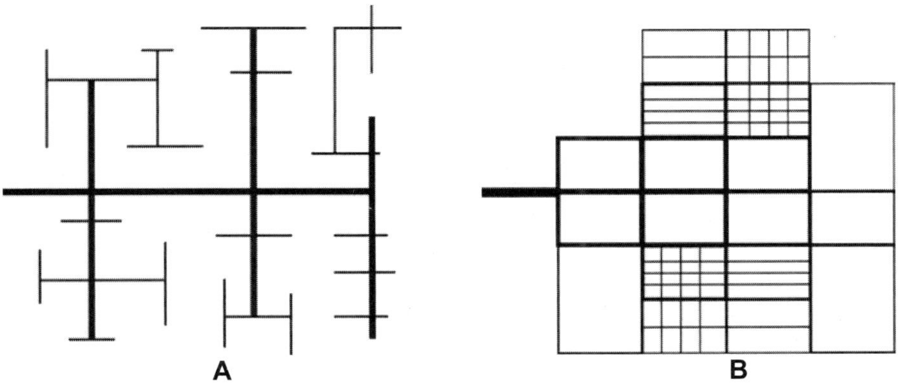

FIGURE 1-3 Two Basic Configurations for Water Distribution Systems. (A) Branched configuration. (B) Looped configuration.

of transmission mains spaced from 1.5 to 2 miles (2,400 to 3,200 m) apart with dual-service mains spaced 3,000 to 4,000 feet (900 to 1,200 m) apart. Service mains should be located in every street.

Storage Tanks and Reservoirs

Storage tanks and reservoirs are used to provide storage capacity to meet fluctuations in demand (or shave off peaks), to provide reserve supply for fire-fighting use and emergency needs, to stabilize pressures in the distribution system, to increase operating convenience and provide flexibility in pumping, to provide water during source or pump failures, and to blend different water sources. The recommended location of a storage tank is just beyond the center of demand in the service area (AWWA, 1998). Elevated tanks are used most frequently, but other types of tanks and reservoirs include in-ground tanks and open or closed reservoirs. Common tank materials include concrete and steel.

An issue that has drawn a great deal of interest is the problem of low water turnover in these facilities resulting in long detention times. Much of the water volume in storage tanks is dedicated to fire protection, and unless utilities properly manage their tanks to control water quality, there can be problems attributable to both water aging and inadequate water mixing. Excessive water age can be conducive to depletion of the disinfectant residual, leading to biofilm growth, other biological changes in the water including nitrification, and the emergence of taste and odor problems. Improper mixing can lead to stratification and large stagnant (dead) zones within the bulk water volume that have depleted disinfectant residual. As discussed later in this report, neither historical designs nor operational procedures have adequately maintained high water quality in storage tanks (Clark et al., 1996).

Security is an important issue with both storage tanks and pumps because of their potential use as a point of entry for deliberate contamination of distribution systems.

Pumps

Pumps are used to impart energy to the water in order to boost it to higher elevations or to increase pressure. Pumps are typically made from steel or cast iron. Most pumps used in distribution systems are centrifugal in nature, in that water from an intake pipe enters the pump through the action of a "spinning impeller" where it is discharged outward between vanes and into the discharge piping. The cost of power for pumping constitutes one of the major operating costs for a water supply.

Valves

The two types of valves generally utilized in a water distribution system are isolation valves (or stop or shutoff valves) and control valves. Isolation valves (typically either gate valves or butterfly valves) are used to isolate sections for maintenance and repair and are located so that the areas isolated will cause a minimum of inconvenience to other service areas. Maintenance of the valves is one of the major activities carried out by a utility. Many utilities have a regular valve-turning program in which a percentage of the valves are opened and closed on a regular basis. It is desirable to turn each valve in the system at least once per year. The implementation of such a program ensures that water can be shut off or diverted when needed, especially during an emergency, and that valves have not been inadvertently closed.

Control valves are used to control the flow or pressure in a distribution system. They are normally sized based on the desired maximum and minimum flow rates, the upstream and downstream pressure differentials, and the flow velocities. Typical types of control valves include pressure-reducing, pressure-sustaining, and pressure-relief valves; flow-control valves; throttling valves; float valves; and check valves.

Most valves are either steel or cast iron, although those found in premise plumbing to allow for easy shut-off in the event of repairs are usually brass. They exist throughout the distribution system and are more widely spaced in the transmission mains compared to the smaller-diameter pipes.

Other appurtenances in a water system include blow-off and air-release/vacuum valves, which are used to flush water mains and release entrained air. On transmission mains, blow-off valves are typically located at every low point, and an air release/vacuum valve at every high point on the main. Blow-off valves are sometimes located near dead ends where water can stagnate or where rust and other debris can accumulate. Care must be taken at these locations to prevent unprotected connections to sanitary or storm sewers.

Hydrants

Hydrants are primarily part of the fire fighting aspect of a water system. Proper design, spacing, and maintenance are needed to insure an adequate flow to satisfy fire-fighting requirements. Fire hydrants are typically exercised and tested annually by water utility or fire department personnel. Fire flow tests are conducted periodically to satisfy the requirements of the Insurance Services Office or as part of a water distribution system calibration program (ISO, 1980). Fire hydrants are installed in areas that are easily accessible by fire fighters and are not obstacles to pedestrians and vehicles. In addition to being used for fire fighting, hydrants are also for routine flushing programs, emergency flushing, preventive flushing, testing and corrective action, and for street cleaning and construction projects (AWWA, 1986).

Infrastructure Design and Operation

The function of a water distribution system is to deliver water to all customers of the system in sufficient quantity for potable drinking water and fire protection purposes, at the appropriate pressure, with minimal loss, of safe and acceptable quality, and as economically as possible. To convey water, pumps must provide working pressures, pipes must carry sufficient water, storage facilities must hold the water, and valves must open and close properly. Indeed, the carrying capacity of a water distribution system is defined as its ability to supply adequate water quantity and maintain adequate pressure (Male and Walski, 1991). Adequate pressure is defined in terms of the minimum and maximum design pressure supplied to customers under specific demand conditions. The maximum pressure is normally in the range of 80 to 100 psi; for example, the Uniform Plumbing Code requires that water pressure not exceed 80 psi (552 kPa) at service connections, unless the service is provided with a pressure-reducing valve. The minimum pressure during peak hours is typically in the range of 40 to 50 psi (276–345 kPa), while the recommended minimum pressure during fire flow is 20 psi (138 kPa).

Residential Drinking Water Provision

Of the 34 billion gallons of water produced daily by public water systems in the United States, approximately 63 percent is used by residential customers for indoor and outdoor purposes. Mayer et al. (1999) evaluated 1,188 homes from 14 cities across six regions of North America and found that 42 percent of annual residential water use was for indoor purposes and 58 percent for outdoor purposes. Outdoor water use varies quite significantly from region to region and includes irrigation. Of the indoor water use, less than 20 percent is for consumption or related activities, as shown below:

Human Consumption or Related Use – 17.1 %......	Faucet use – 15.7 % Dishwasher – 1.4 %
Human Contact Only – 18.5 %........................	Shower – 16.8 % Bath – 1.7 %
Non-Human Ingestion or Contact Uses – 64.3 %...	Toilet – 26.7 % Clothes Washer – 21.7 % Leaks – 13.7 % Other – 2.2 %

Most of the water supplied to residences is used primarily for laundering, showering, lawn watering, flushing toilets, or washing cars, and not for consumption. Nonetheless, except in a few rare circumstances, distribution systems are assumed to be designed and operated to provide water of a quality acceptable for human consumption. Normal household use is generally in the range of 200 gallons per day (757 L per day) with a typical flow rate of 2 to 20 gallons per minute (gpm) [7.57–75.7 L per minute (Lpm)]; fire flow can be orders of magnitude greater than these levels, as discussed below.

Fire Flow Provision

Besides providing drinking water, a major function of most distribution systems is to provide adequate standby fire flow, the standards for which are governed by the National Fire Protection Association (NFPA, 1986). Fire-flow requirements for a single family house vary from 750 to 1,500 gpm (2,839–5,678 Lpm); for multi-family structures the values range from 2,000 to 5,000 gpm (7,570–18,927 Lpm); for commercial structures the values range from 2,000 to 10,000 gpm (7,570–37,854 Lpm), and for industrial structures the values range from 3,000 to over 10,000 gpm (11,356–37,854 Lpm) (AWWA, 1998). The duration for which these fire flows must be sustained normally ranges from three to eight hours.

In order to satisfy this need for adequate standby capacity and pressure, most distribution systems use standpipes, elevated tanks, and large storage reservoirs. Furthermore, the sizing of water mains is partly based on fire protection requirements set by the Insurance Services Office (AWWA, 1986; Von Huben, 1999). (The minimum flow that the water system can sustain for a specific period of time governs its fire protection rating, which then is used to set the fire insurance rates for the communities that are served by the system.) As a consequence, fire-flow governs much of the design of a distribution system, especially for smaller systems. A study conducted by the American Water Works Association Research Foundation confirmed the impact of fire-flow capacity on the operation of, and the water quality in, drinking water networks (Snyder et al., 2002). It found that although the amount of water used for fire fighting is generally a small percentage of the annual water consumed, the required rates of water delivery for fire fighting have a significant and quantifiable impact on the size of water mains, tank storage volumes, water age, and operating and maintenance costs. Generally nearly 75 percent of the capacity of a typical drinking water distribution system is devoted to fire fighting (Walski et al., 2001).

The effect of designing and operating a system to maintain adequate fire flow and redundant capacity is that there are long transit times between the treatment plant and the consumer, which may be detrimental to meeting drinking water MCLs (Clark and Grayman, 1998; Brandt et al., 2004). Snyder et al. (2002) recommended that water systems evaluate existing storage tanks to determine if modification or elimination of the tanks was feasible. Water efficient fire suppression technologies exist that use less water than conventional standards. In particular, the universal application of automatic sprinkler systems provides the most proven method for reducing loss of life and property due to fire, while at the same time providing faster response to the fire and requiring significantly less water than conventional fire-fighting techniques. Snyder et al. (2002) also recommended that the universal application of automatic fire sprinklers be adopted by local jurisdictions for homes as well as in other buildings.

There is a growing recognition that embedded designs in most urban areas have resulted in distribution systems that have long water residence times due to the large amounts of storage required for fire fighting capacity. More than ten years ago, Clark and Grayman (1992) expressed concern that long residence times resulting from excess capacity for fire fighting and other municipal uses would also provide optimum conditions for the formation of DBPs and the regrowth of microorganisms. They hypothesized that eventually the drinking water industry would be in conflict over protecting public health and protecting public safety.

Non-conventional water distribution system designs that might address some of these issues are discussed below including decentralized treatment, dual distribution systems, and an approach that utilizes enhanced treatment to solve distribution system water quality problems. These alternative concepts were not part of the committee's statement of task, such that addressing them extensively is beyond the scope of the report. However, their potential future role in abating the problems discussed above warrants mention here and further consideration by EPA and water utilities.

Decentralized Treatment

Distributed or decentralized treatment systems refer to those in which a centralized treatment plant is augmented with additional treatment units that are located at various key points throughout the distribution system. Usually, the distributed units provide advanced treatment to meet stringent water quality requirements at consumer endpoints that would otherwise be in violation. Distributed units would be located either at the point-of-entry of households, for example, or at a more upstream location from which different water use could be served. This might be at the neighborhood or district level, depending on technological and financial requirements.

How the decentralized treatment concept might be implemented in water systems worldwide is still at a theoretical stage (e.g., Norton, 2006 and Weber, 2002, 2004). Weber's approach involves having distributed networks (Distributed Optimal Technologies Networks or DOT-Nets) in which water supply is optimized by separately treating several components of water *and* wastewater streams using decentralized treatment units. The approach largely views water supply, treatment, and waste disposal as different aspects of the same integrated system. Box 1-1 describes the concepts in detail.

BOX 1-1
Distributed Optimal Technologies Networks

DOT-Net is a decentralized treatment concept in which water supplies are segregated based on uses (or use functions) and levels of quality, to which a qualitative ranking on a scale of 1 to 10 is assigned, with 1 being the best quality and 10 the worst. The use functions include potable water, black water, gray water, various industrial discharges, etc. For example, water extracted from a local surface water source might be given a rank of 6. Following centralized treatment, the water would have a rank of 2. There is then an assumption that this supply will be degraded in distribution systems to a level that is generally not acceptable as potable water (say 3). To address this, advanced treatment technologies such as membranes and super-critical water treatment would be located as satellite systems close to the point of use, producing a water of ranking 1.

This concept hinges upon segregating water into the various use functions and developing and deploying the technology needed to bring about the desired water quality for each function. For example, water for drinking, showering, and cooking would require the highest level of quality and should be treated appropriately using satellite systems and advanced technologies. Advanced technologies exist for the treatment, analysis, and control of personal water including sophisticated electromechanical systems for rapid monitoring and feedback. The existing distribution system would still be used, but would be supplemented with treatment units to treat a portion of the water supply. For example, satellite treatment units may be located in large buildings with a high population density or distributed over neighborhoods.

The concept extends to the waste streams generated by each type of water use, as shown in Figure 1-4. Thus, advanced water treatment would be used not only prior to water delivery, but also upon water disposal but before it is discharged into a centralized collection system. For example if a certain commercial enterprise produced a highly degraded waste stream (with a ranking of 10), a satellite unit could be used to raise the quality to that of the other common waste streams (say 7). Such advanced and other wastewater treatment would be implemented in a manner to eventually resupply the source waters or to

continues

The principal trade-off associated with utilizing such systems is the alternative cost associated with upgrading large centralized treatment facilities and distribution networks. On a per person basis, it is less expensive to build one large treatment system than to build several small ones. In addition to these costs, multiple or new pipe networks are a necessary part of the design framework for these satellite systems. That is, new piping would be needed from the advanced water treatment system into the household (or industry), although it would travel a short distance and would be a small percentage of the total plumbing for the building. It is possible that investing in larger satellite systems with separate piping might offer a cost advantage compared to small satellite systems, based on economies of scale (Norton, 2006). Clearly, there would have to be a policy to avoid social injustice such that decentralized treatment when implemented is affordable to the average user.

BOX 1-1 Continued

produce water for another use function (e.g., recreational or industrial use). Also envisioned is the potential recovery of energy from the treatment of black water as well as some industrial sources.

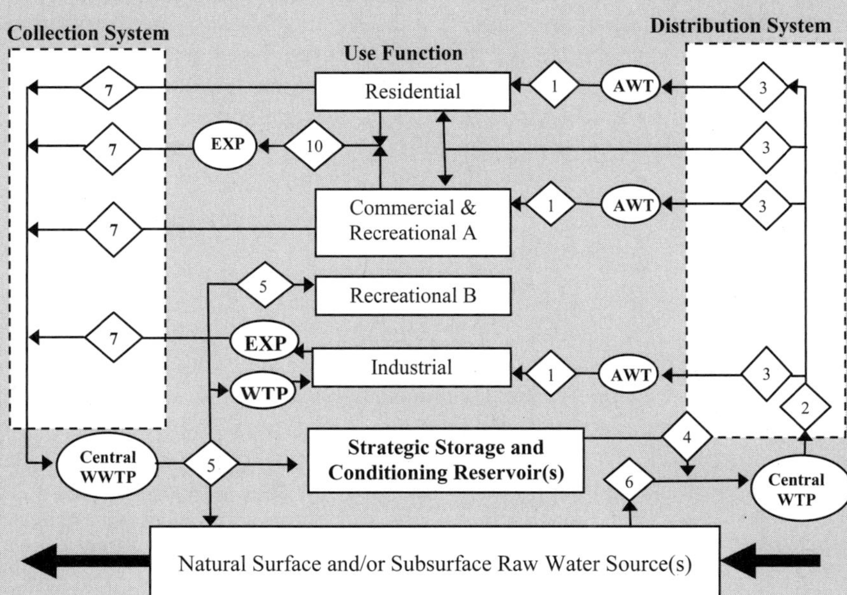

FIGURE 1-4 Distributed Optimal Technologies Networks. SOURCE: Reprinted, with permission, from Weber (2005). © 2005 by Weber.

A second important consideration is the need to monitor the satellite systems, and whose responsibility that monitoring would be. Maintenance activities, such as repair and replacement of a new piping system associated with satellite treatment, would also have to be well planned in order to prevent contamination of the distribution system downstream of the treatment unit. Incorporation of remote control technologies and other monitoring adaptations could reduce the need for human intervention while ensuring that the units operate satisfactorily.

The decentralized treatment concept was tested on a limited basis in the field (from October 1995 to September 1996) by Lyonnaise des Eaux-CIRSEE in the municipality of Dampierre, France. A one-year study was carried out using an ultrafiltration/nanofiltration system to treat water for 121 homes through 13,123 ft (4,000 m) of pipe at an average flow of 22.0 gpm (5 m^3/h) and a peak flow of 44.0 gpm (10 m^3/h). The ultrafiltration/nanofiltration system was fully automatic and monitored by remote control. Results from the study were very satisfactory from a quality perspective, and the cost calculations showed that the system was cost competitive with centralized treatment if production volumes were greater than 5,284,020 gal/year (20,000 m^3/year) (Levi et al., 1997).

A more prospective example is provided by the Las Vegas Valley Water District (LVVWD) and the Southern Nevada Water Authority (SNWA), which serve one of the most rapidly growing areas in the United States (see Box 1-2). Because of concerns over proposed MCLs for DBPs and the compliance framework being established by the Stage 2 D/DBPR, Las Vegas is investigating the application of decentralized or satellite water treatment systems within its distribution network. Currently only about 10 percent of the network is having trouble with compliance but it is anticipated as the system expands, more and more of the network will be out of compliance.

Enhanced Treatment

A third approach to slowing water quality deterioration involves centralized treatment options that can improve the quality of water to such a degree that formation of DBPs and loss of disinfectant residual are minimized. This approach is practiced by the Greater Cincinnati Water Works, which serves a large metropolitan area consisting of urban and suburban areas with potable water and fire flow protection. The distribution system is served by two treatment plants, the largest being the Miller Plant, which has a design capacity of 220 mgd (833,000 m^3/day) with an average water production of 133 mgd (428,000 m^3/day). The Miller Plant (Figure 1-5) has 12 granular activated carbon contactors, each containing 21,000 ft^3 (600 m^3) of GAC. During normal plant operation, between seven and 11 of these contactors are used (in parallel) to process water. Once GAC becomes spent, it is reactivated using an on-site reactivation system. GAC treatment reduces total organic carbon (TOC) levels from an

BOX 1-2
Application of Decentralized Treatment within the Las Vegas Valley Water District

Between 1989 and 2004, Las Vegas grew faster than any other metropolitan area in the U.S. As a result, during this period LVVWD has more than doubled its service area population. In 1989, the service area population was 558,000 but by 2004 it had grown to 1,209,000 (Jacobsen and Kamojjala, 2005).

The LVVWD receives its water on a wholesale basis from the Southern Nevada Water Authority (SNWA), which operates two water treatment plants with a combined design capacity of 900 mgd (3.41 mil m^3 per day). The source of water is Lake Mead. The treatment train at both plants is nearly identical, consisting of ozone and direct filtration. Chlorine is utilized as the final disinfectant. Coagulation dosages are limited and TOC removals through the biologically active filters range from 10 to 30 percent. The distribution system consists of 3,300 miles (5,280 km) of pipe and 29 water storage reservoirs. The system experiences long residence times (in some case greater than a week), resulting in an increase in water temperature as it moves through the system. Consequently it is difficult to maintain chlorine residuals in some parts of the system, necessitating the addition of chlorine at many locations. Currently the system is in compliance with all Safe Drinking Water Act regulations. However, based on distribution system hydraulic modeling estimates of detention time and known formation rates for trihalomethanes and haloacetic acids, it is expected that some areas in the LVVWD will not comply with the DBP MCLs and compliance framework being established by the Stage 2 D/DBPR. In order to meet Stage 2 regulations, the LVVWD/SNWA evaluated several alternatives to change its treatment and/or its residual disinfectant. Advanced oxidation, granular activated carbon (GAC) adsorption, enhanced coagulation (including addition of clarification), and nanofiltration were considered possible changes to treatment that could be helpful. In addition, operational and residual disinfectant changes were considered such as conversion from free chlorine to chloramine and a reduction in distribution system detention time. A more unconventional option considered by LVVWD/SNWA evaluated the potential for targeted or "hot-spot" treatment using several smaller-scale treatment systems that would reduce the concentration of DBPs in those areas of the distribution system that might exceed the MCLs established by the Stage 2 D/DBPR.

Ultimately, the LVVWD/SNWA chose to use the "hot-spot" treatment approach for the following reasons. It would provide a cost-effective approach by only treating water where needed at specific locations, instead of treating water for the entire system. It would reduce residuals production from treatment as compared to intensive organics removal. And it would provide for the continuous use of chlorine and avoid potential nitrification problems. The decentralized treatment options being considered are (1) DBP and natural organic material (NOM) removal by GAC adsorption, (2) DBP and NOM removal by biologically active carbon (BAC), and (3) control of DBP reformation after treatment by GAC and BAC. The American Water Works Association Research Foundation has funded a project that will test the concept of decentralized treatment and its application to the LVVWD (Jacobsen et al., 2005).

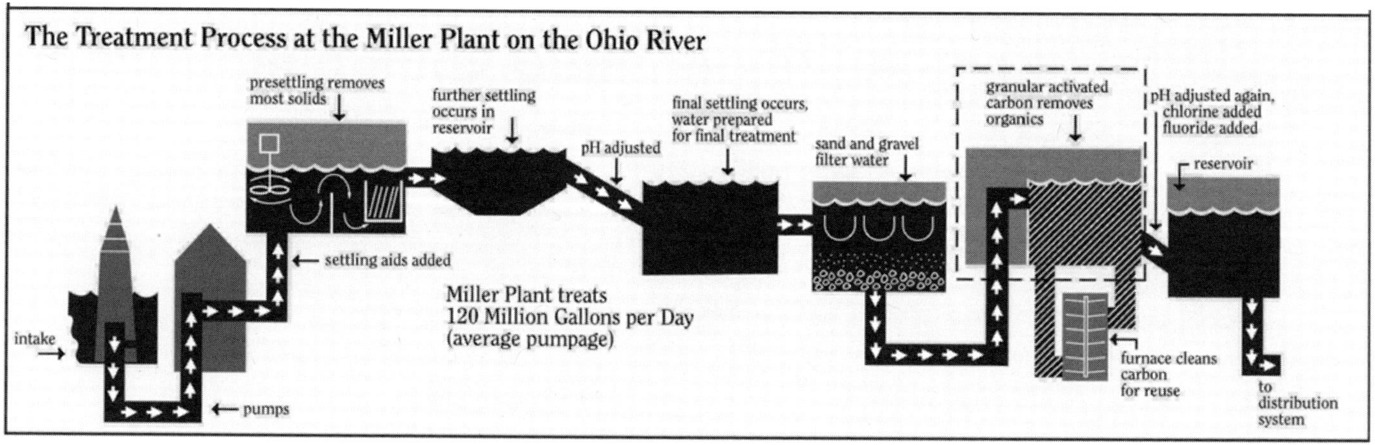

FIGURE 1-5 Schematic of the GCWW Ohio River Treatment Plant (Miller Plant).

annual average of 1.5 mg/L prior to GAC treatment to a combined five-year average of 0.6 mg/L after GAC treatment. The plant is one of the world's largest municipal GAC potable water treatment systems (Moore et al., 2003).

Las Vegas and Cincinnati have chosen distinctly different approaches to meeting and solving the residence time and excess capacity problem. The Las Vegas system is currently conducting studies to explore the possibility of applying treatment at various locations. Decentralized treatment units would be installed at points in the system where DBPs might exceed the Stage 2 D/DBPR. Research is being conducted that would focus primarily on removing the precursor material in the water in order to keep the DBP formation potential below regulated limits. A key aspect of this strategy is to use distribution system models and GIS technology to monitor residence time and DBP formation potential in the system. Cincinnati, on the other hand, has chosen the more traditional but very effective approach of removing DBP precursor material prior to distribution and thereby minimizing the potential formation of DBPs throughout the system. Although the Greater Cincinnati Water Works has a very large distribution system composed of a wide variety of pipe materials, the utility routinely provides water well below the total trihalomethane level of 80 μg/L and the total haloacetic acid level of 60 μg/L at all locations in the system.

Dual Distribution Systems

Another option for design and operation of distribution systems is the creation of dual systems in which separate pipe networks are constructed for potable and nonpotable water. In these types of systems, reclaimed wastewater or water of sub-potable quality may be used for fire fighting and other special purposes such as irrigation of lawns, parks, roadway borders and medians; air conditioning and industrial cooling towers; stack gas scrubbing; industrial processing; toilet and urinal flushing; construction; cleansing and maintenance, including vehicle washing; scenic waters and fountains; and environmental and recreational purposes. The design of these systems differentiates dual systems from most community water supplies, in which one distribution system provides potable water to serve all purposes.

Most dual systems in use today were installed by adding reclaimed water lines alongside (but not connected to) potable water lines already in place. For example, in St. Petersburg, Florida, a reclaimed water distribution system was placed into operation in 1976, and fire protection is provided from both the potable and reclaimed water lines. San Francisco has a nonpotable system, constructed after the 1906 earthquake, that serves the downtown area to augment fire protection. Rouse Hill, Australia was the first community to plan a dual water system with the reclaimed water lines to serve all nonpotable uses, including fire protection, such that the potable water line can have much smaller pipe diameters. Both the potable and nonpotable systems have service reservoirs for meeting diurnal variations in demand, and if a shortage of water for fire protection occurs, potable water can be transferred to the nonpotable system.

In a recent exchange of letters in the Journal of the American Water Works Association, Dr. Dan Okun (Okun, 2005) and Dr. Neil Grigg (Grigg, 2005a) addressed the merits of dual distribution systems for U.S. drinking water utilities, especially given that ingestion and human consumption are minor uses in most urban areas (see above). The argument is that because existing water distribution systems are designed primarily for fire protection, the majority of the distribution system uses pipes that are much larger than would be needed if the water was intended only for personal use. This leads to residence times of weeks in traditional systems versus potentially hours in a system comprised of much smaller pipes. In the absence of smaller sized distribution systems, utilities have had to implement flushing programs and use higher dosages of disinfectants to maintain water quality in distribution systems. This has the unfortunate side effect of increasing DBP formation as well as taste and odor problems, which contribute to the public's perception that the water quality is poor. Furthermore, large pipes are generally cement-lined or unlined ductile iron pipe typically with more than 300 joints per mile. These joints are frequently not water tight, leading to water losses as well as providing an opportunity for external contamination of finished water.

From an engineering perspective it seems intuitively obvious that it is most efficient to satisfy all needs by installing one pipe and to minimize the number of pipe excavations. This philosophy worked well in the early days of water system development. However, it has resulted in water systems with long residence times (and their negative consequences) under normal water use patterns and a major investment in above-ground (pumps and storage tanks) and below-ground (transmission mains, distribution pipes, service connections, etc.) infrastructure. Therefore as suggested in Okun (2005) it may be time to look at alternatives for supplying the various water needs in urban areas such as dual distribution systems. The water reuse aspect of dual systems is particularly attractive in arid sections of the U.S. that otherwise require transportation of large quantities of water into these areas.

Although there are many examples of water reuse in the United States (EPA, 1992), not many of them involve the use of a dual distribution system. The City of St. Petersburg, which operates one of the largest urban reuse systems in the world, provides reclaimed water to more then 7,000 residential homes and businesses. In 1991, the city provided approximately 21 mgd (79,500 m^3/day) of reclaimed water for irrigation needs of individual homes, condominiums, parks, school grounds, and golf courses; cooling tower make-up; and supplemental fire protection. In Irving, Texas, advanced secondary treated wastewater and raw water from the Elk Fork of the Trinity River are used to irrigate golf courses, medians, and greenbelt areas, and to maintain water levels at the Las Colinas Development. The reclaimed water originates from the 11.5 mgd (43,500 m^3/day) Central Regional wastewater treatment plant. A third example is provided in Hilton Head, South Carolina, where about 5 mgd

(18,900 m^3/day) of wastewater is being used for wetlands applications and golf course irrigation. All of the wastewater treatment systems have been upgraded to tertiary systems, and an additional flow rate of the same size as the first is being planned. Perhaps the most famous water reuse operation using a dual distribution system is the Irvine Ranch Water District in Irvine, California (see Box 1-3). There is a recent trend in California toward the use of more dual distribution systems, particularly in new developments, as a result of statutory requirements to use reclaimed water in lieu of domestic water for non-potable uses (California Water Code Section 13550-13551) and because of the need to conserve water to meet increasing local and regional water demands.

The potential advantages of using dual distribution systems include the fact that much smaller volumes of water would need be treated to high standards, which would result in cost savings at the treatment plant if all water supplied were to be treated in this fashion. Another advantage is that flow in the potable line would be expected to be relatively constant compared to a traditional system where large quantities of water would need to be transferred over short time periods (e.g., during fires). The associated flow and pressure changes in a pipe carrying the total water needs for a community are expected to be much greater than in the potable line of a dual distribution system. As discussed later in this document, there is evidence that pressure transients may result in intrusion of contaminated water. Furthermore, use of improved materials in the newer, smaller distribution system would minimize water degradation, loss, and intrusion.

However, the creation of dual distribution systems necessitates the retrofitting of an existing water supply system and reliance on existing pipes to provide non-potable supply obtained from wastewater or other sources. Large costs would be incurred when installing the new, small diameter pipe for potable water, disconnecting the existing system from homes and other users so that it could be used reliably for only nonpotable needs, and other retrofitting measures. These costs can be reduced if a new system is used only for reclaimed water distribution, as was done at Irvine Ranch, but this of course would not decrease the extent of quality degradation now experienced in existing systems. It is also critical to differentiate between full and partial adoption of dual distribution systems, the latter of which has occurred in several cities. For example, if a new nonpotable line is installed alongside an existing potable line, the non-potable line can draw demand away from the potable line, thereby increasing its detention time and aggravating water quality deterioration in the potable line. Furthermore, if the potable system is still used for fire flow, which generally governs pipe sizing, many of the advantages of the dual system will not be realized.

Dual systems may be most advantageous in new communities where neither type of distribution system currently exists. New communities could better optimize their systems because both types of piping systems could be built simultaneously. The cost savings from the need to treat a much smaller portion of the total water to a higher quality could partially offset the costs of constructing two

BOX 1-3
Irvine Ranch Water District

The Irvine Ranch Water District is one of the first water districts in the United States to practice wastewater reuse. It serves 316,287 people over 133 square miles (344.5 km^2), making it about a quarter the size of the Los Angeles Department of Water and Power. There are 85,500 domestic connections and 3,700 recycled connections. As of 2003, there are 1,075 miles (1,730 m) of pipe for the potable system. As part of the potable system, there are 28 above-ground and below-ground storage tanks that range in volume from 0.75 million gallons (0.0028 million m^3) to 16 million gallons (0.061 million m^3) and have a total storage capacity of 131.75 million gallons (0.50 million m^3). Much of the District's infrastructure is below grade due to aesthetic considerations.

The most unusual aspect of the District system is the recycled (reclaimed) water network. There are 350 miles (563.2 km) of reclaimed water lines compared to 1,075 miles (1729.7 km) of potable network lines. Domestic water tanks sit side by side with reclaimed water tanks. The recycled water is used only for toilet flushing in a few high-rise buildings, for cooling towers, for landscape irrigation especially at golf courses and condominium complexes, for food crops, and by one carpet manufacturer. Recycled water for toilet flushing is not used in residences, only in businesses. Recycled water itself is tertiary treated wastewater. It meets all of the water quality standards for drinking water, but it is high in salt. Interestingly, in the summer the recycled water has a much lower retention time in the distribution system than the potable water because of greater demand for the recycled water for landscaping. However, when the demand for recycled water is less than the input from WWTPs, the recycled water is put it in long-term storage. Indeed, one of the reasons dual systems were installed in high rises and other buildings was to make demand for recycled water more level throughout the year. There are no hydrants on the recycled system, so the reclaimed water is not used for fighting fires, and the pipe sizes in the recycled system are generally smaller than in the potable system. Chloramine provides residual disinfection in the potable system but chlorine is used in the recycled system (as mandated by California regulations). The SCADA system, which consists of 6,000 sampling points, provides minute-by-minute monitoring of chlorine residuals in the recycled system.

The potable system is required to meet all SDWA regulatory requirements such as the TCR, SWTR, D/DBPR, LCR, and source water monitoring on the imported water sources and the well water. Special purpose monitoring includes a nitrification action plan that requires tank sampling. For the recycled system, however, there are no specific monitoring objectives required by regulations because the NPDES permit has been met at the end of the WWTP. Internal requirements include bi-monthly sampling of conductivity, turbidity, color, pH, chlorine residual, total coliform, and fecal coliform and total suspended solids (at special locations). The water uses for the recycled water are very specific, and it is the goal of the utility to make sure the water is of an acceptable quality for those uses. Domestic potable water costs are 64 cents per thousand gallons for domestic water and 59 cents per thousand gallons for recycled water.

As might be expected, Irvine Ranch has a very extensive cross-connection control (CCC) program. There are approximately 13,000 CCC devices in place throughout the system. The District conducts an annual cross-connection shut-down test for the recycled irrigation water, and only one cross connection has been found in the last 10 years. For backflow prevention, a reduced pressure principle assembly at the meter is used, as required by the state of CA. Additional devices are installed if found to be needed.

SOURCE: Johannessen et al. (2005).

systems. Clearly, better understanding the technological potential and economic consequences of dual distribution systems is an important research goal.

***

Non-traditional options for drinking water provision present many unanswered questions but few case studies from which to gather information. The primary concerns include determining their economic feasibility and the existence of unknown costs, developing a plan for transition and implementation (which are expected to be very significant undertakings in existing communities), and maintenance of quality assurance and quality control in systems that would be potentially much more complicated than the current system. Furthermore, it is not clear how alternative distribution system designs will affect water security, an important consideration since September 11, 2001. The potential for cross connections or misuse of water supplies of lesser quality is greatly increased in dual distribution systems and decentralized treatment. Larger-scale questions involve potential social inequities and the extent to which nontraditional approaches will transfer costs to the consumer. These issues will have to be considered carefully in communities that decide to adopt these new designs for water provision.

The previous discussion raises a number of research issues, some of which are already noted. With regard to the influence of fire fighting requirements, distribution systems are frequently designed to supply water to meet maximum day demand and fire flow requirements simultaneously. This affects minimum pipe diameters, minimum system pressures (under maximum day plus fire flow demand), fire hydrant spacing, valve placement, and water storage. Generally, agencies that set fire flow requirements are not concerned about water quality while drinking water utilities must be concerned about both quality and fire flow capacity. It will be important to better evaluate the effectiveness of alternative fire suppression technologies including automatic sprinkler systems in a wide range of building types, including residences. Such systems have rarely been evaluated for their positive and negative features with respect to water quality. Furthermore, if fire suppression technologies were improved, it might be possible to rely on smaller sized pipes in distribution systems, as is being tested in Europe (Snyder et al., 2002), rather than moving to dual distribution systems.

If alternatives such as satellite systems and dual systems are not used, continued efforts will be required to upgrade existing distribution systems and to treat water to acceptable levels of quality, so that quality does not deteriorate during distribution. The balance of this report is focused on traditional distribution system design, in which water originates from a centralized treatment plant or well and is then distributed through one pipe network to consumers. Nonetheless, many of the report recommendations are relevant even if an alternative distribution system design is used.

Water System Diversity

Water utilities in the United States vary greatly in size, ownership, and type of operation. The SDWA defines public water systems as consisting of community water supply systems; transient, non-community water supply systems; and non-transient, non-community water supply systems. A community water supply system serves year-round residents and ranges in size from those that serve as few as 25 people to those that serve several million. A transient, non-community water supply system serves areas such as campgrounds or gas stations where people do not remain for long periods of time. A non-transient, non-community water supply system serves primarily non-residential customers but must serve at least 25 of the same people for at least six months of the year (such as schools, hospitals, and factories that have their own water supply). There are 159,796 water systems in the United States that meet the federal definition of a public water system (EPA, 2005b). Thirty-three (33) percent (52,838) of these systems are categorized as community water supply systems, 55 percent are categorized as transient, noncommunity water supplies, and 12 percent (19,375) are non-transient, non-community water systems (EPA, 2005b). Overall, public water systems serve 297 million residential and commercial customers. Although the vast majority (98 percent) of systems serves less than 10,000 people, almost three quarters of all Americans get their water from community water supplies serving more than 10,000 people (EPA, 2005b). Not all water supplies deliver water directly to consumers, but rather deliver water to other supplies. Community water supply systems are defined as "consecutive systems" if they receive their water from another community water supply through one or more interconnections (Fujiwara et al., 1995).

Some utilities rely primarily on surface water supplies while others rely primarily on groundwater. Surface water is the primary source of 22 percent of the community water supply systems, while groundwater is used by 78 percent of community water supply systems. Of the non-community water supply systems (both transient and non-transient), 97 percent are served by groundwater. Many systems serve communities using multiple sources of supply such as a combination of groundwater and/or surface water sources. This is important because in a grid/looped system, the mixing of water from different sources can have a detrimental influence on water quality, including taste and odor, in the distribution system (Clark et al., 1988, 1991a,b).

Some utilities, like the one operating in New York City, own large areas of the watersheds from which their water source is derived, while other utilities depend on water pumped directly from major rivers like the Mississippi River or the Ohio River, and therefore own little if any watershed land. The SDWA was amended in 1986 and again in 1996 to emphasize source water protection in order to prevent microbial contaminants from entering drinking water supplies (Borst et al., 2001). Owning or controlling its watershed provides an opportunity for a drinking water utility to exercise increased control of its source water quality (Peckenham et al., 2005).

The water supply industry in the United States has a long history of local government control over operation and financial management, with varying degrees of oversight and regulation by state and federal government. Water supply systems serving cities and towns are generally administered by departments of municipalities or counties (public systems) or by investor owned companies (private systems). Public systems are predominately owned by local municipal governments, and they serve approximately 78 percent of the total population that uses community water supplies. Approximately 82 percent of urban water systems (those serving more than 50,000 persons) are publicly owned. There are about 33,000 privately owned water systems that serve the remaining 22 percent of people served by community water systems. Private systems are usually investor-owned in the larger population size categories but can include many small systems as part of one large organization. In the small- and medium-sized categories, the privately owned systems tend to be owned by homeowners associations or developers. Finally, there are several classifications of state chartered public corporations, quasi-governmental units, and municipally owned systems that operate differently than traditional public and private systems. These systems include special districts, independent non-political boards, and state chartered corporations.

Infrastructure Viability over the Long Term

The extent of water distribution pipes in the United States is estimated to be a total length of 980,000 miles (1.6 x 10^6 km), which is being replaced at an estimated rate of once every 200 years (Grigg, 2005b). Rates of repair and rehabilitation have not been estimated. There is a large range in the type and age of the pipes that make up water distribution systems. The oldest cast iron pipes from the late 19th century are typically described as having an expected average useful lifespan of about 120 years because of the pipe wall thickness (AWWA, 2001; AWWSC, 2002). In the 1920s the manufacture of iron pipes changed to improve pipe strength, but the changes also produced a thinner wall. These pipes have an expected average life of about 100 years. Pipe manufacturing continued to evolve in the 1950s and 1960s with the introduction of ductile iron pipe that is stronger than cast iron and more resistant to corrosion. Polyvinyl chloride (PVC) pipes were introduced in the 1970s and high-density polyethylene in the 1990s. Both of these are very resistant to corrosion but they do not have the strength of ductile iron. Post-World War II pipes tend to have an expected average life of 75 years (AWWA, 2001; AWWSC, 2002).

In the 20th century, most of the water systems and distribution pipes were relatively new and well within their expected lifespan. However, as is obvious from the above paragraph and recent reports (AWWA, 2001; AWWSC, 2002), these different types of pipes, installed during different time periods, will all be reaching the end of their expected life spans in the next 30 years. Indeed, an estimated 26 percent of the distribution pipe in the country is unlined and in poor condition. For example, an analysis of main breaks at one large Midwestern water utility that kept careful records of distribution system management documented a sharp increase in the annual number of main breaks from 1970 (approximately 250 breaks per year) to 1989 (approximately 2,200 breaks per year) (AWWSC, 2002). Thus, the water industry is entering an era where it must make substantial investments in pipe repair and replacement. As shown in Figure 1-6, an EPA report on water infrastructure needs (EPA, 2002c) predicted that transmission and distribution replacement rates will rise to 2.0 percent per year by 2040 in order to adequately maintain the water infrastructure, which is about four times the current replacement rate according to Grigg (2005b).

These data on the aging of the nation's infrastructure suggest that utilities will have to engage in regular and proactive infrastructure assessment and replacement in order to avoid a future characterized by more frequent failures, which might overwhelm the water industry's capability to react effectively (Beecher, 2002). Although the public health significance of increasingly frequent pipe failures is unknown given the variability in utility response to such events, it is reasonable to assume that the likelihood of external distribution system contamination events will increase in parallel with infrastructure failure rates.

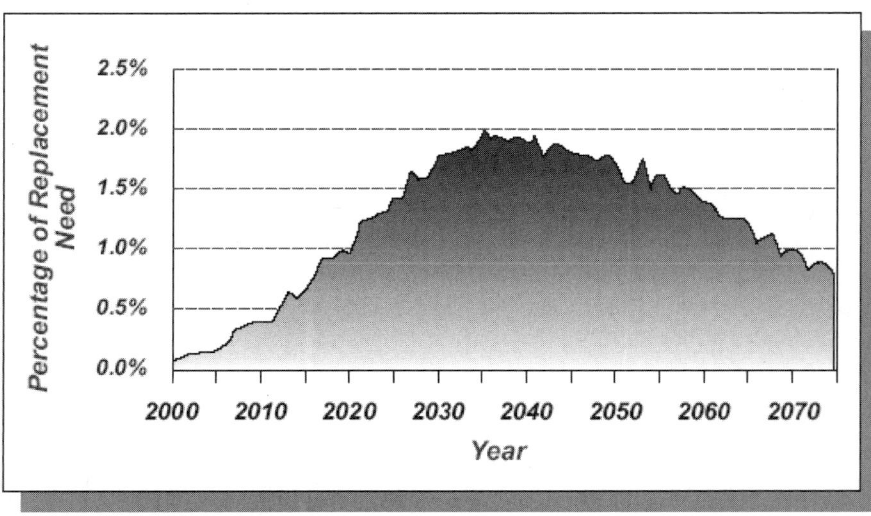

FIGURE 1-6 Projected annual replacement needs for transmission lines and distribution mains, 2000–2075. SOURCE: EPA (2002c).

DISTRIBUTION SYSTEM INTEGRITY

Many factors affect both the quantity and quality of water in distribution systems. As discussed in detail in Appendix A, events both internal and external to the distribution system can degrade water quality, leading to violation of water quality standards and possible public health risks. Corrosion and leaching of pipe materials, growth of biofilms and nitrifying microorganisms, and the formation of DBPs are events internal to the distribution system that are potentially detrimental. Furthermore, most are exacerbated by increased water age within the distribution system. External contamination can enter the distribution system through infrastructure breaks, leaks, and cross connections as a result of faulty construction, backflow, and pressure transients. Repair and replacement activities as well as permeable pipe materials also present routes for exposing the distribution system to external contamination. All of these events act to compromise the integrity of the distribution system.

For the purposes of this report, distribution system integrity is defined as having three basic components: (1) physical integrity, which refers to the maintenance of a physical barrier between the distribution system interior and the external environment, (2) hydraulic integrity, which refers to the maintenance of a desirable water flow, water pressure, and water age, taking both potable drinking water and fire flow provision into account, and (3) water quality integrity, which refers to the maintenance of finished water quality via prevention of internally derived contamination. This division is important because the three types of integrity have different causes of their loss, different consequences once they are lost, different methods for detecting and preventing a loss, and different remedies for regaining integrity. Factors important in maintaining the physical integrity of a distribution system include the maintenance of the distribution system components, such as the protection of pipes and joints against internal and external corrosion and the presence of devices to prevent cross-connections and backflow. Hydraulic integrity depends on, for example, proper system operation to minimize residence time and on preventing the encrustation and tuberculation of corrosion products and biofilms on the pipe walls that increase hydraulic roughness and decrease effective diameter. Maintaining water quality integrity in the face of internal contamination can involve control of nitrifying organisms and biofilms via changes in disinfection practices.

In addition to the distinctions mentioned above, there are also commonalities between the three types of integrity. All three are subject to system specificity, in that they are dependent on such site-specific factors as local water quality, types of materials present, area served, and population density. Furthermore, certain events involve the loss of more than one type of integrity—for example, backflow due to backsiphonage involves the loss of both hydraulic and physical integrity. Materials quality is important for both physical and water quality integrity. In order for a law or regulation to adequately address distribution system integrity—one of the options being considered during revision of the TCR—it must encompass physical, hydraulic, and water quality integrity.

IMPETUS FOR THE STUDY AND REPORT ROADMAP

Water supply systems have historically been designed for efficiency in water delivery to points of use, hydraulic reliability, and fire protection, while most regulatory mandates have been focused on enforcing water quality standards at the treatment plant. Ideally, there should be no change in the quality of treated water from the time it leaves the treatment plant until the time it is consumed, but in reality substantial changes may occur as a result of complex physical, chemical, and biological reactions. Distribution systems are the final barrier to the degradation of treated water quality, and maintaining the integrity of these systems is vital to ensuring that the water is safe for consumption.

The sections above have discussed the aging of the nation's water infrastructure and the continuing contribution of distribution systems to public health risks from drinking water. For the last five years, EPA has engaged experts and stakeholders in a series of meetings on the topic of distribution systems, with the goal of defining the extent of the problem and considering how it can be addressed during revisions to the TCR. As part of this effort, EPA led in the creation of nine white papers that summarized the state-of-the-art of research and knowledge in the area of drinking water distribution systems:

- Cross-Connections and Backflow (EPA, 2002a)
- Intrusion of Contaminants from Pressure Transients (LeChevallier et al., 2002)
- Nitrification (AWWA and EES, Inc., 2002e)
- Permeation and Leaching (AWWA and EES Inc., 2002a)
- Microbial Growth and Biofilms (EPA, 2002b)
- New or Repaired Water Mains (AWWA and EES Inc., 2002e)
- Finished Water Storage (AWWA and EES, Inc., 2002c)
- Water Age (AWWA and EES, Inc., 2002b)
- Deteriorating Buried Infrastructure (AWWSC, 2002)

Additional activities are ongoing, including consideration of a revision of the TCR to provide a more comprehensive approach for addressing the integrity of the distribution system. To assist in this process, EPA requested that the National Academies' Water Science and Technology Board conduct a study of water quality issues associated with public water supply distribution systems and their potential risks to consumers. An expert committee was formed in October 2004 with the following statement of task:

1) Identify trends relevant to the deterioration of drinking water in water supply distribution systems, as background and based on available information.
2) Identify and prioritize issues of greatest concern for distribution systems based on review of published material.

3) Focusing on the highest priority issues as revealed by task #2, (a) evaluate different approaches for characterization of public health risks posed by water quality deteriorating events or conditions that may occur in public water supply distribution systems; and (b) identify and evaluate the effectiveness of relevant existing codes and regulations and identify general actions, strategies, performance measures, and policies that could be considered by water utilities and other stakeholders to reduce the risks posed by water-quality deteriorating events or conditions. Case studies, either at state or utility level, where distribution system control programs (e.g., Hazard Analysis and Critical Control Point System, cross connection control, etc.) have been successfully designed and implemented will be identified and recommendations will be presented in their context.

4) Identify advances in detection, monitoring and modeling, analytical methods, information needs and technologies, research and development opportunities, and communication strategies that will enable the water supply industry and other stakeholders to further reduce risks associated with public water supply distribution systems.

The NRC committee addressed tasks one and two in its first report (NRC, 2005), which is included as Appendix A to this report. The following trends were identified as relevant to the deterioration of water quality in distribution systems:

- The aging distribution system infrastructure, including increasing numbers of main breaks and pipe replacement.
- Decreasing numbers of waterborne outbreaks reported per year since 1982, but an increasing percentage attributable to distribution system issues.
- Increasing host susceptibility to infection and disease in the U.S. population.
- Increasing use of bottled water and point-of-use treatment devices.

It was recommended in NRC (2005) that EPA consider these trends as it revises the TCR to encompass distribution system integrity. The committee was made aware of another important trend subsequent to the release of NRC (2005)—population shifts and how they have affected water demand. Older industrial cities in the northeast and Midwest United States no longer have industries that use high volumes of water, and they have also experienced major population shifts from the inner city to the suburbs. As a consequence, the utilities have an overcapacity to produce water, mainly in the form of oversized mains, at central locations, while needing to provide water to suburbs at greater distances from the treatment plant. Both factors can contribute to problems associated with high water residence times in the distribution system.

As part of its second task, the NRC committee prioritized the issues that are the subject of the nine EPA white papers, and it identified several significant issues that were overlooked in previous reports. The highest priority issues were those that have a recognized health risk based on clear epidemiological and surveillance data. These include cross connections and backflow; contamination during installation, rehabilitation, and repair of water mains and appurtenances; improperly maintained and operated storage facilities; and control of water quality in premise plumbing.

This report focuses on the committee's third and fourth tasks and makes recommendations to EPA regarding new directions and priorities to consider. All of the issues discussed in NRC (2005) are presented here, but considerably more information is presented on the higher priority issues when recommending detection, mitigation, and remediation strategies for distribution systems. The report is intended to inform decision makers within EPA, public water utilities, other government agencies and the private sector about potential options for managing distribution systems.

It should be pointed out that this report is premised on the assumption that water entering the distribution system has undergone adequate treatment. [As recognized in the SDWA, adequate treatment is a function of the quality of source water. For example, some lower quality source waters may require filtration to achieve a product entering the distribution system that is of the same quality (and hence poses the same risk) as a cleaner source water that was treated only with disinfection.] There is not, therefore, an in-depth discussion of drinking water treatment in the report except where it is pertinent to mitigating the risks of degraded water quality in the distribution system. For example, if the lack of disinfectant residual in the distribution system is identified as a risk, the options for mitigating that risk must first consider whether the root cause is inadequate treatment (e.g., insufficient reduction in disinfectant demand), or causes attributable to the distribution system (e.g., excessive water age in storage facilities). It should also be noted that deliberate acts of distribution system contamination are not considered, at the request of the study sponsor.

Chapter 2 reviews the legal and regulatory environment in which distribution systems are designed, operated, and monitored, including federal, state, and local regulations. The limitations and possibilities associated with non-regulatory approaches are also mentioned. Chapter 3 presents the three primary approaches for assessing the public health risk of contaminated distribution systems, focusing on short-term acute risks from microbial pathogens.

Chapters 4, 5, and 6 consider the physical, hydraulic, and water quality integrity of distribution systems, respectively. For each type of integrity, the chapters consider what causes its loss, the consequences if it is lost, and how to detect, maintain, and recover the type of integrity. In most cases, the events that compromise distribution system integrity are discussed only once, in the earliest chapter to which they are relevant. Many of the common themes from these chapters are brought together in Chapter 7, which presents a holistic framework for distribution system management, highlighting those activities felt to be of greatest importance to reducing public health risks. Areas where emerging science and technology can play a role are discussed, including real-time, on-line monitoring and modeling. The report concludes in Chapter 8 by considering the importance of premise plumbing to overall water quality at the tap, the need for additional monitoring of premise plumbing, and the need for greater involvement by regulatory agencies in exercising authority over premise plumbing. Premise plumbing is an issue not generally considered to be the responsibility of drinking water utilities, but there is growing interest—in terms of public health protection—about the role of premise plumbing in contributing to water quality degradation.

REFERENCES

American Water Works Association (AWWA). 1974. Water distribution research and applied development needs. J. Amer. Water Works Assoc. 6:385–390.

AWWA. 1986. Introduction to Water Distribution Principles and Practices of Water Supply Operations. Denver, CO: AWWA.

AWWA. 1998. AWWA Manual M31: Distribution system requirements for fire protection. Denver, CO: AWWA.

AWWA. 2001. Reinvesting in Drinking Water Structure: Dawn of the Replacement Era. Denver, CO: AWWA.

AWWA. 2003. Water Stats 2002 Distribution Survey CD-ROM. Denver, CO: AWWA.

AWWA and EES, Inc. 2002a. Permeation and leaching. Available on-line at *http://www.epa.gov/safewater/tcr/pdf/permleach.pdf*. Accessed May 4, 2006.

AWWA and EES, Inc. 2002b. Effects of water age on distribution system water quality. *http://www.epa.gov/safewater/tcr/pdf/waterage.pdf*. Accessed May 4, 2006.

AWWA and EES, Inc. 2002c. Finished water storage facilities. Available on-line at *http://www.epa.gov/safewater/tcr/pdf/storage.pdf*. Accessed May 4, 2006.

AWWA and EES, Inc. 2002e. New or repaired water mains. Available on-line at *http://www.epa.gov/safewater/tcr/pdf/maincontam.pdf*. Accessed May 4, 2006.

American Water Works Service Co., Inc. (AWWSC). 2002. Deteriorating buried infrastructure management challenges and strategies. Available on-line at *http://www.epa.gov/safewter/tcr/pdf/infrastructure.pdf*. Accessed May 4, 2006.

Baker, M. H. 1948. The quest for pure water. The American Water Works Association. Lancaster, PA: Lancaster Press.

Beecher, J. A. 2002. The infrastructure gap: myth, reality, and strategies. *In:* Assessing the Future: Water Utility Infrastructure Management. D. M. Hughes (ed.). Denver, CO: AWWA.

Berger, P. S., R. M. Clark, and D. J. Reasoner. 2000. Water, Drinking. *In:* Encyclopedia of Microbiology 4:898–912.

Blackburn, B. G., G. F. Craun, J. S. Yoder, V. Hill, R. L. Calderon, N. Chen, S. H. Lee, D. A. Levy, and M. J. Beach. 2004. Surveillance for waterborne-disease outbreaks associated with drinking water—United States, 2001–2002. MMWR 53(SS-8):23–45.

Borst, M., M. Krudner, L. O'Shea, J. M. Perdek, D. Reasoner, and M. D. Royer. 2001. Source water protection: its role in controlling disinfection by-products (DBPs) and microbial contaminants. *In:* Controlling Disinfection By-Products and Microbial Contaminants in Drinking Water. R. M. Clark and B. K. Boutin (eds.). EPA/600/R-01/110. Washington, DC: EPA Office of Research and Development.

Brandt, M., J. Clement, J. Powell, R. Casey, D. Holt, N. Harris, and C. T. Ta. 2004. Managing Distribution Retention Time to Improve Water Quality-Phase I. Denver CO: AwwaRF.

Clark, R. M. 1978. The Safe Drinking Water Act: implications for planning. Pp. 117–137 *In:* Municipal Water Systems—The Challenge for Urban Resources Management. D. Holtz and S. Sebastian (eds.). Bloomington, IN: Indiana University Press.

Clark, R. M., and D. L. Tippen. 1990. Water Supply. Pp. 5.173–5.220 *In:* Standard Handbook of Environmental Engineering. R. A. Corbitt (ed.). New York: McGraw-Hill Publishing Company.

Clark, R. M., and W. M. Grayman. 1992. Distribution system quality: a trade-off between public health and public safety. J. Amer. Water Works Assoc. 84(7):18.

Clark, R. M., and W. M. Grayman. 1998. Modeling water quality in drinking water distribution systems. Denver, CO: AWWA.

Clark, R. M., E. E. Geldreich, K. R. Fox, E. W. Rice, C. H. Johnson, J. A. Goodrich, J.A. Barnick, and F. Abdesaken. 1996. Tracking a *Salmonella* Serovar *Typhimurium* Outbreak in Gideon, Missouri: role of contaminant propagation modeling. Journal of Water Supply Research and Technology–Aqua 45(4):171–183.

Clark, R. M., J. A. Goodrich, and J.C. Ireland. 1984. Cost and benefit of drinking water treatment. Journal of Environmental Systems 14(1):1–30.

Clark, R. M., W. M. Grayman, and R. M. Males. 1988. Contaminant propagation in distribution systems. Jour. of Environ. Eng. ASCE 114(2):929–943.

Clark, R. M., W. M. Grayman, and J. A. Goodrich. 1991a. Water quality modeling: its regulatory implications. *In:* Proceedings of the AwwaRF/EPA Conference on Water Quality Modeling in Distribution Systems, Cincinnati, OH.

Clark, R. M., W. M. Grayman, J. A. Goodrich, R. A. Deininger, and A. F. Hess. 1991b. Field testing of distribution water quality models. J. Amer. Water Works Assoc. 83(7):67–75.

Clark, R. M., G. S. Rizzo, J. A. Belknap, and C. Cochrane. 1999. Water quality and the replacement of drinking water infrastructure: the Washington, DC case study. Journal of Water Supply Research and Technology – Aqua 48(3):106–114.

Clark, R. M., W. M. Grayman, S. G. Buchberger, Y. Lee, and D. J. Hartman. 2004. Drinking water distribution systems: an overview. Pp. 4.1–4.49 *In:* Water Supply Systems Security. L. W. Mays (ed.). New York: McGraw-Hill.

Craun, G. F., and R. Calderon. 2001. Waterborne disease outbreaks caused by distribution system deficiencies. J. Amer. Water Works Assoc. 93(9):64–75.

Edwards, M., D. Bosch, G. V. Loganathan, and A. M. Dietrich. 2003. The Future Challenge of Controlling Distribution System Water Quality and Protecting Plumbing Infrastructure: Focusing on Consumers. Proceedings of the IWA Leading Edge Conference in Noordwijk, Netherlands.

Environmental Protection Agency (EPA). 1991. Water Conservation Guidelines. Washington, DC: EPA.

EPA. 1992. Guidelines for Water Reuse. EPA/625/R-92/004. Washington, DC: EPA.

EPA. 2000. National primary drinking water regulations for lead and copper: final rule. Federal Register 65:1950. Washington, DC: EPA.

EPA. 2002a. 2000 Community water system survey. EPA 815-R-02-005A. Washington, DC: EPA Office of Water.

EPA. 2002b. Potential contamination due to cross-connections and backflow and the associated health risks: an issues paper. Available on-line at *http://www.epa.gov/safewater/tcr/pdf/ccrwhite.pdf*. Accessed May 8, 2006.

EPA. 2002c. The clean water and drinking water infrastructure gap analysis. Washington, DC: EPA.

EPA. 2005a. Drinking water infrastructure needs survey. EPA 816-R-05-001. Washington, DC: EPA Office of Water.

EPA. 2005b. Factoids: drinking water and ground water statistics for 2003. EPA 816-K-05-001. Washington, DC: EPA Office of Water.

Fujiwara, M., J. M. Manwaring, and R. M. Clark. 1995. Drinking water in Japan and the United States: conference objectives. *In:* Drinking water quality management. R. M. Clark and D. A. Clark (eds.). Lancaster, PA: Technomic Publishing Company Inc.

Grigg, N. S. 2005a. Letter to the editor: designing future water distribution systems. J. Amer. Water Works Assoc. 97(6):99–101.

Grigg, N. S. 2005b. Assessment and renewal of water distribution systems. J. Amer. Water Works Assoc. 97(2):58–68.

Grindler, B. J. 1967. Water and water rights: a treatise on the laws of water and allied problems: eastern, western, federal. Vol 3. Indianapolis, IN: The Allan Smith Company.

Hanke, S. H. 1972. Pricing Urban Water. Pp. 283–306 *In*: Public Prices for Public Products. S. Mushkin (ed.). Washington, DC: The Urban Institute.

Insurance Services Office. 1980. Fire suppression rating schedule. New York: Insurance Services Office.

Jacobsen, L. 2005. Las Vegas Valley Water District. April 18, 2005. Presented to the NRC Committee on Public Water Supply Distribution Systems. Washington, DC.

Jacobsen, L., and S. Kamojjala. 2005. Full System Models and GIS Integration. *In:* Proceedings of the AWWA Annual Conference and Exposition, San Francisco, CA.

Jacobsen, L., S. Kamojjala, and M. Fang. 2005. Integrating hydraulic and water quality models with other utility systems: a case study. *In:* Proceedings of the AWWA Information Management and Technology Conference, Denver, CO.

Johannessen, J., C. Kinner, and M. Velardes. 2005. Dual Distribution Systems: The Irvine Ranch Water District Experience. January 13, 2005. Presented to the NRC Committee on Public Water Supply Distribution Systems. Irvine, CA.

Kirmeyer, G., W. Richards, and C. D. Smith. 1994. An assessment of water distribution systems and associated research needs. Denver, CO: AwwaRF.

LeChevallier, M., R. Gullick, and M. Karim. 2002. The potential for health risks from intrusion of contaminants into the distribution system from pressure transients. Draft Distribution System White Paper. Washington, DC: EPA.

Lee, S. H., D. A. Levy, G. F. Craun, M. J. Beach, and R. L. Calderon. 2002. Surveillance for waterborne-disease outbreaks in the United States, 1999–2000. MMWR 51(No. SS-8):1–49.

Levi, Y., S. Pernettes, O. Wable, and L. Kiene. 1997. Demonstration unit of satellite treatment in distribution system using ultrafiltration and nanofiltration. Pp. 581–595 *In*: Proceedings of the AWWA Conference on Membrane Technology Conference. New Orleans, LA.

Mayer, P., W. B. DeOreo, E. M. Opitz, J. C. Kiefer, W. Y. Davis, B. Dziegielewski, and J. O. Nelson. 1999. Residential End Uses of Water. Denver, CO: AwwaRF.

Male, J. W., and T. M. Walski. 1991. Water Distribution Systems: A Troubleshooting Manual. Chelsea, MI: Lewis Publishers, Inc.

Moore, B. C., F. S. Cannon, D. H. Metz, and J. DeMarco. 2003. GAC pore structure in Cincinnati during full-scale treatment/reactivation. J. Amer. Water Works Assoc. 95(2):103–118.

National Fire Protection Association (NFPA). 1986. Fire Protection Handbook, 16th edition. A. E. Cote and J. L. Linville (eds.). Batterymarch Park Quincy, MA: National Fire Protection Association.

National Research Council (NRC). 2005. Public Water Supply Distribution Systems: Assessing and Reducing Risks, First Report. Washington, DC: The National Academies Press.

Norton, J. W., Jr. 2006. Cost advantages of implementing distributed treatment technologies for reduction of water-borne risk factors. Pp. 78-105 *In:* Financial and Technical Feasibility of Distributed Advanced Technology Water Treatment Systems. Ph.D. Dissertation. University of Michigan, Department of Environmental and Water Resources Engineering.

Okun, D. A. 2005. Letter to the editor: designing future water distribution systems. J. Amer. Water Works Assoc. (6):99–101.

Panguluri, S., W. M. Grayman, and R. M. Clark. 2005. Distribution system water quality report: a guide to the assessment and management of drinking water quality in distribution systems. Cincinnati, OH: EPA Office of Research and Development.

Peckenham, J. M., C. V. Schmitt, J. L. McNelly, and A. L. Tolman. 2005. Linking water quality to the watershed: developing tools for source water protection. J. Amer. Water Works Assoc. 97(9):62–69.

Pierson, G., K. Martel, A. Hill, G. Burlingame, and A. Godfree. 2001. Methods to prevent microbiological contamination associated with main rehabilitation and replacement. Denver, CO: AwwaRF.

Snyder, J. K., A. K. Deb, F. M. Grablutz, S. B. McCammon, W. M. Grayman, R. M. Clark, D. A. Okun, S. M. Tyler, and D. Savic. 2002. Impacts of fire flow on distribution system water quality, design and operation. Denver, CO: AwwaRF.

Von Huben, H. (Tech. Ed). 1999. Water distribution operator training handbook, 2nd edition. Denver, CO: AWWA.

Walski, T. M., D. V. Chase, and D. A. Savic. 2001. Water Distribution Modeling, 1st Edition. Waterbury, CT: Haestad Press.

Weber, Jr., W. 2002. Distributed optimal technology networks: a concept and strategy for potable water sustainability. Water Science and Technology 46(6–7):241–246.

Weber, Jr., W. 2004. Optimal uses of advanced technologies for water and wastewater treatment in urban environments. Water Science and Technology: Water Supply 4(1):7–12.

Weber, Jr., W. 2005. Distributed Systems. April 18, 2005. Presented to the NRC Committee on Public Water Supply Distribution Systems. Washington, DC.

2
Regulations, Non-regulatory Approaches, and their Limitations

This chapter provides an overview of the existing regulatory framework as well as non-regulatory approaches that are intended to protect drinking water quality within water distribution systems. Included is a discussion of federal and state statutes and regulations and local codes, along with their limitations. In addition, several non-regulatory programs are described that are intended to complement existing regulations.

REGULATORY ENVIRONMENT

Federal and state statutes and regulations along with local codes are used to establish requirements intended to protect the drinking water quality within distribution systems. The federal Safe Drinking Water Act (SDWA) is the vehicle used nationally to address drinking water quality issues. Prior to the passage of the SDWA, federal involvement in water supply had been limited to development of large multi-purpose water projects and regulation of water quality with respect to interstate carriers. After passage of the SDWA, the federal government became involved in developing national drinking water regulations pursuant to the new law and in conducting research to support these regulations. States implement the federal mandates but also utilize their own statutory and regulatory requirements to protect drinking water quality. For example, the states play a significant role in oversight functions ranging from licensing of water treatment plant operators to the approval of new sources of supply and the approval of new treatment facility design. Local agencies such as health departments, environmental health programs, and building departments implement codes and ordinances that address water distribution systems, most often that portion of the infrastructure not controlled by public water systems. This section provides an overview of the various statutory and regulatory approaches that apply to distribution systems.

Safe Drinking Water Act

The SDWA (Public Law 93-523), enacted in 1974 and amended in 1986 (Public Law 99-339), 1988 (Public Law 100-572), and 1996 (Public Law 104-182), provides the statutory bases by which public water systems are regulated.

Pursuant to the SDWA, the U.S. Environmental Protection Agency (EPA) is mandated to establish regulations for drinking water in the form of either maximum contaminant levels (MCL) or maximum contaminant level goals (MCLGs). MCLs are water quality standards that must be met by utilities and are enforced by state or federal agencies. Unlike MCLs, MCLGs are non-enforceable and are set at a level at which no known or anticipated adverse human health effects occur. Where it is not economically or technologically feasible to ascertain the level of a contaminant, a treatment technique is prescribed by EPA in lieu of establishing an MCL. For example, because the viable concentration of *Giardia lamblia* is difficult to measure, it has been established that if water is treated at a given pH, temperature, and chlorine concentration for a specified length of time (all of which are verified by the water utility), a fixed level of *Giardia* inactivation will take place.

The SDWA also provides EPA with the authority to delegate the implementation of the SDWA requirements to the states through the process of primacy. Forty-nine (49) of the 50 states have accepted primacy, with Wyoming being the exception. The SDWA applies to public water systems, which can be publicly or privately owned. Public water systems are defined as providing drinking water to at least 25 people or 15 service connections for at least 60 days per year. As mentioned in Chapter 1, there are approximately 160,000 public water systems in the United States, providing water to more than 290 million people.

Currently, 51 organic chemicals, 16 inorganic chemicals, seven disinfectants and disinfection byproducts (DBPs), four radionuclides, and coliform bacteria are monitored for compliance with the SDWA (EPA, 2005a). Standards for most contaminants are required to be met at the point of entry to the distribution system, such that the SDWA does not directly address *distribution system* contamination for most compounds. Despite these spatial restrictions, the SDWA does provide EPA with the authority to regulate contaminants within distribution systems—an authority that EPA has used to promulgate several regulations that address distribution system water quality including the Total Coliform Rule (TCR), the Lead and Copper Rule (LCR), the Surface Water Treatment Rule (SWTR), and the Disinfectants/Disinfection Byproducts Rule (D/DBPR).

The 1996 amendments to the SDWA mandated that EPA conduct research to strengthen the scientific foundation for standards that limit public exposure to drinking water contaminants. Specific requirements were given for research on waterborne pathogens such as *Cryptosporidium* and Norovirus, DBPs, arsenic, and other harmful substances in drinking water. EPA was also directed to conduct studies to identify and characterize population groups, such as children, that may be at greater risk from exposure to contaminants in drinking water than is the general population. In response to that mandate EPA has developed a Multi-Year Plan that describes drinking water research program activities and plans for fiscal years 2003–2010 (see Box 2-1).

BOX 2-1
EPA Multi-Year Plan for Drinking Water

The Multi-Year Plan establishes three long-term goals:

1. By 2010, develop scientifically sound data and approaches to assess and manage risks to human health posed by exposure to regulated waterborne pathogens and chemicals, including those addressed by the Arsenic, M/DBP, and Six-Year Review Rules.

2. By 2010, develop new data, innovative tools, and improved technologies to support decision making by the EPA Office of Water on the Contaminant Candidate List and other regulatory issues, and to support implementation of rules by states, local authorities, and water utilities.

3. By 2009, provide data, tools, and technologies to support management decisions by the EPA Office of Water, state, local authorities, and utilities to protect source water and the quality of water in the distribution system.

Some of the tasks in the Multi-Year Plan related to distribution systems include:

- Collect data to assess the stability of arsenic in water distribution systems.
- Prepare a report on chlorine and chloramines to control biofilms in model distribution systems.
- Prepare a report on the mechanisms and kinetics of chloramine loss and DBP formation in distribution systems. This work includes the modeling of n-nitrosodimethylamine formation.
- Prepare a report on the effect of oxidizing conditions on metal releases, corrosion rate, and scale properties of distribution system materials.
- Prepare a report on biofilm formation rates in pilot-scale distribution systems.
- Report on the characterization and prediction of scale formation (including aluminum) in distribution systems.
- Prepare a report on the detection of opportunistic pathogens (*E. coli, Aeromonas, Mycobacterium*) in biofilms using molecular detection techniques.
- Collect data on the treatment conditions which may enhance the solubilization of arsenic-containing iron oxides within the distribution system.
- Prepare a report on the link between the distribution system and *Mycobacterium avium* complex (MAC) found in clinical cases.
- Prepare a report on characterization of drinking water distribution system biofilm microbial populations using molecular detection methods.
- Prepare a report on corrosion chemistry relationships and treatment approaches.
- Prepare a report on the impact of change from conventional treatment of surface water to alternative treatment (membrane) on biofilm growth in water distribution systems in support of regulation development.
- Improve methods for rapid detection of water quality changes.
- Conduct leaching studies to characterize organotin concentrations in distribution systems.

SOURCE: EPA (2003a).

Associated Federal Regulations

There are several federal regulations that are designed to address specific distribution system water quality issues, although none of these regulations deal wholly with the integrity of distribution systems as defined in Chapter 1. The following provides a brief description of each of these regulations.

National Interim Primary Drinking Water Regulations

Following the passage of the SDWA, EPA adopted the National Interim Primary Drinking Water Regulations (NIPDWR) on December 24, 1975 and on July 9, 1976. The NIPDWR established the first national standards for drinking water quality. These standards included limits for ten inorganic chemicals, six organic pesticides, turbidity, and five radionuclides. In addition, the NIPDWR established standards for microbiological contamination based on total coliform organisms.

Total Coliform Rule

The primary purpose of the TCR is to ensure public health protection from microbial contamination of drinking water, and it applies to all public water systems. It is the only regulation that is intended to measure the microbiological quality of water within that part of the distribution system controlled by the public water supply. In 1989 EPA promulgated the TCR as a revision to the existing regulation that required public water systems to monitor for coliform organisms in the distribution system. The TCR changed the concept of monitoring for coliform organisms from one based on measuring the concentration of coliforms to determining the presence or absence of coliforms. In addition, the TCR established an MCL based on the presence or absence of total coliforms, modified monitoring requirements including testing for fecal coliforms or *E. coli,* required the use of a sample siting plan, and also required sanitary surveys for water systems collecting fewer than five samples per month. The MCL for total coliforms is as follows:

- For a system serving more than 33,000 people and collecting more than 40 samples per month, a non-acute violation occurs when more than 5.0 percent of the samples collected during the month are total coliform positive.

- For systems serving 33,000 people or less and collecting less than 40 samples per month, a non-acute violation occurs when more than one sample is total coliform positive in a given month.

- Any fecal coliform positive repeat sample, *E. coli* positive repeat sample, or any total coliform positive repeat sample following a fecal coliform or *E. coli* positive routine sample constitutes an acute violation of the MCL for total coliforms.

The sampling frequency ranges from one sample per month for water systems serving 25 people to 480 samples per month for the largest of water systems serving greater than 3,960,000 people (40 CFR 141.21 & 141.63). Sampling locations, identified in the sample siting plan, are required to be representative of water throughout the distribution system, including all pressure zones and areas supplied by each water source and distribution reservoir.

Trihalomethane Rule

In 1979 EPA promulgated a rule that established a drinking water standard for trihalomethanes (THMs), a group of chemicals produced as a consequence of chlorine disinfection. These chemicals are regulated because of the concern over their potential carcinogenic risk. The drinking water standard set at 0.10 mg/L addressed the total concentration of four specific THMs: chloroform, dichlorobromomethane, dibromochloromethane, and bromoform. This rule was the first to regulate the chemical quality of drinking water in the distribution system. The rule affected public water systems serving greater than 10,000 people because EPA was concerned that smaller systems would not have sufficient expertise available to deal with elevated levels of THMs without compromising microbiological safety. Water systems were required to sample quarterly at a minimum of four points in the distribution system and determine the average concentration of the four sample points. Compliance with the standard was based on the running average of any four consecutive quarterly results (EPA, 1979).

Surface Water Treatment Rule

On June 29, 1989, the EPA published the SWTR in response to Congress' mandate to require systems that draw their water from surface water sources (rivers, lakes, and reservoirs) and groundwater under the influence of surface water to filter, where appropriate, and to disinfect their water before distribution. The SWTR seeks to reduce the occurrence of unsafe levels of disease-causing microbes, including viruses, *Legionella* bacteria, and the protozoan *Giardia lamblia*. The SWTR requires water systems that filter to meet specific turbidity limits, and it assumes that this will achieve reductions in *Giardia lamblia* cysts (99.9 per cent) and viruses (99.99 per cent). Also, water systems are required to continuously monitor the residual disinfection concentration entering the distribution system, except those serving less than 3,300 people, which are allowed to collect grab samples. Furthermore, water systems (both filtered and unfiltered) are required to ensure a residual disinfectant concentration of not less than 0.2 mg/L entering the distribution system and to maintain a detectable residual disinfectant concentration in the distribution system measured as total chlorine, combined chlorine, or chlorine dioxide. The use of the heterotrophic bacteria plate count (HPC) is allowed as a surrogate for a detectable disinfectant in the distribution system provided that the concentration of heterotrophic bacteria is less than or equal to 500 colony forming units/milliliter (EPA, 1989). Samples for measuring residual disinfectant concentrations or heterotrophic bacteria must be taken at the same locations in the distribution system and at the same time as samples collected for total coliforms.

Lead and Copper Rule

The LCR was published in June 1991 and is intended to address the concern over chronic exposure of young children to lead in drinking water, the lead being principally from the leaching of the chemical from premise plumbing, fixtures, solder, and flux, and acute effects from copper. Indeed, since June 19, 1986, the use of solder and flux with more than 0.2 percent lead and the use of pipes and pipe fittings with more than 8.0 percent lead in the installation or repair of any public water system or plumbing in residential or non-residential facilities has been prohibited. States are required to enforce these requirements through state or local codes.

Unlike the TCR, which is intended to assess water quality that is representative of the entire distribution system in a dynamic or flowing state, the LCR is predicated on assessing water quality that represents worst case conditions. The LCR established monitoring requirements for tap water at "primary" locations—homes that contain lead pipes or copper pipes with lead solder installed after 1982. These homes were generally identified through a review of permits and records in the files of the building department(s) that indicate the plumbing materials installed within publicly and privately owned structures connected to the distribution system and the material composition of the service connections. The number of required samples depends on the size of the water system. Samples are collected from interior taps where water is typically drawn for consumption and after the tap has been left unused in a static state for a minimum of six hours. Table 2-1 describes the standard and reduced monitoring requirements of the LCR.

The LCR also established requirements for corrosion control treatment, source water treatment, lead service line replacement, and public education. The LCR establishes "action levels" in lieu of MCLs. The action level for lead was established at 0.015 mg/L while the action level for copper was set at 1.3 mg/L. An action level is exceeded when greater than 10 percent of samples collected from the sample pool contain lead levels above 0.015 mg/L or copper levels above 1.3 mg/L. Water systems exceeding the respective action level are

TABLE 2-1 Standard and Reduced Monitoring Requirements of the Lead and Copper Rule

System size (number of people served)	Standard monitoring requirements (number of sites)	Reduced monitoring requirements* (number of sites)
100,000	100	50
10,001 to 100,000	60	30
3,301 to 10,000	40	20
501 to 3,300	20	10
101 to 500	10	5
< 100	5	5

*Utilities can reduce the number of sampling sites and the frequency of monitoring from the required semi-annual frequency to a lesser frequency if their water system meets the following conditions:

Reduce to Annual monitoring if:

- the system serves less than 50,000 people and the lead and copper levels are less than the action level for two consecutive six-month monitoring periods or,
- the system meets Optimal Water Quality Parameter (OWQP) specifications for two consecutive six-month monitoring periods

Reduce to Triennial Monitoring if:

- the system serves more than 50,000 people and the lead and copper levels are less than the action level for three consecutive years or,
- the system meets OWQP specifications for three consecutive years of monitoring or,
- the system has 90th percentile lead levels less than 0.005 mg/L and 90th percentile copper levels less than 0.65 mg/L for two consecutive six-month monitoring periods or,
- The system has demonstrated optimized corrosion control

Reduce to Monitoring once every nine years if:

- the system serves less than 3,300 people, the distribution system, the service lines, and the premise plumbing are free of lead-containing and copper-containing materials and,
- the system has 90th percentile lead levels less than 0.005 mg/L and 90th percentile copper levels less than 0.65 mg/L for one six-month monitoring period.

required to install corrosion control treatment and conduct lead service line replacement and mandatory lead education.

Information Collection Rule

In May 1996, EPA promulgated the Information Collection Rule (ICR), which established monitoring and data reporting requirements for large public water systems including surface water systems serving at least 100,000 people and groundwater systems serving at least 50,000. The rule was intended to provide EPA with information on the occurrence in drinking water of (1) DBPs and (2) disease-causing microbes including *Cryptosporidium* (EPA, 1996). EPA used the information generated by the rule to develop new regulations for disinfectants and DBPs (EPA, 2006a).

Operator Certification

Pursuant to the SDWA amendments of 1996, EPA in cooperation with the states was directed to issue guidelines specifying minimum standards for certification and recertification of the water treatment and distribution system operators of all public water systems. The guidelines were required to take into account the size and complexity of the system, existing state programs, and other factors aimed at providing an effective program at reasonable cost to states and public water systems (EPA, 1999). EPA, through grants to the states allocated on the basis of "reasonable costs," was required to reimburse training and certification costs for operators of systems serving 3,300 persons or fewer, including an appropriate per diem for unsalaried operators who had to undergo training as a result of the federal requirement. States are required to adopt and implement a program for the certification of operators of public water systems that meet or are equivalent to the requirements of the EPA guidelines.

Stage 1 Disinfection and Disinfection Byproducts Rule

On December 16, 1998, EPA published the Stage 1 D/DBPR, making more stringent the existing standard for trihalomethanes as well as establishing new standards for disinfectants and other DBPs (EPA, 1998a). The rule, which applies to all public water systems, lowers the existing TTHM standard from 0.10 mg/L to 0.080 mg/L and establishes new standards for five haloacetic acids (HAAs) at 0.060 mg/L, bromate at 0.010 mg/L, and chlorite at 1.0 mg/L. In addition, the Rule establishes limits for disinfectants including chlorine, chloramine, and chlorine dioxide within the distribution system (via Maximum Residual Disinfectant Levels or MRDLs). For chlorine and chloramines, samples for measuring residual disinfectant must be taken at the same locations in the distribution system and at the same time as samples collected for total coliforms. For chlorine dioxide, samples must be taken daily at the entrance to the distribution system. Compliance with the MRDLs for chlorine and chloramines is based on the annual running average of all monthly samples collected, while compliance with the MRDL for chlorine dioxide is based on each daily sample. Finally, the Rule requires enhanced coagulation for certain systems in order to achieve specific reductions of DBP precursor material (as measured by total organic carbon concentrations).

Interim Enhanced Surface Water Treatment Rule

In December 1998, EPA promulgated the Interim Enhanced Surface Water Treatment Rule (IESWTR) that applied to public water systems serving greater than 10,000 people that were subject to the original SWTR. The IESWTR established a requirement for the reduction of *Cryptosporidium* and a more stringent turbidity requirement for filtered water supplies, among other provisions. The IESWTR also requires certain water systems to evaluate their disinfection practices to ensure that there will be no significant reduction in microbial protection as the result of modifying disinfection practices to meet MCLs specified by the Stage 1 D/DBPR. In addition, the IESWTR requires that all finished water storage facilities, for which construction began after February 16, 1999, be covered. EPA further indicated that it would consider whether or not to require the covering of existing reservoirs during the development of subsequent microbial regulations (EPA, 1998b).

Long Term 1 Enhanced Surface Water Treatment Rule

In 2002 EPA promulgated the Long Term 1 Enhanced Surface Water Treatment Rule (LT1ESWTR). The LT1ESWTR applies to public water systems that use surface water or groundwater under the direct influence of surface water and serve fewer than 10,000 persons. The purposes of the LT1ESWTR are to improve control of microbial pathogens, specifically *Cryptosporidium*, in drinking water and to address risk trade-offs with DBPs. The LT1ESWTR requires systems to meet strengthened filtration requirements as well as to calculate benchmark levels of microbial inactivation to ensure that microbial protection is not jeopardized if systems make changes to comply with requirements of the Stage 1 D/DBPR (EPA, 2002a). The only difference between this rule and the IESWTR is the size of the affected community.

Stage 2 Disinfectants and Disinfection Byproducts Rule

On January 4, 2006, EPA adopted the Stage 2 D/DBPR that makes more stringent the previous rule regulating certain DBPs. Under the Stage 1 D/DBPR water systems are allowed to average the DBP sample results from across the distribution system. As a result some customers could be exposed to levels of DBPs that consistently exceeded the MCLs and that might escape detection. The new rule requires that water systems meet the MCLs for THMs and HAAs at each sampling location based on the running annual average of any four consecutive quarterly sample results at that location. The intent of this change is to reduce DBP exposure and provide more equitable health protection and to lower potential cancer, reproductive, and developmental risks (EPA, 2006a).

To determine the locations within the distribution system where the highest levels of THMs and HAAs are expected to occur, the Rule requires water systems to conduct an Initial Distribution System Evaluation. Initial Distribution System Evaluations are studies that evaluate THM and HAA levels at various points within the distribution system. The results from these studies along with existing compliance monitoring information will be used to determine future compliance monitoring locations.

Long Term 2 Enhanced Surface Water Treatment Rule

On January 5, 2006, EPA adopted the Long Term 2 Enhanced Surface Water Treatment Rule (LT2ESWTR). The LT2ESWTR applies to public water systems that use surface water or groundwater under the direct influence of surface water. The purpose of the LT2ESWTR is to reduce disease incidence associated with *Cryptosporidium* and other pathogenic microorganisms in drinking water. The LT2ESWTR supplements existing regulations by targeting additional *Cryptosporidium* treatment requirements to higher risk systems based on actual monitoring data of source water quality.

The LT2ESWTR also contains provisions to mitigate risks from uncovered finished water storage facilities. Water systems with uncovered finished water storage reservoirs are required to cover the reservoir or treat the reservoir discharge to the distribution system to achieve inactivation and/or removal of at least 2-log *Cryptosporidium*, 3-log *Giardia*, and 4-log virus (EPA, 2006b).

Finally, to ensure that systems maintain microbial protection as they take steps to reduce the formation of DBPs the LT2ESWTR requires water systems that proposed to modify their disinfection process to reduce THMs and HAAs to assess the existing levels of disinfection that the system provides. Systems are required to establish a benchmark, which is the system's lowest monthly average microbial inactivation. If the benchmark is more than the required inactivation of 3-log removal for *Giardia* and 4-log removal for viruses, the system may consider decreasing the amount of disinfectant added or the contact time, or altering other disinfection practices to lower THM and HAA levels (EPA, 2006b).

Unregulated Contaminant Monitoring Rule 2

On August 22, 2005, EPA proposed the second of two Unregulated Contaminant Monitoring Rules (UCMR2), which will require monitoring for a list of 26 chemical contaminants suspected to be present in drinking water. The purpose of the UCMR2 is to develop data on the occurrence of these contaminants in drinking water, the size of the population exposed to these contaminants, and the levels of the exposure. This information will be used along with health effects information to determine whether or not drinking water standards should be established for these contaminants. All community water systems and non-transient, non-community water systems serving more than 10,000 people will be required to monitor, while a representative sample of 800 community water systems and non-transient, non-community water systems serving less than 10,000 people will have to carry out monitoring. The monitoring is proposed to begin in 2007.

Unlike the first UCMR (which is not discussed above), the UCMR2 will include contaminants that are considered potential DBPs and for which monitoring will be conducted in the distribution system. These contaminants include the nitrosamines N-nitroso-diethylamine (NDEA), N-nitroso-dimethylamine

(NDMA), N-nitroso-di-n-butylamine (NDBA), N-nitroso-di-n-propylamine (NDPA), N-nitroso-methylethylamine (NMEA) and N-nitroso-pyrrolidine (NPYR). Nitrosamines are considered potential human carcinogens, and NDMA has been shown to form in chlorinated or chloraminated water as a result of disinfection (EPA, 2005b).

Water Security-related Directives and Laws

Although not a new issue, security has become paramount to the water utility industry since the events of September 11, 2001. The potential for natural, accidental, and purposeful contamination of water supply has been present for decades whether in the form of earthquakes, floods, spills of toxic chemicals, or acts of vandalism. For example, in May 1998, President Clinton issued Presidential Directive (PDD) 63 that outlined a policy on critical infrastructure protection, including our nation's water supplies. However, it was not until after September 11, 2001, that the water industry truly focused on the vulnerability of the nation's water supplies to security threats. In recognition of these issues, President Bush signed Public Health Security and Bioterrorism Preparedness and Response Act of 2002 (the "Bioterrorism Act") into law in June 2002 (PL107-188). Under the requirements of the Bioterrorism Act, drinking water utilities are required to prepare vulnerability assessments and emergency response plans for water systems serving at least 3,300 people.

***

Table 2-2 summarizes the key requirement(s) of federal rules and regulations from a distribution system perspective.

State Regulatory Programs

State regulatory programs that address water distribution systems can vary significantly. In general most states have statutory and regulatory requirements that cover (1) design, construction, operation, and maintenance of distribution systems, (2) cross-connection control, and (3) plumbing products certified for use pursuant to American National Standards Institute/ NSF International (ANSI/NSF) standards 60 and 61. Furthermore, most states have adopted a plumbing code that dictates the types of materials that can be used for premise plumbing, although these codes are not generally enforced from a state statutory or regulatory standpoint but rather are implemented at the local county and/or municipal level.

TABLE 2-2 Summary of Regulated Distribution System Requirements

Law/Rule/ Regulation	Key Distribution System Requirements
SDWA	• Established national primary and secondary drinking water regulations (MCLs and MCLGs) • Allowed EPA to establish point of compliance
NIPDWR	• Adopted at the passage of the SDWA and required that representative coliform samples be collected throughout the distribution system
THM Rule	• Established a standard for total THMs of 0.10 mg/L • Compliance based on the annual average of THM levels at all monitoring locations within the distribution system
TCR	• Regulates coliform bacteria, which are used as "surrogate" organisms to indicate whether or not system contamination is occurring • Compliance based on results from representative monitoring locations within the distribution system
SWTR	• Requires that a detectable disinfectant residual be maintained at representative locations in the distribution system • Requires continuous monitoring of disinfectant residual entering the distribution system for water systems serving greater than 3,300 people
LCR	• Requires that lead and copper concentration be below action levels in samples taken at the worst case or highest risk consumer's tap
ICR	• Provides monitoring data to support the interim and long-term enhanced SWTR and Stage 2 DBP rule
1996 SDWAA	• Focused on the role that surface water quality can play in influencing the quality of distributed water • Established requirement for certification of operators of water systems including water distribution system operators
IESWTR	• Enhances protection from pathogens, including *Cryptosporidium,* and tries to prevent increases in microbial risk for large systems while they comply with the Stage 1 D/DBPR • Prohibits the construction of new uncovered finished water storage facilities
Stage 1 D/DBPR	• Lowers the standard for total THMs from 0.10 mg/L to 0.08 mg/L. This standard applies to all community water supplies in the U.S. • Set an MCL for 5 HAAs of 0.06 mg/L.

continues

TABLE 2-2 Continued

Law/ Rule/ Regulation	Key Distribution System Requirements
LT1ESWTR	• Enhances protection from pathogens, including *Cryptosporidium,* and tries to prevent increases in microbial risk for systems serving less than 10,000 people while they comply with the Stage 1 D/DBPR
Stage 2 D/DBPR	• Requires an Initial Distribution System Evaluation (IDSEs) • Compliance based on the locational running annual average of total THM and HAA levels at each monitoring location within the distribution system
LT2ESWTR	• Requires additional *Cryptosporidium* treatment for high risk systems and maintenance of microbial protection while reducing the formation of DBPs • Requires uncovered finished water storage facilities to be covered or the discharge from the finished water storage facilities to the distribution system to be treated to achieve inactivation and/or removal of at least 4-log virus, 3-log *Giardia*, and 2-log *Cryptosporidium*
UCMR2 (Proposed)	• Will require distribution system monitoring for nitrosamines to determine their occurrence as DBPs

Requirements for Design, Construction, Operation, and Maintenance

Using their existing statutory authority, many states have established requirements for the design, construction, operation, and maintenance of distribution systems. This was revealed in a survey of state drinking water programs conducted by the Association of State Drinking Water Administrators (ASDWA) in March 2003. Of the 34 states responding, the majority reported having some requirements for water-main design and construction, storage facilities and pump station design and construction, and distribution system operation and maintenance (ASDWA, 2003). A summary of the responses is provided in Tables 2-3, 2-4, and 2-5, respectively.

There appears to be less consistency between states, however, regarding the individual elements that each state requires be met. For example, most states have requirements for minimum operational pressures and the types of pipes that can be used, while less than half the states have requirements for storage and handling of pipes and distribution system maintenance plans. Only a small number of states have requirements for nitrification control and storage tank water quality monitoring. States also use different approaches for establishing these requirements. In some cases states have established their own requirements, while in others requirements are based on third party standards such as

TABLE 2-3 Summary of Results from the ASDWA Distribution System and Total Coliform Rule Survey: Water Main Design and Construction

Element	Numbers of States		
	Required	*Encouraged*	*Not Addressed*
Minimum pipe diameter (set minimum or size based on flow, number of service connections, etc.)	26	3	5
Design for an operational pressure of at least 20 psi under all flow conditions	32	0	2
Minimum flow velocity through pipes	9	6	19
Maximum flow velocity through pipes	9	8	17
Pipe material	30	2	2
Storage and handling of pipes	16	7	9 (2 NR)
Minimum depth of cover over pipes to prevent freezing and damage	25	7	2
Pressure/leakage testing before placing new mains into service	26	7	1
Disinfection, flushing, and microbial testing before placing new mains into service	29	5	0
Looping of pipes/minimization of dead ends	17	15	2
Proper flushing devices at dead ends	23	9	2
Protection of air-release and air vacuum valves	22	9	1 (1 NR)
Isolation valves at intersections and over lengthy stretches of water main	23	8	3
Separation of water mains and sanitary sewers to protect the water main from contamination	29	4	1
Protection of water main at surface water crossings	21	11	2
Exterior corrosion protection of water mains	14	12	8
Cross connection control/backflow prevention (through the drinking water program)	29	2	3

NR: No Response
Note: State practices may have changed since 2003. This survey is not a complete census of all state drinking water programs, rather it is indicative of the practices of the 34 states that responded to the survey.
SOURCE: Reprinted, with permission, from ASDWA (2003). © 2003 by ASDWA.

TABLE 2-4 Summary of Results from the ASDWA Distribution System and Total Coliform Rule Survey: Storage Facilities and Pump Station Design and Construction

Element	**Numbers of States**		
	Required	*Encouraged*	*Not Addressed*
Standards for tank design and construction	28	5	1
Tanks designed to ensure adequate turnover	15	16	3
Storage tank vents, screens, overflows, and access hatches	30	4	0
Telemetry or other means for controlling/monitoring the storage facility	15	15	4
Provisions for draining the storage facility	22	10	2
Standards for paints and coatings and provisions for testing before placing the storage facility in service	31	3	0
Cathodic protection for storage facilities	15	12	7
Standards for pump station design and construct	26	6	2
Drainage of underground pump stations and valve vaults	22	8	3 (1 NR)
Minimum inlet pressure for in-line booster pumps	25	7	2

Note: State practices may have changed since 2003. This survey is not a complete census of all state drinking water programs, rather it is indicative of the practices of the 34 states that responded to the survey.
SOURCE: Reprinted, with permission, from ASDWA (2003). © 2003 by ASDWA.

those developed by the American Water Works Association (AWWA) or the Ten State Standards (ASDWA, 2003).

Cross-Connection Control Requirements

One of most common means of contaminating distribution systems is through a cross connection. Cross connections occur when a nonpotable water source is connected to a potable water source. Under this condition contaminated water has the potential to flow back into the potable source. Backflow can occur when the pressure in the distribution system is less than the pressure in the nonpotable source, described as backsiphonage. Conditions under which backsiphonage can occur include water main breaks, firefighting demands, and pump failures. Backflow can also occur when there is increased pressure from the nonpotable source that exceeds the pressure in the distribution system, described as backpressure. Backpressure can occur when industrial operations connected to the potable source are exerting higher internal pressure than the pressure in

TABLE 2-5 Summary of Results from the ASDWA Distribution System and Total Coliform Rule Survey: Distribution System Operation and Maintenance

Element	Numbers of States		
	Required	*Encouraged*	*Not Addressed*
Operational pressure ≥ 20 psi under all flow conditions	30	3	1
Distribution system maintenance plan	16	11	7
Routine distribution system flushing, cleaning and/or pigging	11	20	3
Valve and hydrant exercise/ maintenance plan	10	19	5
Telemetry or other means for controlling/monitoring the DS	7	14	13
Unaccounted for water requirements	12	13	9
Disinfection, flushing, testing, and other follow-up action before returning a water main to service after repairs	26	8	0
Tank flushing	5	19	10
Tank inspection and maintenance	13	16	5
Tank cleaning	8	18	8
Provisions for testing before placing the storage facility back in service following cleaning/maintenance	24	8	2
Maintaining a minimum disinfectant residual in groundwater systems (if disinfection is provided)	21	7	6
Storage tank water quality monitoring	5	11	18
Nitrification control	4	7	23
Other water quality monitoring in the distribution system (beyond the SWTR, TCR, and LCR)	17	8	8 (1 NR)

Note: State practices may have changed since 2003. This survey is not a complete census of all state drinking water programs, rather it is indicative of the practices of the 34 states that responded to the survey.

SOURCE: Reprinted, with permission, from ASDWA (2003). © 2003 by ASDWA.

the distribution system or when irrigation systems connected to the potable system are pumping from a separate water source and the pump pressure exceeds the distribution system pressure.

Of 30 states surveyed by ASDWA, the vast majority required some sort of cross-connection control program, either through regulations (23 states) or guidelines, that is administered by the Drinking Water Program or as part of the State's Plumbing Code (ASDWA, 1999). However, these requirements and the authority to implement them vary considerably in terms of how detailed a water system's program must be, the types of systems (community and/or non-community) required to have a program, and the role the states play in implementing and maintaining a program. Some states rely solely on plumbing codes to address cross connections and backflow, which is problematic because plumbing codes, in most cases, do not require testing and follow-up inspections of backflow prevention devices.

A similar assessment of state cross-connection control requirements by EPA (EPA, 2002b), which is summarized in Table 2-6, demonstrates the variability in state requirements. Based on the EPA review, there are 48 states which have some minimum requirement relating to cross connections in their state administrative code or state law (EPA, 2002b). A number of states do not go beyond these minimum requirements or require public water systems to administer any type of cross-connection control program at the local level. These states tend to rely on community water systems to implement cross-connection control programs. In a few cases, states specify that systems which serve a population of a certain size category must implement a cross-connection control program.

There are five primary elements of an effective cross-connection control program. The first is *authority*; effective cross-connection control programs must have the legal authority to implement program requirements. Legislation must provide the authority to: (1) enter premises and inspect facilities to determine hazards; (2) install, repair, and test backflow devices; (3) license inspectors to test assemblies; and (4) terminate water service in case of non-compliance. According to the American Backflow Prevention Association State Program Survey (ABPA, 1999), 16 of 26 states require utilities to have the authority to implement program requirements. However, on average only 55 percent of systems required to have an enforceable program actually have one in place.

The second requirement is to *inspect facilities and test devices*. It is important to conduct site inspections, and the right of entry enables the inspector to identify where a high hazard might exist. The frequency of inspections and testing is typically based on the degree of hazard. A testing program must identify the appropriate standards that a backflow prevention device must meet, and assemblies must be tested by a certified backflow assembly tester. Many states require in regulation some of the critical components that make up a testing program. For example, 35 of 50 states specified a list of design standards that backflow assemblies must meet, and 34 of 50 states stipulated a testing frequency interval for various backflow assemblies in their regulations (EPA, 2002b). A

TABLE 2-6 State Cross-Connection Control Requirements

Requirement	Number of States with Requirement
Does the state have a requirement for the control of cross-connections and/or backflow prevention?	50
Is it specified in the requirement that the system must implement or develop a cross-connection control and/or backflow prevention program?	32
Does the state require authority to implement a local ordinance or rule for cross-connection control and/or backflow prevention?	33
Must the authority cover testing of backflow prevention assemblies?	27
Must the authority cover the use of only licensed or certified backflow assembly testers?	16
Must the authority cover the entry of the premises for the sake of inspecting the premises?	14
Must the authority cover the entry of the premises for the sake of inspecting and/or installing backflow prevention assemblies?	15
Does the state require training, licensing, or certification of backflow prevention assembly testers?	26
Does the state require training, licensing, or certification of backflow prevention assembly and/or device installers?	6
Does the state require training, licensing, or certification of backflow prevention assembly and/or device repairers?	10
Does the state require training, licensing, or certification of cross-connection control inspectors?	19
Does the state require inspection of backflow prevention devices and/or testing of backflow prevention assemblies?	37
Does the state require the system to include recordkeeping as part of cross-connection control?	34
Does the requirement include keeping records of hazard assessment surveys?	11
Does the state require the system to notify the public following the occurrence of a backflow event?	3
Does the state require the local rule or ordinance to allow the system to take enforcement action against customers who do not comply with the cross-connection control and backflow prevention requirements?	23
Does the state conduct periodic reviews of cross-connection control programs?	3
Does the state regulation or plumbing code require public education regarding cross-connection control and/or backflow prevention?	7

SOURCE: EPA (2002b).

fewer number of states included certification specifications for testers in regulation.

A third issue is *training and certification.* The testing of backflow prevention assemblies by a certified tester is necessary to ensure that the assembly is functioning properly and will prevent backflow. The EPA survey revealed that 26 of 50 states require certification of backflow assembly testers (EPA, 2002b). The states often require the tester to pass a proficiency test and written exam to qualify for certification. A smaller number of states expand their training requirements to program managers, installers, and/or repairers. States rely on plumbers for cross-connection control testers/repairers, survey inspectors, and program managers. Twenty-seven (27) percent of the training was conducted by plumber-affiliated organizations, 15 percent by AWWA-affiliated organizations, 12 percent by state agencies, 6 percent by others, and 40 percent did not specify the source of training.

A fourth important element is *record keeping* following inspections and testing. According to the ABPA survey, 17 of 26 states require record keeping, and 10 of 26 states indicated a requirement for water systems to report backflow incidents to the state. Additional details are found in Table 2-7.

Public education is a final critical element. According to the ABPA survey, five of 26 states required public awareness of backflow potential as an element of their cross-connection control program. Public education is usually a function of the local water purveyor which may educate the public through bill inserts and special mailings. States also maintain internet sites that educate consumers about cross-connection control programs and the role they play in protecting the public's drinking water.

TABLE 2-7 ABPA State Survey Results on Record Keeping Requirements

Record Keeping Requirement	Percent of States	
Number of States that require record keeping	(17 of 26)	65%
Records of inventory of backflow assemblies in service	(14 of 26)	53%
Records of reports of routine testing of assemblies	(16 of 26)	61%
Records of hazard assessment surveys	(9 of 26)	34%
Records of enforcement activities	(8 of 26)	30%
Number of States which require annual reporting to the States	(6 of 26)	23%
Number of States which require reporting of backflow incidents	(10 of 26)	38%

SOURCE: Reprinted, with permission, from The American Backflow Prevention Association (ABPA) State Program Survey (1999). © 1999 by ABPA.

At the current time, there is no unified basis from which cross-connection control programs are designed, adopted, and implemented, which is reflected in the immense variability in programs discussed above. EPA has not adopted national cross-connection control program requirements, although the agency has provided guidance on cross-connection control issues for approximately two decades through its Cross-Connection Control Manual. In 2003 EPA published the third edition (EPA, 2003b), which is designed as a tool for health officials, waterworks personnel, plumbers, and any others involved directly or indirectly in water supply distribution systems. It is intended to be used for educational, administrative, and technical reference in conducting cross-connection control programs. Interestingly, the states that have strong cross-connection control programs are generally not in favor of greater EPA involvement because their programs might be compromised. Those states with programs that are lacking, however, could benefit greatly from EPA directives.

An indirect benefit of a cross-connection control program that has an effective inspection aspect is its ability to identify improper customer account information, missing water meters, unauthorized use of water, and illegal connections. This can result in a reduction in lost water and in the generation of more revenue.

Requirements for Drinking Water Products, Components, and Materials

Because of the potential for drinking water products, components, and materials to add contaminants to drinking water, EPA initiated the development of a Drinking Water Additives third party certification program in 1985. The purpose was to establish standards by which products, components, and materials would be tested to ensure that contaminants of health concern would not introduced into drinking water at levels that imposed a risk to the public. The resulting standards—ANSI/NSF Standard 60 and ANSI/NSF Standard 61—were initially adopted by NSF through a consensus standards development process in October 1988. These standards are designed to test products that are added to drinking water (Standard 60) and products, components, and materials that come into contact with drinking water (Standard 61).

ANSI/NSF Standard 61 is the more relevant standard with regards to water distribution systems. Thirty-six (36) states have adopted ANSI/NSF Standard 61 by either statute or regulation and thus require water systems to use only water distribution system products, components, and materials that are certified pursuant to the standard. Eight additional states have policies (but not requirements) that water systems use products, components, and materials that meet the standard (ASDWA, 2004). Standard 61 applies to all distribution system materials (including pipes, valves, coatings, storage tank materials, etc.) as well as to premise plumbing including home water faucets. These standards can be used by water utilities (along with AWWA industry standards) in the specification of materials they purchase or allow to be installed in their systems.

Plumbing Codes

Plumbing codes are used by states, territories, counties, local governments, and any other form of governance which has a responsibility to protect their constituents' health and safety. Plumbing code requirements do not generally apply to the utility-owned portion of public water systems but rather to residential and non-residential property. Accountability in enforcing the codes primarily resides with the inspection entity, though in many states the licensed plumber and design professionals are also held accountable. Once adopted the codes are used by all sectors of the plumbing industry and public, including inspectors/plan reviewers; contractors/masters; journeymen/apprentices; engineers/architects; material, pipe, and product manufacturers; and certification organizations and test labs. Plumbing codes are usually implemented by the "Authority Having Jurisdiction", which can be a state agency, county commission, or local building department. In some cases plumbing codes are implemented by agencies of the federal government such as the Army Corps of Engineers, Air Force, or the Department of Housing and Urban Development (Chaney, 2005).

The major plumbing codes include the Uniform Plumbing Code (UPC), the International Plumbing Code (IPC), and the Southern Building Code Congress International. As indicated in Table 2-8, by 1999 47 states had adopted plumbing codes, with the UPC, developed and maintained by the International Association of Plumbing and Mechanical Officials (IAPMO), being the most commonly used code (14 states) (EPA, 2002b). More recent information indicates that the various codes were amalgamated by the year 2000 into the three codes that are in use today: the UPC, the IPC, developed and maintained by the International Code Council (ICC), and the National Standard Plumbing Code

TABLE 2-8 Plumbing Codes Adopted by the States by 1999

Plumbing Code	Number of States Adopting
Statewide Code	47
No Statewide Code	3
Statewide Codes Adopted	
Uniform Plumbing Code	14
State Code	7
International Plumbing Code	5
National Standard Plumbing Code	4
Southern Building Code Congress International	4
Other	13

SOURCE: EPA (2002b).

(NSPC). NSPC, published by the Plumbing, Heating, and Cooling Contractors National Association, is adopted in New Jersey and some counties of Maryland but is otherwise not used widely. The UPC has now been adopted in approximately 28 states (Chaney, 2005).

The UPC and IPC have different contents and permit different materials and devices. The UPC, for instance, allows for some piping material that is not permitted under the IPC. The IPC permits air admittance valves not permitted in the UPC. Some venting configurations are permitted in one code and not the other. Both the UPC and the IPC include important cross-connection control requirements intended to prevent contamination of the domestic water supply that is internal to the property as well as to the drinking water delivered by the public water system. Both codes also establish minimum requirements for the separation of water and sewer lines as well as requirements for the disinfection of new or repaired potable water systems. Both codes, however, have certain shortcomings. For examples, the UPC does not prohibit the installation of water service or water distribution pipe in soil contaminated with solvents, fuels, organic compounds, or other detrimental material which could cause permeation, corrosion, degradation, or structural failure of the piping material. The UPC does not require that water service and distribution pipe and fittings conform to ANSI/NSF Standard 61, which is intended to prevent the use of materials that will leach contaminants into drinking water at levels that may constitute a health risk. The IPC requires that all cross-connection control devices be inspected annually including devices that cannot be tested and air gaps, while the UPC only requires inspection of testable devices. Inspection of all devices is preferable to ensure that tampering has not occurred. Both the IPC and UPC have established minimum distances between water supply wells and sewage disposal systems. The distances established by the IPC are less conservative and may not provide adequate protection from potential contamination. A comparison of the two codes with regard to the principal requirements within the codes that address water distribution system integrity is contained in Table 2-9.

The major difference between the UPC and IPC is the procedural process by which the codes are maintained. IAPMO uses an American National Standards Institute (ANSI) consensus development process for the UPC, while the ICC uses a government or inspector only process for the IPC. The ICC predominantly consists of building inspectors from three organizations (Building Officials and Code Administrators, Southern Building Code Congress International, and International Conference of Building Officials) that have been widely involved in developing structural and fire codes for years. The ANSI consensus code development and maintenance process used by IAPMO is open to all interested parties, it is balanced to prevent any one sector of the industry from dominating, and it provides for due process (participants have appeal rights to ANSI) (Chaney, 2005). Given the disparities between the codes, and the possible resulting confusion, efforts are underway to combine the UPC and the IPC into a single model code (IAPMO, 2005).

TABLE 2-9 Comparison of UPC and IPC: Requirements for the Protection of Water System Distribution Systems

Element	UPC	IPC
Code Maintenance	ANSI Consensus Process	Inspectors from Specific Process Organizations
Cross-Connection Control Requirements		
Devices		Similar device requirements for degree of hazard, but IPC more detailed regarding type of device and application
Minimum Required Air Gaps		Same requirements except for ¾ inch openings affected by side wall where IPC more restrictive
Protection from Lawn Irrigation Systems	Similar requirements but UPC provides more detail	
Protection from Fire Sprinkler Systems		Similar requirements but IPC is more specific as to requirements for systems not under constant pressure
Inspections and Testing	UPC requires inspections of testable devices only	IPC requires inspection of testable and non-testable devices and air gaps
Additional Distribution System Requirements		
Separation of Water and Sewer Lines	Requires minimum 12 inch vertical separation	Require minimum 12-inch vertical separation but IPC is more restrictive on horizontal clearance where vertical clearance is less than 12 inches
Disinfection of New or Repaired Water Pipe	Flushing with potable water; 50 parts per million (ppm) of chlorine solution/24 hours or 200 ppm for 3 hours; flush to purge chlorine; bacteriological analysis	Flushing with potable water; 50 parts per million (ppm) of chlorine solution/24 hours or 200 ppm for 3 hours; flush to purge chlorine; bacteriological analysis
Identification of Potable and Nonpotable Water Systems	UPC requires color coding of each system	IPC requires color coding or metal tags
Pipe Materials	UPC does not require pipe material to meet ANSI/NSF 61	IPC requires pipe material meet ANSI/NSF Standard 61
Pipe Placement	UPC does not address	IPC prohibits placement of water pipe in soils contaminated with contaminants that could adversely affect the pipe
Water Supply Protection Requirements		
Water Supply Well Protection	UPC requires 50 feet between water supply wells and sewage disposal systems such as septic tanks and 100 feet between water supply wells and disposal fields	IPC requires 25 feet between water supply wells and sewage disposal systems such as septic tanks and 50 feet between water supply wells and disposal fields

Note: Where certain entries are blank, the two codes are similar and the small difference is mentioned for only one of the codes. SOURCES: IPC (2003); UPC (2003); Chaney (2005).

In the United States, plumbing codes are adopted in one of two ways: (1) through statutory adoption which usually occurs through the enactment of legislation or (2) through regulatory adoption which occurs upon the implementation of regulations or procedures. At the state level, codes are usually adopted through a public hearing process that allows interested parties to present testimony (Chaney, 2005).

Although states will adopt the UPC or IPC as their base plumbing code, they may amend the code to address specific issues. In addition, plumbing codes may also be adopted at the local county and municipal level that are at least as stringent as the state plumbing code. For example, in Iowa, the state adopted the UPC as the plumbing code but then amended the UPC to add additional backflow prevention provisions including a requirement that cities with populations of 15,000 or greater enact a backflow prevention program with containment by January 1, 1996. Although local jurisdictions in Iowa must adhere to the provisions of the state plumbing code, these jurisdictions may adopt local ordinances or rules and regulations that provide for higher but not lower standards than those found in the state plumbing code (State of Iowa, 2005). As examples, the City of Des Moines, and Linn County, Iowa have adopted the UPC with some modifications. In the case of Linn County the modifications require the examination, qualification, and licensing of plumbing contractors, plumbers, and the registration of apprentice plumbers (Linn County, 2004). In addition, homeowners are prohibited from carrying out plumbing work on their residence unless they pass the County's homeowners examination.

LIMITATIONS OF REGULATORY PROGRAMS

Existing federal regulations such as the TCR, SWTR, LCR, LT1ESWTR, and the Stage 1 and Stage 2 D/DBP Rules are intended to address only certain aspects of distribution system water quality and are not designed to address the integrity of the distribution system in its totality. Of these regulations, only the TCR may provide some indication of potential problems with distribution system integrity related to microbial contamination. However, the TCR has significant limitations that affect its use as an indicator of distribution system integrity.

TCR sampling requirements are based on water system size and as a result vary widely, from as many as hundreds of samples per month to one sample per month. Each water system is required to develop a sample siting plan that is approved by the state regulatory agency. For larger water systems even a sample siting plan that results in hundreds of samples per month may not adequately cover the myriad of potential points where contamination could occur, such as storage tanks, premise plumbing, and service connections. For smaller systems the sampling is so infrequent that contamination would be easily missed. Although most reported outbreaks associated with distribution systems have occurred in community water systems because of their greater size and complexity, there have been a number of outbreaks associated with noncommunity water systems that have been attributed to deficiencies in the distribution system. In addition to the problems associated with sample locations and the frequency of sampling, TCR monitoring does not provide real-time information. There are inherent delays between sampling and reporting of coliform results that do not allow for sufficient time to recognize a contamination event and to prevent public exposure and disease transmission. (It generally takes about 24 hours to obtain results from the time of sample collection to the completion of coliform analysis using presently available analytical methods.)

The TCR encompasses only microbiological indicators. With the exception of monitoring for disinfectant residuals and DBPs within the distribution system and lead and copper at the customer's tap, existing federal regulations do not address other chemical contaminants within the distribution system. Yet there have been a number of examples of waterborne outbreaks associated with chemical contamination (chlordane, ethylene glycol) of the distribution system as a result of cross connections, contamination of water mains during construction, and contamination of storage facilities (Craun and Calderon., 2001; Blackburn et al., 2004).

Some federal regulations are inherently contradictory to one another, as they relate to distribution integrity and maintenance of water quality, such that water suppliers have found it difficult to be in compliance with both simultaneously. For example, the SWTR and TCR recommend the use of chlorine to minimize risk from microbiological contamination. However, chlorine or other disinfectants interact with naturally occurring organic matter in treated water to form DBPs. As a result many water systems have changed disinfectants (generally from chlorine to chloramine) in order to be in compliance with the MCLs for DBPs in the distribution system. The increased reliance on chloramine can be problematic if close attention is not paid to controlling nitrifying bacteria in the distribution system. Biological nitrification can result in the loss of chloramine residual, which may then present a health threat to the consumer (as discussed in Appendix A). Simultaneous compliance with the D/DBPR and the LCR can also create problems for the maintenance of distribution integrity and water quality. Raising the pH of treated water will assist in controlling corrosion (and hence reduce lead concentrations) but may increase the formation of THMs.

In areas where federal regulations are weak, state regulations and local ordinance contribute to public safety from drinking water contamination. States have adopted requirements that address certain aspects of distribution system integrity. All states appear to have provisions for the control of cross connections and/or backflow prevention, although there is considerable variation in how they are implemented and by whom. The majority of states have established regulations within their drinking water programs requiring cross-connection control programs to be implemented by water systems or local authorities, while some have adopted plumbing codes that included the requirements and others have established only guidelines for cross-connection control programs (ASDWA, 2003; EPA, 2002b). In general, very few states provide dedicated resources for implementing a cross-connection control program but rather incorporate the program activities into the overall public water system supervision program. At best, most states attempt to assess that a water system has an effective cross-connection control program when carrying out a sanitary survey of the water system. However, because sanitary surveys may occur only once every several years, it is difficult to ascertain the level of compliance. A few states track the number of cross-connection control devices that are annually installed and tested while others determine programs effectiveness by the number of backflow incidents reported (ASDWA, 1999).

Although most states have also established requirements for the design, construction, operation, and maintenance of distribution systems, as discussed previously these requirements vary significantly and some states only encourage certain contamination prevention activities while others do not address them at all. For example, some states only encourage the separation of water mains and sanitary sewers to protect the water main from contamination or the disinfection, flushing, testing, and other follow-up actions before returning a water main to service after repairs. Even where states have established extensive requirements, the onus for ensuring implementation is placed on the water system. States do not dedicate resources to routinely oversee that implementation occurs.

Local regulatory programs are implemented through the plumbing code. Because local plumbing codes must be consistent with the provisions of the state plumbing codes, local regulatory programs should have the authorities to address certain distribution system integrity issues including cross-connection control, use of appropriate pipe and other plumbing materials, and separation of water and sewer lines. However, program implementation can vary from one local jurisdiction to another. For example, licensing of plumbing contractors and plumbers is normally part of the local jurisdictions regulatory program. Neither of the two prominent plumbing codes—the UPC and the IPC—address licensing requirements, and there is no national system for licensing of plumbers or plumbing inspectors. There also appears to be no uniformity regarding the training and licensing of personnel who install, maintain, and inspect backflow prevention devices. Yet there are numerous organizations such as AWWA, New England Water Works Association, American Society of Safety Engineers, American Backflow Prevention Association, Backflow Prevention Institute, University of Southern California Foundation for Cross-Connection Control and Hydraulic Research, and IAPMO that offer personnel certifications that address competency.

There also is a significant difference between the approach taken by state drinking water regulatory programs and water systems to ensure high water quality within premises, particularly residential dwellings, versus utility-owned portions of the distribution system. Plumbing codes (UPC and IPC) address requirements for the installation of plumbing fixtures, appurtenances, and backflow prevention devices within premise plumbing where necessary such as to prevent contamination of the public water system (UPC, 2003; IPC, 2003). However, there are no provisions for ongoing inspections or surveillance to ensure that modifications to the premise plumbing by the homeowner will not adversely affect the quality of the drinking water, either within the premise or within the water distribution system. Plumbing codes (UPC and IPC) have also never addressed ongoing water quality within the premise. Provisions for periodic premise inspections to check for cross contamination, to ensure that the integrity of the system is being maintained, and to assess premise water quality could be required by local ordinances, but funding mechanisms would have to be created (Chaney, 2005).

Finally, there is no incentive for homeowners to keep their premise plumbing in compliance with codes. Houses are built to code but many fall out of compliance due to age and as the code changes. In addition there are no organizations that advise homeowners on how to maintain their plumbing systems such as when flushing is necessary, water temperature recommendations, home treatment devices, etc. (Chaney, 2005). A further discussion of issues associated with premise plumbing and possible solutions can be found in Chapter 8.

VOLUNTARY AND NON-REGULATORY PROGRAMS THAT INFLUENCE DISTRIBUTION SYSTEM INTEGRITY

Voluntary and non-regulatory programs exist that are designed to provide public water systems with approaches for maintaining and improving distribution system integrity. There are several objectives of these non-regulatory water quality improvement programs for water supplies, foremost among them being to further protect public health and to engage in risk management efforts beyond what is provided by federal, state, and local regulations and the enforcement system developed for primacy agencies. A related motivation for a utility to implement such programs is to help organize their many activities—i.e., to have a unifying umbrella that encompasses all of the piecemeal requirements of the federal, state, and local regulations. A second important objective of these programs is to increase customer satisfaction, which is based largely on a perception of the quality of service and the cost and quality of the delivered product. One common theme among these programs is their intent to assist utilities in identifying best practices and then affirm that the utility is employing these practices. Examples of best practices include continuing or expanding monitoring of water quality and setting up water quality goals, engagement in plant optimization projects, studies on applicability of emerging technologies, and proactive preparation for upcoming regulations—activities that, along with routine operation, compliance monitoring, and maintenance, are often collectively described in a utility's distribution system management plan (if one exists). Voluntary and non-regulatory programs can also help utilities to improve efficiency, as manifested in responsiveness and cost. Performing services at a low cost is desirable but customers and others require a high level of service. A balance must be achieved to satisfy the expectations of regulators, customers, and owners at a reasonable cost.

Voluntary programs are attractive because although public water systems recognize the need for health and environmental regulations to protect the public, utilities (particularly larger ones) seek the flexibility to undertake activities that will achieve these goals within the broader existing regulatory framework while reducing the need for intensive regulatory oversight. Programs such as voluntary accreditation are being designed that will allow water systems to implement industry best practices that go beyond regulatory requirements to produce a drinking water quality that exceeds the minimum established by law.

Given the need to improve public confidence in drinking water quality, water systems can use the recognition that they receive from implementing these voluntary programs to promote these efforts to their customers. In particular, water systems can communicate how they are achieving their water quality goals along with an increased level of service without the need for a significant increase in cost to their customers. Water systems are also able to demonstrate that the product that they are providing not only exceeds regulatory requirements but competes equally with other sources such as bottled water, vended water, and home treatment devices, at far less cost.

A few select voluntary, non-regulatory programs are described below, including accreditation, Hazard Analysis and Critical Control Points (HACCP) Plans, and Water Safety Plans, that can serve as guides to water utilities that want to improve their distribution system management. Note that the Partnership for Safe Water and QualServe, two voluntary AWWA programs that target drinking water quality, are not discussed because distribution systems are not their primary focus. QualServe uses self-assessment and peer-review methods to identify opportunities for improvement in water and wastewater utility services, while the Partnership for Safe Water focuses on water treatment plant optimization.

Accreditation Standards

Currently, there is no nationwide system that accredits water utilities. However, a voluntary, nationwide accreditation program for all water utilities, including small utilities, is currently under development by AWWA. The basis of the program is to verify the application of standards and best practices that will ensure the delivery of high quality services, exceeding regulatory compliance. The program will be carried out by independent auditors who will verify conformation with the accreditation standards on-site. The goals of the program are not only to improve customer satisfaction, but also to provide a tool for regulatory agencies to use in evaluation of water utilities and to encourage utilities to evolve beyond seeking compliance with existing regulations to seeking the best strategies to protect public health.

The accreditation standards developed so far are water treatment plant operation and management (G100), distribution system operation and management (G200), and source water management and protection (G300). (After piloting the implementation of these standards at both large and small systems in August 2005, other areas of accreditation standards will be developed.) The Distribution System Operation and Management Standards (G200) (AWWA/ANSI, 2004), published in May 2004, are intended to improve distribution systems' water quality and utility's management efficiency by voluntarily adhering to standards that exceed current regulatory requirements and by performing independent audits to verify performance. The standards call for development of water quality sampling plans at prescribed sites in distribution systems. Nitrification control; booster chlorination; internal corrosion monitoring and control; reduction of the formation of DBPs; and color, taste, and odor monitoring and control are defined as programs that should have individual goals and action plans established specifically for each utility. Distribution system management activities listed in the standard include system pressure monitoring, backflow prevention, permeation prevention, water loss minimization, valve exercising and replacement, fire hydrant maintenance and testing, maintenance of coatings and linings, water use metering, external corrosion control, water quality monitoring, and energy management. The verification step of the standard includes providing certain required documents and records. For those utilities that decide to develop a distribution system management plan that meets the AWWA G200 standard, conformance would be verified on a periodic basis. Because G200 provides a comprehensive framework in which a water utility can manage distribution system integrity and it targets those activities felt by the committee to be of highest priority in reducing public health risks, it is further discussed in Chapter 7.

Hazard Analysis and Critical Control Points

Voluntary programs that deal with water quality and management issues from the perspective of risk evaluation and reduction are being adapted to drinking water treatment, operations, and distribution from other branches of the industry. An example is the Hazard Analysis and Critical Control Points (HACCP) program, which was developed by NASA in the 1960s for the U.S. space program, later transferred to food safety, and recently formatted for drinking water quality. The program relies on three steps, which are addressed continuously in a cycle: hazard identification, remediation, and verification. HACCP for the drinking water industry is based around the same seven principles as were developed for NASA and other industries (NASA, 1991; Codex Alimentarius Commission, 1993, 1997; Mucklow, 1997). The HACCP principles are to:

- Identify hazards and control measures
- Identify critical control points
- Establish critical limits
- Identify monitoring procedures

- Establish corrective action procedures
- Verify and validate the HACCP Plan
- Establish record keeping and responsibility

HACCP is a risk management program because utilities use it to first identify and evaluate hazards/risks, and then to establish control systems to minimize the occurrence and effects of incidents that may impact the safety and quality of the water. A water utility can choose to apply HACCP to any one "process"—i.e., watershed protection, treatment, or the distribution system. Some utilities may already have good watershed protection programs and good control over treatment facilities, and so may view the distribution system as a priority. However, because HACCP is a proactive approach to system management that helps the utility to identify "hazards" further upstream, it works quite well as a comprehensive system plan, from source to tap. For maximum benefits, it is important to leave the decision to individual utilities and not be too prescriptive about how to apply HACCP (Friedman et al., 2005).

A recently completed project sponsored by the AWWA Research Foundation (Friedman et al., 2005) describes HACCP pilot studies conducted with three utilities' distribution systems—Greater Cincinnati Water Works, Cincinnati, Ohio; Calgary Water Works, Calgary, Alberta; and the City of Everett, Everett, Washington. Training workshops were held at each utility location to explain HACCP terminology and to initiate development of the utility's HACCP plan. Each participating utility formed a HACCP team to further develop the HACCP plan and to guide its implementation. The goal was for each utility to implement their HACCP plan over a 12-month period during which certain operational and water quality parameters would be monitored. The participating utilities found that the implementation of HACCP to water supply distribution was feasible and practical, but that the time and resource requirements were greater than originally anticipated. The development of the HACCP plan was useful in honing in on the most important risks and process controls for water quality management. Within the 12-month pilot study period, none of the three participating utilities developed a fully implemented HACCP program for certification. A longer period of time and/or a greater resource commitment was likely to be required before the HACCP systems would be considered fully implemented, complete, and certifiable. Box 2-2 describes two other HACCP case studies in detail, for Austin, Texas, and Burwick, Maine.

NSF International provides HACCP certification to water utilities in the United States through its HACCP-9000 registration program. The program consists of third-party verification of utility HACCP plans, combined with a registration with ISO 9000 standards. However, adoption of the HACCP approach need not be tied formally to such administrative programs. HACCP could be an integral part of a utility's distribution system management plan, either in addition to or in lieu of G200 (given the substantial similarities between the two programs). In particular, HACCP is useful for improving a utility's awareness of its existing databases and how it can better manage the information contained within, and for promoting record keeping and reporting. Critics contend that HACCP is little more than properly operating a distribution system. Indeed, there may be little value added in the United States where utilities are relatively heavily regulated compared to other countries where HACCP has been successfully adopted (such as Australia, which has no national water quality standards). However, advocates contend that the part of HACCP that most utilities do not already engage in is checking to verify that actions are working (Martel, 2005). Furthermore, HACCP puts an increased focus on operator training, which can be ignored in the face of so many other competing activities, like compliance monitoring. The program is more likely to be adopted by larger-size utilities because of the need for a larger staff and budget to carry out HACCP.

Nonetheless, there is another practical consideration that makes G200 a more attractive organizing program for distribution systems than HACCP. Programs like HACCP are ideally suited to industries that experience little variation on a day-to-day basis (such as food and beverage processing plants) and are not as easily adapted to the dynamic nature of drinking water distribution systems that may experience changes in water quality depending on season, source of supply, and changing daily demands. Furthermore, unplanned disruptions such as water main breaks require immediate responses in areas that may not be considered critical control points, making it very difficult to proactively control contamination events. Finally, the vast number of locations within a distribution system that could be potential critical control points (presumably every residence where a cross connection exists) argues against the formal adoption of HACCP.

The cost of creating a HACCP plan for a community of 10,000 may be in the range of $10,000, including a day- or two-day-long workshop.

Water Safety Plans

In 1994, the World Health Organization (WHO) adapted the HACCP program through Water Safety Plans, which can be prepared for individual water systems. The WHO's Guidelines for Drinking Water Quality (2004) describe an approach to follow in preparing Water Safety Plans. The approach is to identify, prioritize, and prevent risks arising from hazards associated with distribution of drinking water. The three critical components of a water safety plan are:

- System assessment regarding both the quantity and quality of supplied water
- Identification of control measures
- Management plans describing actions during both normal and extreme conditions and documenting, monitoring, communication, and improvement efforts.

BOX 2-2
HACCP Case Studies

There are few case studies of where HACCP has been applied to distribution system management. One involves a relatively small utility, the South Berwick Water District, in South Berwick, Maine, which serves about 4,000 people. At this utility, a HACCP training workshop was held on June 2003 to assemble the HACCP team, which included the superintendent, foreman, and a service person, as well as outside experts such as an engineer familiar with the South Berwick system, a microbiologist from EPA, a state regulator who was an expert on cross-connection control, and a risk manager from the bottled water industry. As in other cases where HACCP has been applied, assembling a team that has as many people from different cross sections of the water utility as possible is one of the benefits of doing HACCP, but because of the small size of the utility this required outside assistance. The process flow diagram for the entire water system is shown in Figure 2-1.

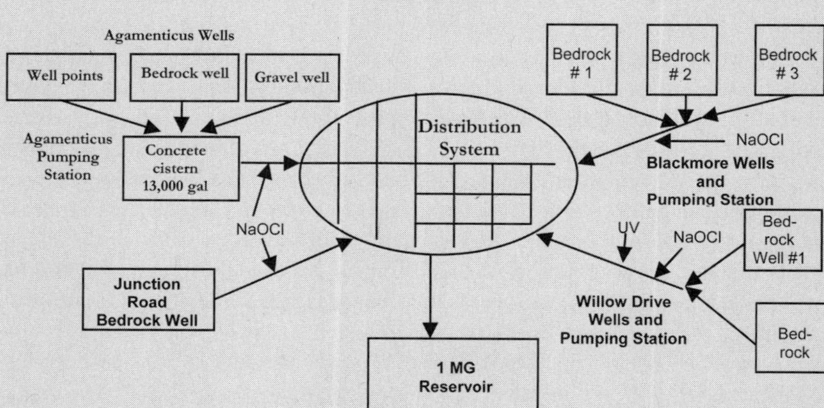

FIGURE 2-1 Process Flow Diagram for the South Berwick Water District. SOURCE: Reprinted, with permission, by Martel (2005).© 2006 by AwwaRF.

Three priority hazards were identified by the HACCP team, two of which involve the distribution system: (1) backflow through unprotected cross connections, (2) long dead-end mains with zero or poor disinfectant residual, and (3) unintentional contamination of shallow well points at the Agamenticus Wellfield. It should be noted that it was very difficult to gather enough information to determine the frequency of occurrence or the severity of these hazards, given the utility's lack of data. For this reason, South Berwick's initial HACCP plan focused on monitoring activities to further characterize these hazards and improve existing control measures. Unfortunately, the HACCP plan was not fully implemented because of a lack of manpower and because of other priorities. With only three full-time employees at the utility, daily system operation and maintenance took priority over HACCP plan implementation. Furthermore, the utility personnel were involved with building a new treatment facility, developing a new rate structure, and addressing local and state political issues. This case study illustrates the need for sufficient manpower to successfully implement a HACCP Plan.

A second case study is from Austin, Texas, a much larger water supply that serves approximately 770,000 people. The interdisciplinary HACCP team consisted primarily of in-house staff: the water quality manager, the water laboratory supervisor, an engineer

continues

/planner, a construction inspector, the cross-connection control supervisor, the Assistant Director of Treatment, the Infrastructure Superintendent, and a state regulator. A HACCP pilot study was conducted from May 2003 to September 2004. The team focused on one pressure zone within the distribution system for the HACCP pilot study (see the flow process diagram below in Figure 2-2):

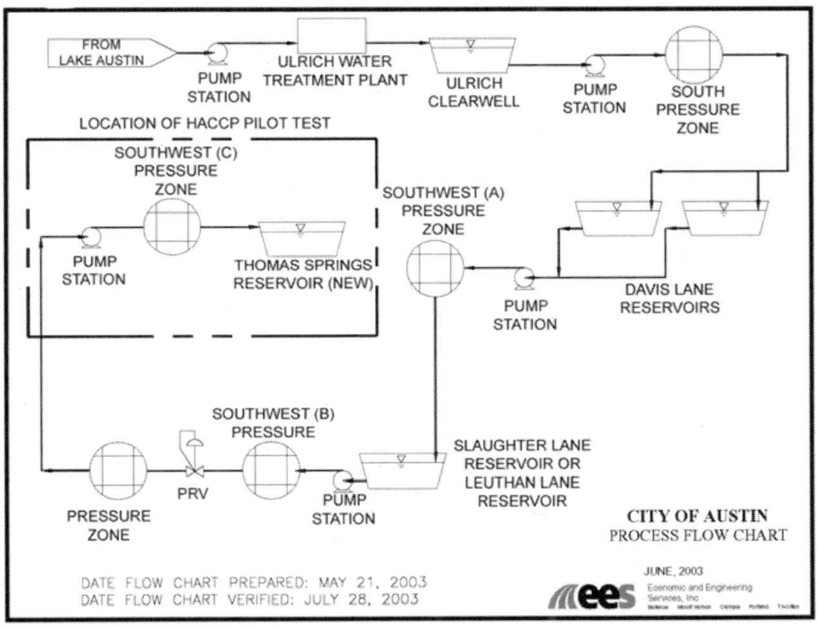

FIGURE 2-2 Flow Process Diagram for the Austin Water Supply. SOURCE: Reprinted, with permission, from Martel (2005).© 2006 by AwwaRF.

Austin's HACCP team identified two high priority hazards: backflow through unprotected cross connections (focusing specifically on irrigation and hydrant vandalism) and contamination from new construction sites (primarily via improper valve turning). Austin found that HACCP is more complex than initially envisioned. Originally, the utility thought that HACCP would involve identifying critical flow paths within the distribution system and monitoring these flow paths more intensively to assure water quality to downstream sites. Instead, by nature of the selected hazards, the measures used to control these hazards focused on operations and maintenance activities rather than water quality monitoring. This approach added layers of complexity to the existing monitoring program. On a positive note, the HACCP approach helped the utility (1) improve understanding of their distribution system hazards; (2) heighten employee awareness of pressure zone boundaries, pressure transients, the need to maintain pressure and to respond quickly to main breaks in small pressure zones; (3) improve awareness of existing databases and monitoring programs; (4) improve data management skills; (5) identify needed improvements to existing databases; and (6) improve reporting procedures for acceptance of new mains.

SOURCE: Martel et al. (2006).

Water safety plans present an affordable risk management tool for all drinking water suppliers, regardless of size. While some critical elements of the plan should be assured by all systems, more costly or time-consuming elements, characterized as not critical, may be added to the plans based on budgetary and staff availability. The most critical elements of the water safety plan documents include system description, water flow diagrams, hazard identification, identification of a team, and contingency plan. Additional items include specification of chemicals and materials, job descriptions for staff responsible for individual operations, corrective actions for deviations, record-keeping procedures, validation data, and incident documentation procedures. Finally, optional elements may include manuals for hygiene, preventive maintenance, and equipment calibration; job descriptions for all staff; training programs and records; documentation of corrective actions, audits, and verification procedures; and consumer complaint policy and procedures.

Clearly, the elements of a Water Safety Plan closely resemble the elements of a HACCP Plan: (1) source-to-tap system assessment; (2) control measures for identified hazards and operational monitoring of control measures; and (3) a management plan that documents the system assessment, control measures, monitoring plan, corrective action procedures to address water quality incidents, communication plan, and supporting programs such as standard operating procedures, employee training, and risk communication. Both HACCP and Water Safety Plans should be used continuously.

A 2004 conference sponsored by NSF International examined a variety of risk management approaches, including HACCP, ISO certification, Water Safety Plans, and Environmental Management Systems. Not only were many commonalities among these programs evident, the distinctions between them were unclear. The conference presented a number of domestic and international case studies where water utilities had utilized one of these risk management systems, but no case studies targeting the distribution system were discussed. Indeed, the choice of the "right" program for any given water utility may present a challenge, specifically because there is no precedence for using these programs for distribution system management, but also because of a lack of coordination between the programs, a lack of tangible benefits beyond what a utility already accomplishes, and inefficient communication to the public about the programs. It is up to an ambitious utility manager and staff to learn about the programs, evaluate their applicability, and select one.

Training for Operators, Inspectors, and Related Personnel

While utilities endeavor to optimize their infrastructure and operate the distribution system to minimize degradation, an integral component not to be ignored are the operators, inspectors, and related personnel charged with running and monitoring the system. Inevitably, the operators and field personnel serve as guardians to minimize degradation in the distribution system and ensure water quality is maintained for the consumer.

Training of distribution system operators was identified as a high priority issue for reducing risk in drinking water distribution systems (NRC, 2005). The need for the continuing and intensive training of operators of distribution systems has increased recently for three reasons. First, as federal and state regulations become increasingly stringent and more complex, they require enhanced skills for proper sample collection and preservation, as well as better understanding of aquatic chemistry and biology for proper implementation and interpretation of results. Second, in many systems the D/DBPR (EPA, 1998a) created a shift in the use of disinfectants in the distribution systems from a relatively simple application of chlorine to the rather complicated application and maintenance of chloramine. Finally, with an increase in the importance of security of drinking water pipes, pumps, reservoirs, and hydrants, there is a corresponding increase in the responsibility of operators to make decisions during perceived security events.

Typically distribution system operators, mechanics, and field crews are well trained in the mechanical aspects of water delivery (such as pipe replacement and repair; pump, valve, and storage facility operation; etc.) and safety. In cases where contractors are used to repair or maintain the infrastructure (for example, many utilities allow certified plumbers to perform the tasks related to backflow prevention and cross-connection control), diligence of construction inspectors in providing oversight is of paramount importance because the contractor may or may not be following standard practices. A case in point regarding the importance of training plumbers is the ban on lead solder implemented in the late 1980s. Because the responsibility for high lead levels in drinking water falls on the utility, many utilities were actively engaged in training plumbers about the dangers of lead from the use of lead solder and about the new requirements of the LCR. This training was critical to reducing the risk of lead exposure from drinking water.

The importance of operator training in protecting public health from contaminated drinking water cannot be overstated. A recent critique of the Walkerton, Ontario Inquiry Report (Hrudey and Walker, 2005) claims that lives could have been saved had operators been properly trained. Failure to perform basic monitoring duties and understand the vulnerability of the system to a contamination event in May 2000 led to more than 2,300 cases of waterborne disease in a system of only 5,000 people. "Water system operators must be able to recognize that the threats to their system contrasted with the system's capability to cope. They have a professional responsibility to ensure deficiencies are identified, made known to management, and effectively remedied. Pending necessary improvements, operators must increase their vigilance and develop contingency plans to cope with periods of stress. Contingency plans should be practiced using simulated incidents before a real crisis develops" states Hrudey. Justice O'Connor who led the multi-million dollar inquiry into the Walkerton tragedy concluded that "Ultimately, the safety of drinking water is protected by effective management systems and operating practices, run by skilled and well-trained staff' (Hrudey and Walker, 2005).

Operator training classes and seminars are offered through industry associations (e.g., AWWA, the National Rural Water Association) and third party contractors. The International Association for Continuing Education and Training (IACET) has recently developed certification for trainers, which is a positive step toward ensuring the quality of instructors who are providing operator training. However, it is well recognized that nationally there is a paucity of adequate training facilities, instructors, and apprentice programs to replace an experienced workforce who will be retiring in the coming decade (Brun, 2006; Eaton, 2006; McCain and Fahrenbruch, 2006; Pomerance and Means, 2006).

As discussed earlier, there are existing EPA guidelines for the certification of treatment plant operators and distribution system operators (EPA, 1999), which have subsequently been implemented by states (leading to state requirements for certification). However, these requirements are not always enforced, particularly on small systems. Stronger enforcement of the distribution system operator certification requirements developed by individual states could be a mechanism to support training and apprentice programs. Also, future regulations need to include mechanisms to fund training and apprentice programs specifically for distribution system operators. Finally, while existing certification exams test generic knowledge, future requirements should ensure that operators understand the system in which they work and are familiar with portions of operating plans that apply to performance of their daily activities.

CONCLUSIONS AND RECOMMENDATIONS

The Total Coliform Rule, the Surface Water Treatment Rule, the Disinfectants/ Disinfection By-Products Rule, and the Lead and Copper Rule are the federal regulations that address water quality within the distribution system, and they do so in a piecemeal fashion. These rules were not intended to address distribution system integrity as defined in Chapter 1, which consists of physical, hydraulic, and water quality integrity. For example, the TCR considers only that microbial contamination indicated by fecal parameters. Nor does the SDWA contemplate federal actions that would address premise plumbing, with the exception of lead in plumbing materials. As a result a more comprehensive approach needs to be taken to ensure that the overall integrity of distribution systems is maintained. The following regulatory recommendations are made.

EPA should work closely with representatives from states, water systems, and local jurisdictions to establish the elements that constitute an acceptable cross-connection control program. Although states, either through drinking water regulations or state plumbing codes, have cross-connection control requirements in place, these requirements are inconsistent amongst states. State oversight of cross-connection control programs varies and is subject to availability of resources. If states expect to maintain primacy over their drinking water programs, they should adopt a cross-connection control program that includes a process for hazard assessment, the selection of appropriate backflow devices, certification and training of backflow device installers, and certification and training of backflow device inspectors. Although tracking compliance by water systems is also an important element, the resource implications of tracking and reporting requirements should be carefully considered. EPA may need to allow use of federal funds for training of backflow prevention device inspectors for small water systems.

Existing plumbing codes should be consolidated into one uniform national code. Although similar with regard to cross-connection control requirements and other premise plumbing protection measures, the two principal plumbing codes that are used nationally, the UPC and the IPC, have different contents and permit different materials and devices. These differences appear to be addressable, recognizing that the two code developing organizations may have other issues that would need to be resolved. In addition to integrating the codes, efforts should be made to ensure more uniform implementation of the plumbing codes. Their implementation can vary significantly between jurisdictions, which can have major impacts on the degree of public health protection afforded to their constituents.

For utilities that desire to operate beyond regulatory requirements, adoption of G200 or an equivalent program is recommended to help utilities develop distribution system management plans. G200 has advantages over other voluntary programs, such as HACCP, in that it is more easily adapted to the dynamic nature of drinking water distribution systems.

More attention should be paid to having adequate facilities, instructors, and apprentice programs to train utility operators, inspectors, foremen, and managers. The need for the continuing and intensive training of operators of distribution systems has increased as a result of more sophisticated federal and state regulations, the shift in the use of disinfectants in the distribution system, and the increase in importance of security of drinking water distribution systems. Recent development of IACET certification for trainers is a positive step toward the quality of instructors providing operator training. Future regulations need to include mechanisms to fund training and apprentice programs.

REFERENCES

American Backflow Prevention Association (ABPA). 1999. American Backflow Prevention Association State Program Survey. Available on-line at: *http://www.abpa.org/originalsite/ABPA_Survey_Report.pdf.* Accessed May 4, 2006.

AWWA/ANSI. 2004. G-200 Distribution Systems Operation and Management. Denver, CO: AWWA.

Association of State Drinking Water Administrators (ASDWA). 1999. Survey of State Cross-Connection Control Programs. September 29, 1999. Washington, DC: ASDWA.

ASDWA. 2003. Summary of Results from the ASDWA Distribution System & TCR Survey, Design and Construct & Operation and Maintenance. Washington, DC: ASDWA.

ASDWA. 2004. Survey of State Adoption of ANSI/NSF Standards 60 and 61. Washington, DC: ASDWA.

Blackburn, B. G., G. F. Craun, J. S. Yoder, V. Hill, R. L. Calderon, N. Chen, S. H. Lee, D. A. Levy, and M. J. Beach. 2004. Surveillance for waterborne-disease outbreaks associated with drinking water—United States, 2001–2002. MMWR 53(SS-8):23–45.

Brun, P. 2006. Is it workforce planning or succession planning? Source 20:6.

Chaney, R. 2005. The Uniform Plumbing Code: development, maintenance and administration as a pathway to reducing risk. April 18, 2005. Presented to the NRC Committee on Public Water Supply Distribution Systems. Washington, DC.

Codex Alimentarius Commission. 1993. Guidelines for the Application of the Hazard Analysis Critical Control Point (HACCP) System, CAC/GL 18-1993. Rome, Italy: Codex Alimentarius Commission and the FAO/WHO Food Standards Program, Food and Agriculture Organization of the United Nations and World Health Organization.

Codex Alimentarius Commission. 1997. Guidelines for the Application of the HACCP System. Rome, Italy: Codex Alimentarius Commission and the FAO/WHO Food Standards Program, Food and Agriculture Organization of the United Nations and World Health Organization.

Craun, G. F., and R. L. Calderon. 2001. Waterborne disease outbreaks caused by distribution system deficiencies. J. Amer. Water Works Assoc. 93(9):64–75.

Eaton, G. 2006. San Diego County Water Authority prepares for the future. Source 20:14–15.

Environmental Protection Agency (EPA). 1979. National Interim Primary Drinking Water Regulations for the Control of Trihalomethanes in Drinking Water, Final Rule. Federal Register 44:68641.

EPA. 1989. National Primary Drinking Water Regulations: Filtration, Disinfection, Turbidity, *Giardia lamblia*, Viruses, *Legionella*, and Heterotrophic Bacteria; Final Rule (SWTR). Federal Register 54:27486.

EPA. 1991. National Primary Drinking Water Regulation: Lead and Copper Rule, Final Rule. Federal Register 56:26460.

EPA. 1996. National Primary Drinking Water Regulations: Monitoring Requirements for Public Drinking Water Supplies; Final Rule. Federal Register 61:24353.

EPA. 1998a. National Primary Drinking Water Regulations: Disinfectants and Disinfection Byproducts, Final Rule. Federal Register 63:69389.

EPA. 1998b. National Primary Drinking Water Regulations; Interim Enhanced Surface Water Treatment Rule; Final Rule. Federal Register 63:69477.

EPA. 1999. Final guidelines for the certification and recertification of the operators of community and nontransient noncommunity public water systems. Federal Register 64:5915–5921.

EPA. 2002a. National Primary Drinking Water Regulations; Long Term 1 Enhanced Surface Water Treatment Rule, Final Rule. Federal Register 67:1811.

EPA. 2002b. Potential contamination due to cross-connections and backflow and the associated health risks, an issue paper. Washington, DC: EPA Office of Ground Water and Drinking Water.

EPA. 2003a. Drinking Water Research Program, Multi-Year Plan, 2003. Washington, DC: EPA Office of Research and Development. Available on-line at: *http://www.epa.gov/osp/myp/dw.pdf.* Accessed May 4, 2006.

EPA. 2003b. Cross-Connection Control Manual. Washington, DC: EPA.

EPA. 2005a. FACTOIDS: Drinking Water and Ground Water Statistics for 2004. Washington, DC: EPA.

EPA. 2005b. Unregulated Contaminant Monitoring Regulation (UCMR) for Public Water Systems Revisions. Federal Register 70:49093.

EPA. 2006a. National Primary Drinking Water Regulations: Stage 2 Disinfectants and Disinfection Byproducts Rule; National Primary and Secondary Drinking Water Regulations, Final Rule. Federal Register 71:387.

EPA. 2006b. National Primary Drinking Water Regulations: Long Term 2 Enhanced Surface Water Treatment Rule, Final Rule. Federal Register 71:653.

Friedman, M., G. Kirmeyer, G. Pierson, S. Harrison, K. Martel, A. Sandvig, and A. Hanson. 2005. Development of distribution system water quality optimization plans. Denver, CO: AwwaRF.

Hrudey, S. E., and R. Walker. 2005. Walkerton—5 years later tragedy could have been prevented. OpFlow 31:1–7.

International Association of Plumbing and Mechanical Officials (IAPMO). 2003. Uniform Plumbing Code, 2003 edition. Ontario, CA: IAPMO.

IAPMO. 2005. Can we make it work? Available on-line at *http://www.iapmo.org /iapmo/news/code-release.html.* Accessed April 26, 2006.

International Code Council. 2003. International Plumbing Code, 2003 Edition. Falls Church, VA: International Code Council.

Linn County. 2004. Linn County Plumbing Regulations. Available on-line at *http://www.linncountyauditor.org/Ordinances/Plumbing--%5B5%5D.pdf.* Accessed April 26, 2006.

MacPhee, M. J. (ed.). 2005. Distribution system water quality challenges in the 21st century: a strategic guide. Denver, CO: AWWA.

Martel, K. 2005. HACCP Applied to Distribution Systems. January 13, 2005. Presented to the NRC Committee on Public Water Supply Distribution Systems. Washington, DC.

Martel, K., G. Kirmeyer, A. Hanson, M. Stevens, J. Mullenger, and D. Deere. 2006. Application of HACCP for Distribution System Protection. Denver, CO: AwwaRF.

McCain, K., and M. Fahrenbruch. 2006. Succession planning: the babies and boomers. Source 20:16–17.

Mucklow, R. 1997. Where did HACCP come from? *In:* Heads Up for HACCP. National Meat Association. Available on-line at *http://www.nmaonline.org/files/ headsup12-1.htm.* Accessed May 4, 2006.

National Aeronautics and Space Administration (NASA). 1991. A dividend in food safety. Spinoff 1991. NASA Technical Report ID 20020086314. Washington, DC: NASA.

National Research Council (NRC). 2005. Public Water Supply Distribution Systems: Assessing and Reducing Risks, First Report. Washington, DC: National Academies Press.

Pomerance, H., and E. G. Means. 2006. Succession planning: leveraging the inevitable. Source 20:10–13.

State of Iowa. 2005. State Plumbing Code. Available on-line at *http://www.legis.state.ia.us/Rules/2002/iac/641iac/64125/64125.pdf.* Accessed April 26, 2006.

World Health Organization (WHO). 2004. Guidelines for drinking water quality, third edition. Available on-line at *http://www.who.int/water_sanitation_health/dwq/gdwq3/en/.* Accessed April 26, 2006. Geneva, Switzerland: WHO.

3
Public Health Risk from Distribution System Contamination

One of the most challenging facets of reducing the risk of contaminated distribution systems is being able to quantify the existing risk. This is made complicated not only by the plethora of factors that can constitute public health risks, including a diversity of microbial pathogens and chemical compounds, but also by the varying response that a given individual will have when exposed to those factors. This chapter describes three primary mechanisms used to assess the acute public health risk of distribution system contamination, the limitations of these methods, and what conclusions can be derived from currently available data.

INTRODUCTION TO RISK

The process of risk assessment involves determining the likelihood and severity of different adverse impacts given exposure of a population to a hazard. Risk analysis includes the process of risk assessment, as well as risk management activities to decide what an acceptable risk level is and to take actions to reduce risk (NRC, 1983). Risk assessment requires the activities of hazard identification, exposure assessment, and dose-response (or exposure-response) assessment. Hazard identification is the determination of what adverse agents might be present and what adverse impacts they might cause. Exposure assessment is the quantitative determination of the levels of contaminants (in the case of environmental exposures) individuals may consume/inhale/contact over a specific time period. Dose-response assessment is the quantitative determination of the likelihood of an individual having a particular adverse effect from a given exposure. Alternatively, this can be viewed as the proportion of persons in a population who are expected to have the adverse effect were they to have the particular exposure.

Various federal agencies, including the U. S. Environmental Protection Agency (EPA), have developed specific guidelines and procedures for performing risk assessment, particularly for carcinogens and for substances that result in non-carcinogenic toxic effects. In the case of infectious agents (which are frequently the concern in drinking water), methodologies are at a developmental stage.

One of the goals of performing risk assessment within a regulatory framework is to develop regulatory guidance or standards (or decide not to undertake such action) based on the results. This process, which is part of risk analysis, requires additional considerations such as cost and equity. Under the Safe Drinking Water Act, EPA is required to set a maximum contaminant level goal (MCLG) for certain contaminants that is absolutely protective against all adverse health effects, given available risk assessment information. For most contaminants with MCLGs, a regulatory level is then established—a maximum contaminant level (MCL)—or a treatment technique is required, both of which incorporate considerations of feasibility (see Box 3-1).

In determining a regulatory level such as an MCL, implicitly or explicitly the acceptable residual risk (after the implementation of any interventions) must be decided upon. The empirical evidence is that, for human carcinogens, EPA has regarded a window of residual lifetime risk of 1/1,000,000 to 1/10,000 to be acceptable (see Box 3-2 for an explanation of the origins of this value and its extension to infectious agents). In other words, a residual risk resulting in no more than 1 extra cancer in the lifetime of a population of 10,000 to 1,000,000 persons is regarded as being acceptable.

Risks from Drinking Water

Drinking water can serve as a transmission vehicle for a variety of hazardous agents: enteric microbial pathogens from human or animal fecal contamination (e.g., noroviruses, *E. coli* O157:H7, *Cryptosporidium*), aquatic microorganisms that can cause harmful infections in humans (e.g., nontuberculous mycobacteria, *Legionella*), toxins from aquatic microorganisms (such as cyanobacteria), and several classes of chemical contaminants (organic chemicals such as benzene, polychlorinated biphenyls, and various pesticides; inorganic chemicals such as arsenic and nitrates; metals such as lead and copper; disinfection byproducts or DBPs such as trihalomethanes; and radioactive compounds).

Contaminants in drinking water can produce adverse effects in humans due to multiple routes of exposure. In addition to risk from ingestion, exposure can also occur from inhalation and dermal routes. For example, inhalation of droplets containing respiratory pathogens (such as *Legionella* or *Mycobacterium*) can result in illness. It is known that DBPs present in drinking water may volatilize resulting in inhalation risk, and these compounds (and likely other organics) may also be transported through the skin (after bathing or showering) into the bloodstream (Jo et al., 1990). Reaction of disinfectants in potable water with other materials in the household may also result in indoor air exposure of contaminants; for example Shepard et al. (1996) reported on release of volatile organics in indoor washing machines. Thus, multiple routes of exposure need to be considered when assessing the risk presented by contaminated distribution systems. It should be noted, however, that the report will not consider such indirect routes of exposure as (1) the loss of pressure and subsequent inadequate fire protection, (2) loss of water for hospitals and dialysis centers, and (3) leaks in household plumbing that lead to toxic mold growth.

BOX 3-1
U.S. Code, Title 42(6A)(XIIB)§300g-1
(Safe Drinking Water Act as Amended)

(A) *Maximum contaminant level goals.* Each maximum contaminant level goal established under this subsection shall be set at the level at which no known or anticipated adverse effects on the health of persons occur and which allows an adequate margin of safety.

(B) *Maximum contaminant levels.* Except as provided in paragraphs (5) and (6)[1], each national primary drinking water regulation for a contaminant for which a maximum contaminant level goal is established under this subsection shall specify a maximum contaminant level for such contaminant which is as close to the maximum contaminant level goal as is feasible.

It has been recognized for some years that consumers face risk from multiple hazards, and that action to reduce the risk from one hazard may increase the risk from other hazards given the same exposure. There are prominent examples of this phenomenon in the drinking water arena that have greatly complicated efforts to reduce overall risk from distribution systems. Havelaar et al. (2000) assessed the relative changes in risk from switching to ozone treatment of drinking water in the Netherlands. In this case, there was a projected reduction in risk from waterborne infectious disease (such as *Cryptosporidium*) while there was a projected increase in risk from DBP formation (the primary one examined was bromate). To compare the net change in overall risk, it is necessary to place the multiple risks (with their different endpoints in terms of disease severity) on the same scale. Havelaar et al. (2000) did this comparison using the methodology of disability adjusted life years (DALY's). In this approach, the severity of an adverse health effect is quantitatively weighted by an index (disability weight) reflecting the proportional degradation in health (a weight of 0 is reflective of absence of an effect, while a weight of 1 is reflected in total impairment); the integral of the years of diminished functioning multiplied by the disability weight is summed with the reduction in lifespan due to premature mortality to get the aggregate impact to a population. In principle, using such an approach one can optimize for the overall net reduction in risk, considering competing hazards. It is noted that the DALY framework has not been adopted for U.S. regulatory practice and remains controversial for a number of technical and policy reasons (including age equity) (Anand and Hanson, 1997).

When risk is assessed for chemical or microbial exposure, it should be considered that not all segments of the population are at the same degree of risk. This may be due to differences in exposure in terms of either consumption (Gerba et al., 1996) or in concentrations (due to heterogeneity in the environ-

[1] Paragraph (5) allows departure upwards from setting the MCL as close to the MCLG as feasible if doing so would result in an increase in risk from other contaminants, or would interfere with the performance of processes used to address other contaminants. Paragraph (6) allows departure upward from the "as close as feasible" criterion in certain circumstances if the benefits would not justify the cost of compliance at that standard.

BOX 3-2
Origin of the 1/10,000 Acceptable Risk Level for Carcinogens and Infectious Agents

EPA has been at the forefront of the issue of acceptable risk in virtually all of its programmatic areas, primarily as the result of court challenges to its regulations. In response to the 1987 Section 112 Clean Air Act decision (*Natural Resources Defense Council vs. U.S. Environmental Protection Agency 824 F. 2nd 1146 [1987]*), EPA decided it would base its regulatory decisions on quantitative risk assessments using the general policy that a lifetime added cancer risk for the most exposed person of 1 in 10,000 (1×10^{-4}) might constitute acceptable risk and that the margin of safety required by statute and reinforced by the court should reduce the risk for the greatest number of persons to an added lifetime risk of no more than 1 in 1 million (1×10^{-6}). However, EPA (along with the courts) has not viewed "safe" as the equivalent of risk-free and has determined that standards should protect against significant public health risks (EPA 49 Fed. Reg. 8386 [1984]; Rodricks et al. 1987; Industrial Union Department, AFL-CIO v. American Petroleum Institute et al. 448 U.S. 607 [1980]). EPA has repeatedly rejected the opinion that it can establish a universal (i.e., brightline) acceptable risk that should never be exceeded under any circumstances, and they maintain that guidance provided under one statute might have little relevance to others because of differing program goals. In practical terms, EPA almost never regulates at a theoretical risk below 1×10^{-6} (de minimis) and almost always regulates at a theoretical risk below 1×10^{-4} (de manifestis)" (NRC, 2004).

Policy with respect to acceptable levels of risk from exposure to infectious agents is less well developed than for chemical carcinogens. However, in framing the Surface Water Treatment Rule (Federal Register, June 29, 1989, page 27486), the rule for reduction of risk from *Giardia* and viruses was set to achieve a residual estimated risk of infection below 1/10,000 per year. This number derived from the then average waterborne illness rate associated with reported waterborne outbreaks (Regli et al., 1991). However it is now recognized that the waterborne illness rate is substantially greater than this value—due to underreporting of outbreaks, as well as to substantial endemic illness. The use of infection rather than illness as an endpoint was intended to compensate for secondary cases and also for presumed heightened infectivity amongst sensitive subpopulations.

The use by EPA of an acceptable risk window for microorganisms in the 10^{-6} to 10^{-4} range as one factor in setting standards continues. As recently as the promulgation of the Long Term 2 Enhanced Surface Water Treatment Rule (Federal Register, January 5, 2006), EPA has stated: "EPA and Advisory Committee deliberations focused on mean source water *Cryptosporidium* concentrations in the range of 0.01–0.1 oocysts/L as threshold levels for requiring additional treatment...these levels are estimated to result in an annual infection risk in the range of 1.7×10^{-4} – 6×10^{-3} ... for a treatment plant achieving 3-log *Cryptosporidium* removal (the treatment efficiency estimated for conventional plants under existing regulations)."

ment, e.g., in the distributed water), or to intrinsic differences in susceptibility (Balbus et al., 2000). Unfortunately, our ability to assess quantitative differences in intrinsic susceptibility remains poor, and therefore protection of susceptible subpopulations often relies upon the imposition of safety factors.

Methods for Characterizing Human Health Risk

Characterization of human health risks may be performed using an epidemiological approach or using a risk assessment approach. These methods are complementary and have different strengths and limitations, and each has been used for assessment of drinking water risks in various applications. Epidemiological approaches study the relationship between exposures and disease in actual populations and are descriptive, correlational, or analytic. In the descriptive study, population surveys or systematic disease surveillance (monitoring) describe disease patterns by various factors such as age, seasonality, and geographic location. Correlational (also called "ecologic") studies collect population level data on disease rates and exposures and look for correlations. Analytical studies (whether experimental or observational) are those in which individual-level data is collected and the investigator tests a formal hypothesis about the association between exposure and disease.

Risk assessment methods, on the other hand, follow the hazard identification, dose-response assessment, exposure assessment, and risk characterization paradigm noted above. Frequently, but not always, the dose-response assessment is based upon extrapolation from results of trials in animals (although results from human exposure may be used where available—for example, in human feeding trials of infectious agents or from studies in populations exposed in occupational or other settings to particular agents of concern).

Epidemiological studies have the advantage of involving human populations, often experiencing the exposure of interest and representing a range of variability in susceptibility and behavior. However to detect a small increase in risk from the baseline, epidemiological studies require very large sample sizes, and thus considerable expense and effort. Epidemiological studies cannot provide direct information on the potential for risk reduction from a proposed change in treatment practice that has not yet been implemented since by definition there is not yet human exposure to conditions expected from the proposed change. However, epidemiological studies can be designed to measure the direct impact of a treatment intervention after it has been implemented. This is very powerful tool and it has provided the evidence base that changes in water treatment have had a positive impact on community health. For example, the recent meta-analysis by Fewtrell and Colford (2004) demonstrates the body of evidence linking improvements in community and household water quality to health.

Risk assessment approaches have the advantage of being flexible in their application to potential (but not yet experienced) situations. A risk assessment can be performed even when the projected risk from a particular exposure or change of exposure is very small. They have the disadvantage of requiring extensive measurement or modeling to ascertain exposure, and also of the need for dose-response studies. Often these dose-response studies are in animals or at higher doses, thereby requiring extrapolation with respect to dose (via a formal mathematical dose-response curve) and/or between species. Generally, whether animal or human data are used to establish the dose-response relationship, the range in variability in susceptibility is small (compared to a full human population) and therefore some margin of safety may need to be explicitly used to account for more susceptible subpopulations.

This chapter discusses what is known about the human health risks that derive from contamination of the distribution system, relying on three primary approaches: risk assessment methods that utilize data on pathogen occurrence measurements, outbreak surveillance data, and epidemiology studies. A special section is devoted to *Legionella*, for which all three types of activities have occurred, leading to greater understanding of the risks inherent from growth of this organism in distribution systems. Because the impetus for this study was revision of the Total Coliform Rule, the report focuses primarily on acute risks from microbial contamination of the distribution system. However, there are short- and long-term risks from chemicals that merit mention (particularly DBPs—lead and copper were outside the scope of the study). DBP concentrations in the distribution system can vary significantly depending on water residence time, the types of disinfectants used, and biological and chemical reactions, among many other factors (see Chapter 6). The concentrations of trihalomethanes in finished water tend to increase with increasing water age, while certain haloacetic acids tend to decrease in concentration over time (see Chapter 6; Arbuckle et al., 2002). A number of epidemiologic studies have examined the health significance of DBP exposure and have reported significantly increased risks of bladder, rectal, and/or colon cancers in some populations (King et al., 1996; Koivusalo et al., 1997; Doyle et al., 1997; Cantor et al., 1998; Yang et al., 1998; King et al., 2000) as well as adverse reproductive outcomes (Waller et al., 1998; Dodds et al., 1999; Klotz and Pyrch, 1999; King et al., 2000). However, determining and classifying DBP exposure in these studies has been extremely challenging and has made it difficult to interpret the findings of these studies (Arbuckle et al., 2002, Weinberg et al., 2006). Furthermore, the contribution of distribution systems to the reported risk, as opposed to drinking water treatment or other processes, has not been elucidated. Because epidemiological studies of DBP exposure have been extensively reviewed by others (Boorman et al., 1999; Nieuwenhuijsen et al., 2000; Graves et al., 2001), they are not reviewed here.

EVIDENCE FROM PATHOGEN OCCURRENCE MEASUREMENTS

The risk assessment approach relies on being able to measure or predict (e.g., by modeling) the concentration of an etiologic agent in the water supply. Certain microbial pathogens are indicative of distribution system contamination stemming from both internal and external sources. These include bacteria known to form biofilms—a physiological state in which organisms attach to and grow on a surface (Characklis and Marshall, 1990)—and bacteria that indicate an external contamination event such as intrusion. In distribution systems, the interior pipe walls, storage tanks, sediments, and other surfaces in contact with finished water are colonized by bacteria, which can survive, grow, and detach depending on local conditions. Other types of bacteria (such as coliforms) as well as enteric viruses and protozoa (Quignon et al., 1997; Piriou et al., 2000) are also found in biofilms. However, their presence can also be attributable to an external contamination event or break through of the treatment barrier.

The microbiology of distribution systems can be influenced by a variety of factors (e.g., poor quality source water, inadequate treatment, unsanitary activity, backflow). Given this report's assumption of adequate treatment, a discussion of all source water microbes and those that would be eliminated during treatment is not warranted. Furthermore, virtually any microorganism in close enough proximity to a vulnerable part of the distribution system (e.g., a cross connection, main break, or leak) could enter during an external contamination event. Control of these events—see Chapters 4 and 5—is important for reducing the risks of not only microbial pathogens but also chemicals that might enter distribution system. Because the complexity of microbes from such diverse sources is beyond the scope of this report, the following section focuses on those organisms most likely to *indicate* either internal or external contamination of the distribution system.

The Microbiology of Bulk Water

The microbiology of distribution systems essentially consists of two different environments—microorganisms in the bulk water column and those in biofilms attached to the surfaces of pipes, sediments, and other materials. Microorganisms in the bulk water column originate from either the source water, from bacterial growth within the treatment process (e.g., within the treatment filters), from biofilms within the distribution system, or from recontamination of the water from cross connections, intrusion, pipe breaks, or other external sources.

Heterotrophic Bacteria

Heterotrophic bacteria (a broad classification that takes into account all bacteria that utilize organic carbon) are commonly found in the bulk water of distribution systems because they readily form biofilms in such systems. They are measured by using heterotrophic plate counts (HPC). Heterotrophs have traditionally been divided into two primary groups based on their cell wall characteristics—Gram-negative and Gram-positive.

The presence of a disinfectant residual in drinking water has a tremendous selective effect, particularly on Gram-negative bacteria, which are relatively sensitive to inactivation by disinfectants. Identification of bacteria using fatty acid analysis (Norton and LeChevallier, 2000) showed that chlorination resulted in a rapid shift from predominately Gram-negative bacteria (97 percent) in the raw water to mostly Gram-positive organisms (98 percent) in the chlorinated water (see Table 3-1). Bacteria in the raw water were diverse, with *Acinetobacter*

TABLE 3-1 Bacterial Populations Isolated from the Water Column During Treatment

Bacterial Identification	Percentage of Population in Raw Water	Percentage of Population in Ozone Contactor	Percentage of Population in Filter Effluent	Percentage of Population in Distribution System Influent
Gram Negative				
Acidovorax spp.	2		4	7
Acinetobacter spp.	29	6		
Alcaligenes spp.	12	2	1	
Alteromonas spp.	2			
Comamonas spp.	1		3	
Enterobacter spp.	2		5	
Flavobacterium spp.	2		5	
Hydrogenophaga spp.	8	3	1	
Klebsiella spp.	10	1	3	
Methylobacterium spp.	1		2	
Pseudomonas spp.	14	53	22	
Rhodobacter spp.	2	1		
Sphingomonas spp.	2	2	19	
Stenotrophomonas spp.	2	1	2	
Xanthobacter spp.	3			
Others*	2	1	5	
Gram Positive				
Bacillus spp.				7
Nocardia spp.	1	3	7	53
Rhodococcus spp.		16	4	
Staphylococcus spp.	1	1		
Others*	1	1	1	
Unidentified	3	9	16	33

* Includes organisms isolated from only one site at a frequency of 1%. 100 isolates were identified from each site.
SOURCE: Adapted from Norton and LeChevallier (2000).

spp., *Pseudomonas* spp., and *Klebsiella* spp. predominate among the 20 genera identified. Ozonation of the raw water reduced the microbial diversity to 13 genera, dominated by *Pseudomonas* spp. and *Rhodococcus* spp. However, following biologically active granular activated carbon filtration, 19 genera were identified in the filter effluent, the majority of which (63 percent) matched isolates observed in the raw water. The predominant genera were *Pseudomonas* spp. and *Sphingomonas* spp., which are known to grow attached to the carbon fines of the filter while utilizing natural organic compounds found in the aquatic environment. Final chlorination of the filtered water resulted in a shift to *Nocardia* spp. as the water entered the pipe system. *Nocardia* spp. possess characteristic fatty acids that are closely related to *Rhodococcus*, *Mycobacterium*, and *Corynebacterium*. Its partially acid-fast cell wall and possession of the catalase enzyme, which breaks down hydrogen peroxide, are important factors that enable the organism to survive disinfection. Other Gram-positive bacteria found in chlorinated drinking water include *Bacillus* and *Staphylococcus* spp. *Bacillus* spp. form environmentally resistant spores that can withstand prolonged contact with chlorine. Some strains of *Bacillus* and *Staphylococcus aureus* can produce toxins when contaminated water is used in food preparation (LeChevallier and Seidler, 1980).

Treated drinking water will include a mixture of Gram-negative and Gram-positive bacteria. In the absence of a disinfectant residual, Gram-negative bacteria will out grow Gram-positive bacteria and dominate the bacterial population. These organisms typically include *Pseudomonas*, *Acinetobacter*, *Flavobacterium*, and *Sphingomonas* spp. For the most part, these organisms have limited public health significance, except for *Pseudomonas aeruginosa,* which is a possible opportunistic pathogen in drinking water and in the biofilms of water systems. It is known to colonize point-of-use carbon filters in drinking water systems (de Victoria and Galvan, 2001; Chaidez and Gerba, 2004). *Pseudomonas aeruginosa* is of concern in bathing waters, especially in swimming pools and spas, where skin infections may result due to exposure. In the case of drinking water, there are a few studies that suggest a relationship between the presence of this organism in the water and disease. In one hospital setting, five of 17 patients with a *Pseudomonas* infection carried a genotype also detected in the tap water (Trautmann et al., 2001). In another outbreak of pediatric *P. aeruginosa* urinary tract infections, two isolates had genotypes similar to those in the water. The outbreak was resolved when the taps in the unit were changed (Ferroni et al., 1998).

Despite these specific incidences, a workgroup recently convened by the World Health Organization (WHO) to address this issue concluded that HPC bacteria were not associated with any adverse health effect (Bartram et al., 2003). "Some epidemiological studies have been conducted into the relationship between HPC exposures from drinking water and human health effects. Other studies relevant to this issue include case studies, especially in clinical situations, and compromised animal challenges using heterotrophic bacteria obtained from drinking water distribution systems. The available body of evidence supports the conclusion that, in the absence of fecal contamination, there is no direct relationship between HPC values in ingested water and human health effects in the population at large. This conclusion is also supported indirectly by evidence from exposures to HPC in foodstuffs where there is no evidence for health effects link in the absence of pathogen contamination. There are a small number of studies that have examined possible links between HPC bacteria and non-intestinal outcomes in general populations. The conclusions of these studies do not support a [health] relationship" (WHO, 2002).

One of the difficulties in interpreting the significance of HPC data is that test methods involve a wide variety of conditions that lead to a wide range of quantitative and qualitative results. For this reason, the EPA has not yet issued a health-based standard. However, the Surface Water Treatment Rule requires that distribution system locations without a detectable disinfectant residual maintain HPC levels at or below 500 colony forming units (CFU)/mL in at least 95 percent of the samples each month (EPA, 1989).

Coliform Bacteria. Total coliform bacteria (a subset of Gram-negative bacteria) are used primarily as a measure of water treatment effectiveness and can occasionally be found in distribution systems. The origins of total coliform bacteria include untreated surface water and groundwater, vegetation, soils, insects, and animal and human fecal material. Typical coliform bacteria found in drinking water systems include *Klebsiella pneumoniae*, *Enterobacter aerogenes, Enterobacter cloacae*, and *Citrobacter freundii*. Other typical species and genera are shown in Table 3-2. Although most coliforms are not pathogenic, they can indicate the potential presence of fecal pathogens and thus in the absence of more specific data may be used as a surrogate measure of public health risk. Indeed, the presence of coliforms is the distribution system is usually interpreted to indicate an external contamination event, such as injured organism passage through treatment barriers or introduction via water line breaks, cross connections, or uncovered or poorly maintained finished water storage facilities (Geldreich et al., 1992; Clark et al., 1996). However, biofilms within distribution systems can support the growth and release of coliforms, even when physical integrity (i.e., breaches in the treatment plant or distribution system) and disinfectant residual have been maintained (Characklis, 1988; Haudidier et al., 1988; Smith et al., 1990), such that their presence may not necessarily indicate a recent external contamination event. Coliform regrowth in the distribution system is more likely during the summer months when temperatures are closer to the optimum growth temperatures of these bacteria.

Thermotolerant coliforms (capable of growth at 44.5 °C), also termed "fecal coliforms" have a higher association with fecal pollution than total coliforms. And *Escherichia coli* is considered to be even more directly related to fecal pollution as it is commonly found in the intestinal track of warm-blooded animals. Although most fecal coliform and *E. coli* strains are not pathogenic, some strains are invasive for intestinal cells and can produce heat-labile or heat-stable toxins (AWWA, 1999). *E. coli* and most of the thermotolerant coliforms do not grow in biofilms, although they most likely can be trapped and retained within biofilms.

TABLE 3-2 Coliform Isolates Typically Found in Drinking Water

Citrobacter	Enterobacter	Escherichia	Klebsiella
C. freundii	*E. aerogenes*	*E. coli*	*K. pneumonia*
C. diversus	*E. agglomerans*		*K. oxytoca*
	E. cloacae		*K. rhinoscleromatis*
			K. ozaena

SOURCE: Adapted from Geldreich and LeChevallier (1999).

Aeromonas. *Aeromonas* spp. are Gram-negative bacteria found in fresh and salt water and cause a wide variety of human infections including septicemia, wound infections, meningitis, pneumonia, respiratory infections, hemolytic uremic syndrome, and gastroenteritis (Carnahan and Altwegg, 1996; Alavandi et al., 1999). The ability of these microorganisms to grow at low temperatures and low nutrient conditions are important in their occurrence in drinking water supplies. Through the Unregulated Contaminant Monitoring Rule (see Chapter 2), EPA examined the occurrence of *Aeromonas* spp. in 308 drinking water systems and found detectable concentrations in 2.6 percent of 5,060 samples and in 13.6 percent of the systems. In a 16-month study conducted on the presence of *A. hydrophila* in drinking water in Indiana, 7.7 percent of the biofilm samples were positive for *A. hydrophila* (Chauret et al., 2001). The health significance of detecting aeromonads in drinking water is not well understood. Some countries (such as the Netherlands) have set standards for aeromonads in drinking water leaving the treatment plant (< 20 CFU/200 mL) and in the distribution system (< 200 CFU/100 mL).

Mycobacteria. Organisms of the genus *Mycobacteria* are also found in drinking water. Of particular concern is the MAC, or *Mycobacterium avium* complex. Studies have detected *M. avium* complex organisms in drinking water distribution systems with concentrations ranging between 0.08 and 45,000 CFU/mL (Haas et al., 1983; duMoulin and Stottmeir, 1986; Carson et al., 1988; duMoulin et al., 1988; Fischeder et al., 1991; von Reyn et al., 1993; Glover et al., 1994; von Reyn et al., 1994; Covert et al., 1999). *M. avium* are resistant to disinfectants, especially free chlorine (Taylor et al., 2000). Indeed, it is postulated that they may in fact be selected for in distribution systems as a result of their resistance to chlorine (Collins et al., 1984; Schulze-Robbecke and Fischeder, 1989; Briganti and Wacker, 1995). However, there is also evidence that MAC are susceptible to chlorine dioxide and chloramine (Vaerewijck et al., 2005).

Falkinham et al. (2001) examined eight, well characterized drinking water systems and reported that 20 percent of the water isolates and 64 percent of the biofilm isolates were identified as *M. avium* or *M. intracellulare*. Additionally, 8 percent of the water isolates were identified as *M. kansasii*. Most of these isolates were detected in raw water samples, with *M. avium* complex organisms detected in five of six surface water sites ranging from 6 to 35 percent of the organisms isolated. *M. avium* complex organisms were not detected in any plant or well effluent sample, but were occasionally detected at low levels (< 1 CFU/mL) in drinking water systems. However, *M. avium* and *M. intracellulare* were recovered frequently from drinking water biofilm samples, indicating that *M. avium* levels were increasing in the distribution system. Increases in *M. avium* levels in drinking water were correlated to levels of AOC ($r^2 = 0.65$, $p = 0.029$) and BDOC ($r^2 = 0.64$, $p = 0.031$) (Falkinham et al., 2001; LeChevallier, 2004).

The greatest increase in *M. avium* complex infections have been with acquired immunodeficiency syndrome (AIDS) patients; approximately 25 to 50 percent of these patients suffer debilitating and life-threatening MAC infections (Horsburgh, 1991; Nightingale et al., 1992), although the availability of highly active antiretroviral therapy has reduced the incidence of MAC in AIDS patients in recent years. Members of the MAC are known opportunistic pathogens, with symptoms of pulmonary infection mimicking that of *M. tuberculosis* (Wolinsky, 1979). The organism infects the gastrointestinal or pulmonary tract, suggesting that food or water may be important routes of transmission for AIDS patients (Singh and Lu, 1994). It should be pointed out that epidemiology studies have not yet identified drinking water as a risk factor for MAC, except perhaps in hospital water systems.

Free-Living Protozoa

Of the genera of protozoa present in distribution systems, *Acanthamoeba*, *Hartmanella* and *Naegleria* are known to feed on bacteria and biofilms by grazing. Previous research has shown that all coliforms as well as bacterial pathogens and opportunistic pathogens may be ingested by protozoa. Ingested bacteria, if not digested, may survive within the protozoa and be protected from residual disinfectant. The survival of *Legionella* has been the subject of numerous reports in the literature with regards to its increased resistance to disinfectants while in the intracellular state (Levy, 1990).

Of the eucaryotes mentioned above, two are known to be pathogenic—*Naegleria* spp. and *Acanthamoeba*. These are usually associated with recreational rather than drinking waters, although *Acanthamoeba* was included as part of the first Contaminant Candidate List (EPA, 1998) as an opportunistic pathogen affecting contact lens wearers. Previous studies have shown that these organisms are usually found at the source. However, cysts have also been isolated from drinking water distribution systems in France (Jacquemin et al., 1981; Geldreich, 1996).

Routine monitoring for free-living protozoa is rarely done. Isolation and identification of these organisms are accomplished only when there is evidence for disease outbreak or when research studies are being conducted. As interest in the ability for protozoa to harbor bacterial pathogens increases, it is probable that more effort will be expended in determining their presence in distribution systems, including premise plumbing.

Fungi

Although many fungi have been found in drinking water systems, their levels are typically low and the organisms have not been directly associated with disease (Kelley et al., 2003). The origin of fungi in drinking water systems has not been well characterized, but it is assumed that they come from environmental sources including surface water and groundwater, soils, and vegetation. The four most frequently occurring genera of filamentous fungi isolated from chlorinated and unchlorinated distribution systems in southern California were *Penicillium, Sporocybe, Acremonium,* and *Paecilomyces* (Nagy and Olson, 1982). *Aspergillus fumigatus* was the predominant species detected in the distribution system water supplies in Finland (Niemi et al., 1982). A variety of fungi (*Cephalosporium* sp., *Verticillium* sp., *Trichodorma sporulosum*, *Nectria veridescens, Phoma* sp., and *Phialophora* sp.) were identified from water service mains in England (Bays et al., 1970; Dott and Waschko-Dransmann, 1981). Outside of specialized research studies, potable water supplies are not routinely tested for fungi.

The Microbiology of Distribution System Biofilms

Biofilms in drinking water pipe networks contain all of the organisms mentioned above that are found in bulk distribution system water, as well as others. The microbial composition of any given pipe segment can be highly variable, and in most cases is poorly, if ever, characterized. The pipe surface itself can influence the composition and activity of biofilm populations. Studies have shown that biofilms developed more quickly on iron pipe surfaces than on plastic PVC pipes, despite the fact that adequate corrosion control was applied, that the water was biologically treated to reduce AOC levels, and that chlorine residuals were consistently maintained (Haas et al., 1983; Camper, 1996).

In addition to influencing the development of biofilms, the pipe surface has also been shown to affect the composition of the microbial communities present within the biofilm (Figure 3-1). Iron pipes supported a more diverse microbial population than did PVC pipes (Norton and LeChevallier, 2000). Undoubtedly part of the reason that certain bacteria associate with certain pipe types is because materials may leach compounds that support bacterial growth. For example, pipe gaskets and elastic sealants (containing polyamide and silicone) can be a source of nutrients for bacterial proliferation. Colbourne et al. (1984) reported that *Legionella* were associated with certain rubber gaskets. Organisms associated with joint-packing materials include populations of *Pseudomonas aeruginosa, Chromobacter spp., Enterobacter aerogenes,* and *Klebsiella pneumoniae* (Schoenen, 1986; Geldreich and LeChevallier, 1999). Coating compounds for storage reservoirs and standpipes can contribute organic polymers and solvents that may support regrowth of heterotrophic bacteria (Schoenen, 1986; Thofern et al., 1987). Liner materials may contain bitumen, chlorinated rubber, epoxy resin, or tar-epoxy resin combinations that can support bacterial regrowth (Schoenen, 1986). PVC pipes and coating materials may leach stabilizers that can result in bacterial growth. Studies performed in the United Kingdom reported that coliform isolations were four times higher when samples were collected from plastic taps than from metallic faucets (cited in Geldreich and

LeChevallier, 1999). The purpose of these studies was not to indicate that certain pipe materials are preferred over another, but to demonstrate the importance of considering the type of materials that come into contact with potable water. Although procedures are available to evaluate the growth stimulation potential of different materials (Bellen et al., 1993), these tests are not applied in the United States by ANSI/NSF.

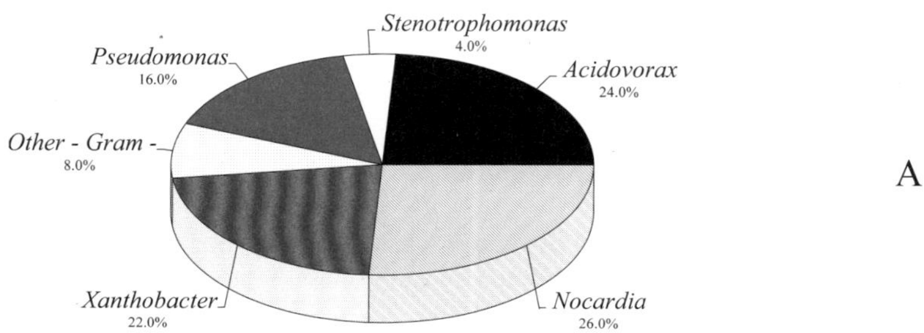

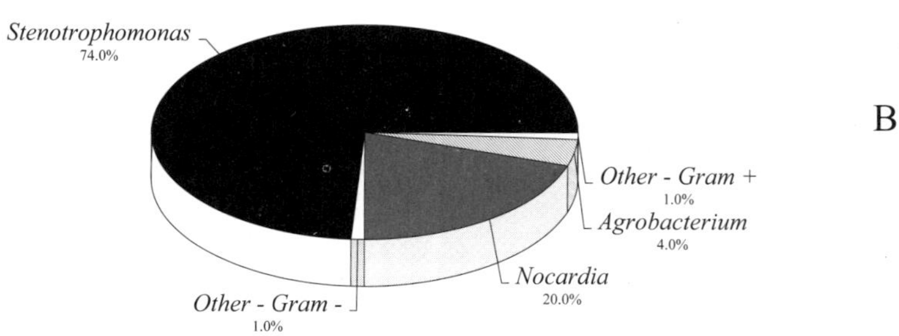

FIGURE 3-1 Microbial populations isolated from iron pipe (A) or PVC (B) surfaces. SOURCE: Adapted from Norton and LeChevallier (2000).

***

For both bulk drinking water and biofilms, the identification of microorganisms typically relies on culturing bacteria from potable supplies, which has important limitations. Culture methods do not detect all microbes that may exist in water, such that only a fraction of viable organisms is recovered (Amann et al., 1995). In addition, most culture methods only detect relatively rapidly growing heterotrophic bacteria, and slowly growing organisms, fastidious or autotrophic organisms, and anaerobes are generally not examined. Diagnostic kits are unreliable for many heterotrophic bacteria because the methodology often requires the analyst to perform a Gram stain, which is difficult because of the slow growth and acid-fast or partially acid-fast nature of bacteria surviving in disinfected drinking water.

An alternative method includes fatty acid profiling. As shown above, this approach can be used to identify organisms from drinking water (Norton and LeChevallier, 2000) but in this study the organisms were cultured prior to identification and therefore the limitations associated with culturing are still present. Additionally, for identification, the lipid profile must match an established profile in a database; these databases are predominated by medical (and not environmental) organisms. The use of fatty acid profiles was further developed by Smith et al. (2000) who used biofilm samples without prior culturing to demonstrate that predominantly Gram-negative bacteria were present, but no further identification was accomplished. A similar approach was taken by Keinanen et al. (2004) who compared profiles from two drinking water systems and showed that they differed, but again, no identifications were obtained. Although fatty acid profiling has been used in these studies to provide some insight on microbial ecology, the limitations associated with the method preclude it from extensive use in characterizing mixed microbial communities.

Molecular methods offer the promise of a more complete determination of the microbiology of drinking water (see Chapter 6 for details). DNA extraction coupled with polymerase chain reaction (PCR) amplification can be used to identify waterborne microbes (Amann et al., 1990, 1995). These procedures can be combined with quantitative real-time PCR, fluorescence in-situ hybridization, or flow cytometry to provide quantitative assessments of bacterial populations. However, careful quality assurance is necessary to ensure complete extraction and recovery of environmental DNA. Martiny et al. (2003) utilized terminal restriction fragment length polymorphisms to identify members of a biofilm consortium over a three-year time period. In this study, several organisms were identified (*Pseudomonas, Sphingomonas, Aquabacterium, Nitrospira, Planctomyces, Acidobacterium*) but for the majority of the peaks no sequence match could be made.

It is telling that there is very little published information about the microbial ecology of distribution systems. At this point in time, the detection methods are expensive, are time consuming, require optimization for specific conditions, and are appropriate only for the research laboratory. As a consequence, there is a lack of information about the types, numbers, and activities of microorganisms in drinking water. It is also unknown how the ecology of the main distribution system is related to that in premise plumbing, how the populations vary between distribution systems in different locations, and how the populations respond to water quality changes within a distribution system. This translates into a lack of understanding about whether organisms of potential public health concern may be present in water systems and further complicates the ability to assess risk due to their presence.

As mandated by the Safe Drinking Water Act, the EPA has issued a second Contaminant Candidate List that includes 10 microbes (or microbial products) for potential future regulation (EPA, 2004) (see Table 3-3). For most of these microbes, methods do not exist for routine testing of drinking water supplies, and basic research is needed on their occurrence, survival, and importance in potable water. Where the current list includes organisms that are not discussed above, they are considered to be of primary concern in untreated or inadequately treated source waters and not in distribution systems, such that a more detailed discussion is beyond the scope of the report.

It can be hard to determine whether the detection of frank or opportunistic pathogens in drinking water poses an unacceptable risk. In addition to the monitoring techniques being difficult, time-consuming, expensive, and of poor sensitivity, the methods do not detect specific virulence determinants, such that many environmental isolates (e.g., *E. coli, Aeromonas, Legionella*, etc.) are indistinguishable from their clinical strains. Therefore even when monitoring for potentially pathogenic organisms is done, the public health significance of the results is often in question. Furthermore, there is insufficient supporting information (in terms of occurrence data for exposure assessment, dose-response data, health effects, and models that can predict pathogen occurrence for different distribution system contamination scenarios such as contamination via cross connections, main breaks, or intrusion) to conduct a risk assessment for many waterborne microbes. For all these reasons, measurement of the microbe itself is

TABLE 3-3 Contaminant Candidate List Microbes

Bacteria	*Mycobacterium avium*
	Helicobacter
	Aeromonas
Viruses	Caliciviruses
	Echovirus
	Coxsackieviruses
	Adenovirus
Protozoa	*Microsporidium*
Toxins	Cyanobacterial toxins

typically insufficient to make a public health determination. Until better monitoring methods, pathogen occurrence models, dose-response data, and risk assessment data are available, pathogen occurrence measurements are best used in conjunction with other supporting data on health outcomes. Such supporting data could include enhanced or syndromic surveillance in communities, as well as the use of microbial or chemical indicators of potential contamination.

EVIDENCE FROM OUTBREAK DATA

Most information on the risks of waterborne disease in the United States comes from surveillance and investigation of waterborne disease outbreaks. A passive voluntary surveillance system for waterborne disease outbreaks started in 1971 and is a collaboration between the Centers for Disease Control and Prevention (CDC), the EPA, and state and regional epidemiologists. This surveillance system includes outbreaks associated with both drinking and recreational water, and outbreaks due to both microbial and chemical agents. The objectives of the surveillance system are to (1) characterize the epidemiology of waterborne disease outbreaks, (2) identify the etiologic agents that cause the outbreaks, (3) determine the risk factors that contributed to the outbreak, (4) inform and train public health personnel to detect and investigate waterborne disease outbreaks, and (5) collaborate with local, regional, national and international agencies on strategies to prevent waterborne diseases (Stanwell-Smith et al., 2003).

From 1971 through 2002, 764 drinking water outbreaks have been reported through this surveillance system. Although this is believed to be an underestimate of the true number of outbreaks that occurred during this period, the information collected in this surveillance system has been extremely valuable for improving our understanding of the agents that cause waterborne disease and the risk factors involved in waterborne disease outbreaks. The data collected in this surveillance system includes:

- Type of exposure (drinking water or recreational water)
- Location and date of outbreak
- Actual or estimated number of persons exposed, ill, hospitalized, dead
- Symptoms, incubation period, duration of illness
- Etiologic agent
- Epidemiological data (attack rate, relative risk or odds ratio)
- Clinical laboratory data (results of fecal and serology tests)
- Type of water system
    - Community, non-community, or individual homeowner drinking water supply
    - Swimming pool, hot tub, water park, or lake for recreational water

- Environmental data (results of water analyses, sanitary survey, water plant inspection)
- Factors contributing to contamination of water

The surveillance data are summarized in biannual reports (*Morbidity and Mortality Weekly Report Surveillance Summaries*) that are published by the CDC and distributed to public health authorities and practitioners throughout the country. The information is also available on the Internet at *http://www.cdc.gov/mmwr*. These reports (Herwaldt et al., 1991; Moore et al., 1993; Kramer et al., 1996; Levy et al., 1998; Barwick et al., 2000; Lee et al., 2002a; Blackburn et al., 2004) indicate three main trends:

1. The overall number of reported waterborne disease outbreaks associated with drinking water is declining from a peak of over 50 reported outbreaks in 1980 to eight reported outbreaks in 2002.
2. For a substantial portion of drinking water outbreaks, the pathogen is not identified and the outbreaks are classified as "acute gastrointestinal illness of unknown etiology" (AGI). From 1986 through 2002, approximately 41 percent of the over 250 outbreaks reported during this period were classified as AGI, and this proportion varies by reporting period from a peak of 68 percent in 1991–1992 to 17 percent in 1993–1994. Overall, *Giardia* and *Cryptosporidium* are the most commonly reported etiologic agents of waterborne disease when a pathogen is identified and are associated with about 20 percent of reported outbreaks associated with drinking water since the mid-1980s. However, with the recent addition of *Legionella* outbreaks to the surveillance system, *Legionella* is now the single most common cause of outbreaks involving drinking water (as discussed below).
3. Most drinking water outbreaks involve groundwater systems, especially untreated groundwater systems. Forty (40) percent of the 25 drinking water outbreaks reported between 2001 and 2002 involved untreated groundwater systems (Blackburn et al., 2004).

Declining Number of Drinking Water Outbreaks

Since the mid-1980s, the number of waterborne outbreaks has declined (Figure 3-2). The reason for the decrease is largely attributed to the promulgation of more stringent drinking water regulations, including the Surface Water Treatment Rule, the Total Coliform Rule, and others. In addition, many water utilities have made voluntary improvements, such as the Partnership for Safe Water program to reduce the risk of waterborne cryptosporidiosis. The Partnership program entails a comprehensive evaluation of treatment practices with a focus on achieving filtered drinking water turbidities less than 0.1 nephelometric turbidity units (NTU). The number of reported outbreaks began to decrease sharply beginning with the 1985–1986 reporting period; this was attributable

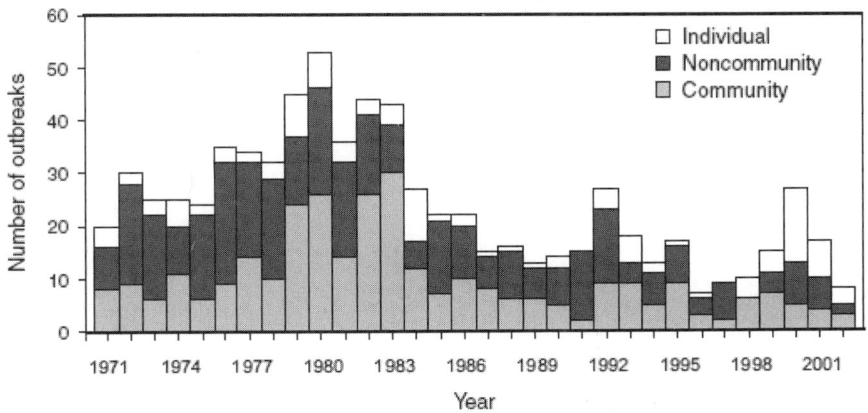

FIGURE 3-2 Number of drinking water disease outbreaks in the United States, 1971–2002. Individual—private or individual water systems (9 percent of U.S. population or 24 million users); Community—systems that serve > 25 users year round (91 percent of U.S. population or 243 million users); Noncommunity—systems that serve < 25 users and transient water systems such as restaurants, highway rest areas, parks (millions of users yearly). SOURCE: Blackburn et al. (2004).

primarily to fewer community and noncommunity outbreaks. With the institution and enforcement of better regulations that chiefly affect these types of water systems (particularly community systems), a marked drop in the number of outbreaks was seen. In contrast, the increase in outbreaks reported during 1999–2000 was attributable primarily to individual homeowner systems, which affect fewer persons, are less regulated, or are more subject to changes in surveillance and reporting. In 2001–2002, individual homeowner systems comprised 40 percent of the waterborne outbreaks (Figure 3-3).

Etiologic Agents Associated With Drinking Water Outbreaks

The agents responsible for waterborne disease outbreaks were predominantly undefined, microbial (parasitic, bacterial, or viral), or chemical. Indeed, surveillance data on waterborne disease outbreaks associated with drinking water in the United States from 2001 to 2002 indicate that almost 30 percent of reported outbreaks were due to bacterial agents, 16 percent were due to protozoa, 16 percent were due to viral agents, 16 percent were due to chemical contaminants, and 23 percent had an unidentified etiology. Figure 3-4 shows the

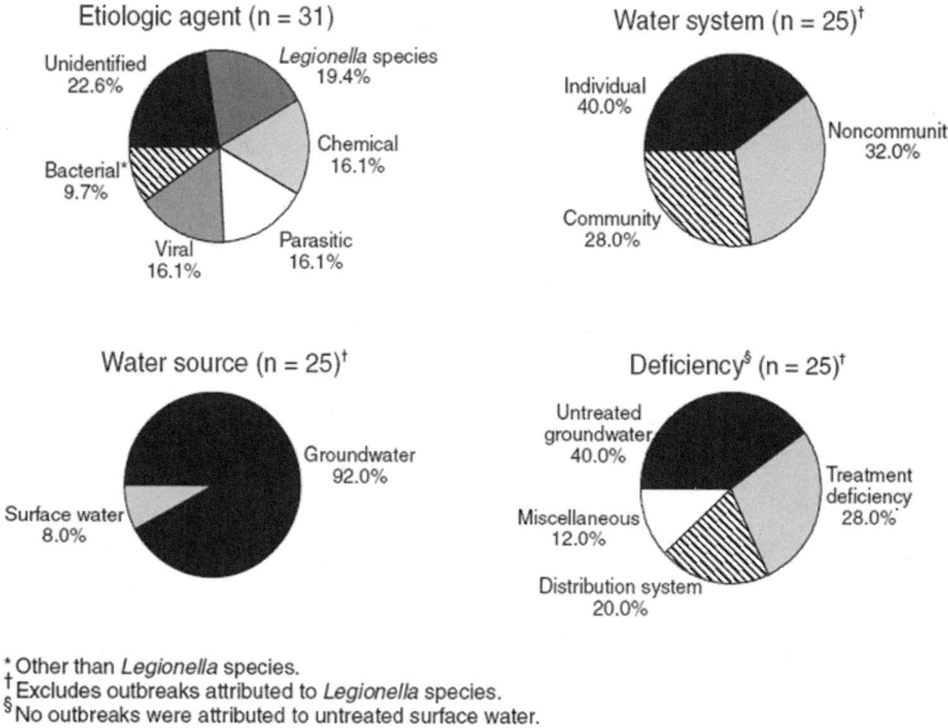

FIGURE 3-3 Waterborne outbreaks by etiological agent, water system, water source, and deficiency—United States, 2001–2002. SOURCE: Blackburn et al. (2004).

etiology of waterborne disease outbreaks over time. The large number of waterborne disease outbreaks associated with protozoa in the early 1980s was mostly caused by *Giardia* and was greatly reduced by the implementation of the Surface Water Treatment Rule in 1989 (Barwick et al., 2000). Relatively few outbreaks due to viruses have been reported, in part because of the difficulty of the detection methodologies for these organisms. However, the number of reported viral outbreaks has increased significantly since 1999 with the development of better diagnostic techniques for noroviruses. Nine of the 15 drinking water outbreaks associated with noroviruses that have been reported since 1986 occurred between 1999 and 2002 (Herwaldt et al., 1991; Moore et al., 1993; Kramer et al., 1996; Levy et al., 1998; Barwick et al., 2000; Lee et al., 2002a; Blackburn et al., 2004).

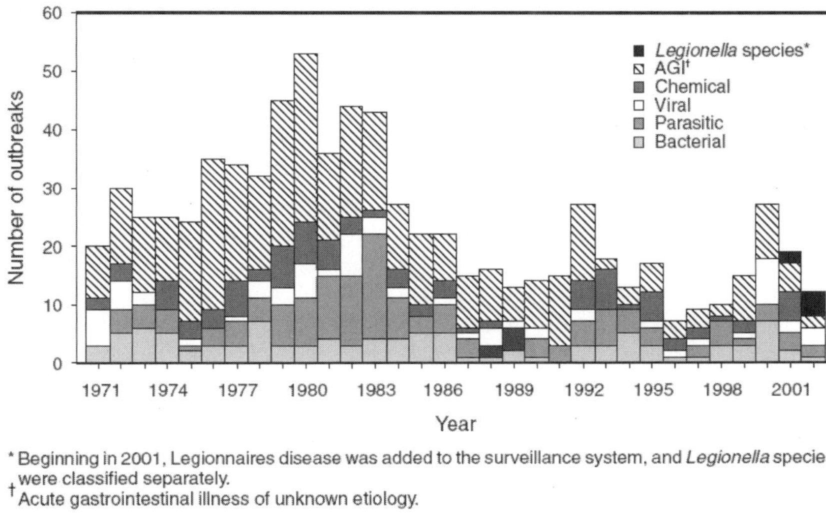

FIGURE 3-4 Agents responsible for waterborne outbreaks. SOURCE: Blackburn et al. (2004).

Over the past 30 years, there has been a wide range of chemical agents associated with drinking water outbreaks, including arsenic, benzene, chlordane, chlorine, chromate, copper, cutting oil, developer fluid, ethyl acrylate, ethylene glycol, fluoride, fuel oil, furadan, lead, leaded gasoline, lubricating oil, kerosene, nitrate, nitrite, phenol, polychlorinated biphenyls, selenium, sodium hydroxide, toluene, xylene, and unidentified herbicides. From 1993 through 2002, most drinking water outbreaks associated with chemical agents have been due to copper (eight outbreaks, usually related to premise plumbing) followed by nitrates/nitrites (six outbreaks, usually related to contamination of groundwater) (Kramer et al., 1996; Levy et al., 1998; Barwick et al., 2000; Lee et al., 2002a; Blackburn et al., 2004).

Outbreaks Associated With Groundwater Systems

In recent years, as treatment of surface water supplies has improved, waterborne outbreaks have increasingly involved groundwater supplies (Figure 3-3). There is increasing recognition that many groundwater supplies have microbial contamination, yet the use of untreated groundwater continues in many small communities and by individual homeowners. A survey of 448 wells in 35 states reported that 31 percent of the sites were positive for at least one virus, and enterovirus RNA was detected in approximately 15 percent, rotavirus RNA in 14 percent, and hepatitis A virus RNA in 7 percent of the wells by reverse-transcription-polymerase chain reaction (RT-PCR) (Abbaszadegan et al., 2003). Fout et al. (2003) examined 321 samples from 29 groundwater sites by RT-PCR and reported that 72 percent of the sites were virus positive. Borchardt et al. (2004) collected monthly samples from four municipal wells in one city in Wisconsin for a 12-month period and detected enteric viruses by RT-PCR in 50 percent of the samples. Two studies in Ontario, Canada examined the relationship between *E. coli* in well water and acute gastrointestinal illness in households using the water for drinking (Raina et al., 1999, Strauss et al., 2001). In the first study of 181 households with untreated well water, water samples were collected five times during the one-year study, and *E. coli* was detected in 20 percent of the household wells. The second study included 235 households in four rural communities (Strauss et al., 2001) and reported that 20 percent of the households had at least one water sample that exceeded the national standards for total coliforms or *E. coli*.

Outbreaks Associated With Distribution Systems

Among the seven outbreaks associated with community water systems in 2001–2002, four (57.1 percent) were related to problems in the water distribution system. Preliminary results from the 2003–2004 surveillance report indicate that distribution systems were associated with 38 percent of the outbreaks associated with drinking water systems during this period (Liang et al., 2006). Other epidemiological and outbreak investigations conducted in the last five years suggest that a substantial proportion of waterborne disease outbreaks, both microbial and chemical, is attributable to problems within distribution systems (Craun and Calderon, 2001; Blackburn et al., 2004) (see Figure 1-1). Craun and Calderon (2001) examined causes of reported waterborne outbreaks from 1971 to 1998 and noted that, in community water systems, 30 percent of 294 outbreaks were associated with distribution system deficiencies, causing an average of 194 illnesses per outbreak. Distribution system contamination was observed to be the single most important cause of outbreaks in community water systems over that time period.

The reason for the apparent increase in the proportion of outbreaks associated with water distribution systems is not entirely clear. Outbreaks associated with distribution system deficiencies have been reported since the surveillance system was started. However, there may be more attention focused on the distribution system now that there are fewer outbreaks associated with inadequate treatment of surface water. Also, better outbreak investigations and reporting systems in some states may result in increased recognition and reporting of all the risk factors contributing to the outbreak, including problems with the distribution system that may have been overlooked in the past. Although waterborne disease outbreaks in general are still under-reported, the surveillance system has become more mature, and outbreak investigations and analyses are becoming more sophisticated.

The CDC surveillance system for waterborne disease outbreaks attempts to collect information on outbreaks and their contributing causes. For example, from 1981 to 1998, the CDC documented 57 waterborne outbreaks related to cross-connections, resulting in 9,734 detected and reported illnesses (Craun and Calderon, 2001). Contamination from cross-connections and backsiphonage were found to cause the majority of the outbreaks associated with distribution systems (51 percent), compared with contamination of water mains following breaks (39 percent) and contamination of storage facilities (the remaining 10 percent). A separate compilation by the EPA of backflow events revealed many more incidents of backflow and resulting outbreaks—a total of 459 incidents resulting in 12,093 illnesses from backflow events from 1970 to 2001 (EPA, 2002). The situation may be of even greater concern because incidents involving premise plumbing are even less recognized.

Most reported outbreaks associated with distribution systems occur in community water systems because of their greater size and complexity. For example, from 1999 to 2002 there were 18 reported outbreaks in community water systems, and nine (50 percent) of these were related to problems in the water distribution system (Lee et al., 2002b; Blackburn et al., 2004). However, there have been a number of reported outbreaks associated with noncommunity water systems that have been attributed to deficiencies in the distribution system. Finally, the magnitude and severity of reported outbreaks associated with distribution systems vary, with an average about almost 200 illnesses per outbreak (Craun and Calderon, 2001) and a total of 13 deaths.

The Extent of Underestimation

The number of identified waterborne disease outbreaks is considered an underestimate because not all outbreaks are recognized, investigated, or reported to health authorities (Blackburn et al., 2004). For example, outbreaks occurring in national parks, tribal lands, or military bases might not be reported to state or local authorities. Factors influencing whether a waterborne outbreak is recognized include awareness of the outbreak, availability of laboratory testing, and resources available for surveillance and investigation of outbreaks. The detection and investigation of waterborne outbreaks is primarily the responsibility of the local, state, and territorial public health departments with varying resources and capacities. Differences in the capacity of local and state public health agencies and laboratories to detect an outbreak might result in reporting and surveillance bias, such that the states with the majority of outbreaks might not be the states with the majority of waterborne disease. Outbreaks are more likely to be recognized when they involve acute illnesses with symptoms requiring medical treatment, or when sensitive laboratory diagnostic methods are readily available. These and other limitations are discussed below.

Underreporting of Outbreaks Involving Individual Homeowner Systems

Although the surveillance system has always included outbreaks associated with individual homeowner water systems, it is likely that most sporadic cases and small clusters of waterborne disease associated with individual homeowner water systems are not recognized or reported because small numbers of people are involved. Furthermore, a cluster of cases of gastroenteritis within a single household may easily be attributed to food contamination or person-to-person transmission, such that the possibility of waterborne transmission may not be considered or investigated. From 1971 to 1980, 37 (11.6 percent) of the 320 reported drinking water outbreaks were associated with individual homeowner systems, and most of these outbreaks involved chemical agents when an etiologic agent was identified (Craun, 1986). From 1993 to 2002, 41 (28.7 percent) of the 143 reported drinking water outbreaks were associated with individual homeowner water systems, suggesting that there may be increased recognition and reporting of these smaller outbreaks in the past ten years of surveillance.

Underreporting of Outbreaks Involving Premise Plumbing

Outbreaks associated with premise plumbing are not specifically identified in the CDC surveillance reports. Adverse health effects associated with premise plumbing problems are less likely to be recognized and reported in this surveillance system, especially if they occur within a single household. However, a number of outbreaks associated with drinking water have been reported from public building settings such as schools, restaurants, churches, factories, and apartment buildings. Some of these outbreaks were due to contamination of a private well that serves the building. Other outbreaks in public buildings were classified as due to distribution system deficiencies and appeared to involve cross-connections and/or backsiphonage problems. Examples of the latter type of outbreak include:

- an outbreak of copper poisoning in the early 1980s that occurred when "backsiphonage of corrosive water containing carbon dioxide from a soda-mixing dispenser caused copper to be leached from piping in a building (Craun, 1986);
- a norovirus outbreak in 1995 at a high school in Wisconsin that affected 148 persons. The school was connected to the community water supply. However, water in the school became contaminated from backsiphonage of water from hoses submerged in a flooded football field (Levy et al., 1998);
- a chemical outbreak in 1995, in which 13 persons in a healthcare facility in Iowa became ill after drinking water that was contaminated with concentrated liquid soap. A valve on the water supply hose to the soap dispenser had been left open and allowed the soap to enter the water supply in the building.

Although the building had vacuum breakers to prevent backsiphonage, these were installed incorrectly at the soap dispensers (Levy et al., 1998);

- a chemical outbreak in 1999, in which four residents of an apartment building in Florida had acute gastroenteritis that was attributed to unidentified chemical poisoning. A cross-connection was discovered between their drinking water and an improper toilet flush-valve. Residents of the apartment had noticed on several occasions that their tap water was blue before the onset of illness (Lee et al., 2002a).
- a small waterborne disease outbreak at a middle school in Florida in 2001 due to a cross-connection between the air conditioning unit and the potable water supply. A maintenance worker used the potable water system to dilute the ethylene glycol solution in the chiller unit. The higher water pressure in the chiller unit forced the diluted ethylene glycol into the school's water supply and pink-colored water was observed in the school bathrooms. Three students became ill with gastrointestinal symptoms (Blackburn et al., 2004).

Underreporting of Outbreaks Involving Chemical Agents

From 1971 to 1980, 38 (11.9 percent) of the 320 reported drinking water outbreaks were attributed to chemical agents (Craun, 1986), and from 1993 to 2002, 25 (17.5 percent) of the 143 reported drinking water outbreaks were attributed to chemical agents (Kramer et al., 1996; Levy et al., 1998; Barwick et al., 2000; Lee et al., 2002a; Blackburn et al., 2004). The CDC believes that waterborne chemical poisonings are underreported for many reasons. First, most of these are probably due to copper and lead leaching from plumbing in private residences and affect relatively few people and are consequently unlikely to be recognized by public health authorities. Furthermore, exposure to chemicals in drinking water can often cause non-specific symptoms that may not be recognized as chemical poisoning or may not be linked to a specific chemical. The detection, investigation, and reporting of waterborne disease outbreaks linked to chemical exposures are not as well established as the methods for dealing with outbreaks associated with infectious agents. Finally, many physicians may have difficulty recognizing and diagnosing chemical poisonings unless they have had additional training in this area (Barwick et al., 2000).

Revisions of the CDC Waterborne Disease Outbreak Surveillance System

The CDC is making several changes to its waterborne disease outbreak surveillance system that are relevant to better understanding the role of distribution systems, including premise plumbing. Previously, the risk factors or deficiencies that contributed to a waterborne disease outbreak were classified as: (1) use of untreated surface water, (2) use of untreated groundwater, (3) treatment deficiency, (4) distribution system problem, or (5) miscellaneous. The 2003–2004

MMWR Surveillance Summary will use a new and detailed classification system for risk factors that contributed to the outbreak, and it will distinguish between deficiencies before or after entry into a building or home. This distinction is important because drinking water before it enters a building is usually managed by the water utility and subject to EPA drinking water regulations. However, drinking water problems that occur after entry into a building, such as those due to *Legionella* colonization in premise plumbing, cross-connections, point-of-use devices, or drink mix machines, may not be the responsibility of the water utility or regulated by EPA (lead and copper are an exception—see Chapter 2). Preliminary results from the surveillance system for 2003–2004 indicate that 48 percent of the outbreaks associated with drinking water were associated with deficiencies in source water, water treatment, and the distribution system and 52 percent of the outbreaks were due to deficiencies after the point of entry. In this latter group of outbreaks, approximately 47 percent involved *Legionella* and 35 percent involved chemical agents (including copper) (Liang et al., 2006). In addition, the surveillance system will now report all the identified deficiencies that contributed to the waterborne disease outbreak rather than reporting only the primary deficiency. Finally, CDC is moving toward a web-based system for reporting outbreaks and developing a public access database on waterborne disease outbreaks that will allow investigators to examine and analyze these data.

EPIDEMIOLOGY STUDIES

Three basic epidemiological study designs can be used to assess the public health risk of contaminated water supplies (Steenland and Moe, 2005): descriptive, correlational or ecological, and analytic. In the *descriptive* study, population surveys or systematic disease surveillance describe disease patterns by various factors such as age, seasonality, and geographic location. These studies do not test a formal hypothesis about the relation between a specific exposure (or risk factor) and disease, but they can help identify specific populations or geographic regions for further study. This category includes the systematic surveillance of outbreaks discussed in the previous section as well as endemic cases. Surveillance systems are useful for showing trends in the causes and risk factors of waterborne disease, but they are not very sensitive and cannot serve as a rapid warning system of a water-related health problem in a specific community because of reporting delays. In addition to the waterborne disease outbreak surveillance system, there is also a national system of notifiable diseases in the United States that mandates that health care providers report specific infections, including a number of potentially waterborne infections such as cholera, cryptosporidiosis, *E. coli* O157:H7, giardiasis, hepatitis A virus, legionellosis, poliomyelitis, salmonellosis, shigellosis, tularemia, and typhoid fever. Like the outbreak surveillance system, the surveillance for notifiable diseases is a voluntary passive surveillance system with low sensitivity and reporting delays. Finally, the descriptive framework has been used in the Foodnet surveillance program to assess occurrence of common gastroenteric illnesses in the population and gather information on the prevalence of various risk factors for diarrheal disease (such as food consumption habits, water consumption habits, and recreational water contact). Although the notifiable disease surveillance system and the Foodnet program provide valuable data on disease occurrence, they provide no information on what proportion of these diseases are related to drinking water.

Correlational or ecologic studies collect population level data on disease rates and exposures and look for correlations. For example, bladder cancer rates in cities with chlorinated surface water can be compared to cities with chlorinated groundwater to see if there may be a correlation between chlorination of surface water, formation of DBPs, and bladder cancer. However, these studies do not collect information on individual risk factors or confounders that may be related to risk of disease, such as smoking. Correlational studies do not test a formal hypothesis and are considered weaker than studies that collect individual-level data. But they can provide valuable information for generating hypotheses. Time-series studies are another example of correlational studies and have been used to examine the relationship between changes in water quality indicators (such as turbidity) and disease rates in the population served by the water supply (such as emergency department visits for gastroenteritis) (Schwartz et al., 1997). These studies have the advantage of comparing the same population at different points in time (thus controlling for confounding) so that only the variables that change are those that are being studied—i.e., water quality and disease rates.

Analytical studies are those in which individual-level data are collected, and the investigator tests a formal hypothesis about the association between exposure and disease. Analytical studies can be *experimental*, such as a clinical trial where some households are given bottled water to drink and other households are asked to drink tap water, and then disease rates between the two study groups are compared to determine the risk of disease attributable to drinking water. In these clinical trials, study participants are randomly assigned to a study group in order to ensure that other potential risk factors for disease are equally distributed among the study groups. An example of this design is the study of Colford et al. (2002) in which home water purification devices were installed in the homes of a test group of study participants and the control group consisted of homes in which "sham" devices were installed. Both groups kept health diaries to record symptoms of gastroenteritis and other health effects. At the end of the observation period, incident rates of disease were compared as a ratio, e.g., diarrhea episodes per person-year in the "exposed group" (those with the sham device) divided by diarrhea episodes per person-year in the "unexposed group" (those with additional purification).

Other analytical studies can be *observational* or natural experiments, where the investigator examines disease rates over time in study groups that have different exposures. Observational studies can use a cohort design, case-control design, or cross-sectional design. In the cohort design, all study participants are disease-free at the beginning of the study and disease rates over time are compared between study participants who are exposed to various risk factors vs. those who are not exposed. This design allows the consideration of multiple health outcomes and can be either prospective or retrospective. The cohort design is useful for rare exposures because the study deliberately recruits a cohort of individuals who are more likely to become exposed because of their occupation or geographic location. An example of this is the study of Frost et al. (2005), who assessed the illness rate of cryptosporidiosis and the presence of antibodies to *Cryptosporidium* in two populations (one exposed to surface water and one to groundwater). They concluded that populations receiving surface-derived water had higher antibody prevalence (but not higher illness rate) than individuals receiving groundwater. Cohort studies are not well suited for rare diseases because the purpose of this study design is to compare how frequently the disease occurs in the exposed group vs. the unexposed group. If the disease is rare, then a very large cohort must be recruited in order to make a meaningful comparison.

Case-control studies are often used to study rare diseases and start with recruiting a group of individuals with the disease of interest (cases) and another group of individuals without the disease (controls). The study individuals are then queried about their past exposure to the specific risk factors of interest. In a case-control study, the measure of association is the "risk odds ratio" which compares the odds of exposure to a specific risk factor among the cases to the odds of exposure among the controls. In contrast to the cohort study, a case-control study can look at only one health outcome but can examine multiple risk factors. An example of the case-control design is the study of Steinmaus et al. (2003) who examined associations of risk factors with bladder cancer in the western U.S. This study found no association of bladder cancer with daily arsenic ingestion in drinking water below 80 μg/day and found some association in smokers at ingestions of greater than 200 μg/d of arsenic.

Cross-sectional studies are similar to ecologic studies in that exposure rates and disease rates are measured at the same time. However, cross-sectional studies collect individual-level data whereas ecologic studies collect population-level data. Seroprevalence surveys are a form of cross-sectional study where, for example, prevalence of antibodies to *Cryptosporidium* can be measured in populations served by different types of water supplies. The use of epidemiological methods to study health risks associated with drinking water has been reviewed by Savitz and Moe (1997).

Descriptive Studies of Endemic Waterborne Disease

The risk of endemic waterborne disease (sporadic cases) is difficult to estimate, although various authors have made educated guesses. Bennett et al. (1987) estimated that the incidence of waterborne disease in the United States was 940,000 cases per year and resulted in 900 deaths. Although the purpose of the study was to rank the importance of various disease categories (water ranked next to the last, above zoonotic diseases) and to define the opportunities for prevention, the study has been criticized as little more than an exercise in guess work. Morris and Levin (1995) used incidence rates for enteric diseases and prevalence rates for specific groups of pathogens detected in water to give waterborne infectious disease estimates of 7.1 million mild infections, 560,000 cases of moderate or severe illness, and 1,200 deaths annually in the United States. The authors concluded, however, that available data were inadequate to refine the estimates.

Recent data on the incidence of diarrheal disease in the U.S. is available from the FoodNet population-based surveillance system (managed by the CDC). The disease estimates from the FoodNet system are based on telephone surveys that used random-digit-dialing and interviewed one individual per household to recall their occurrence of diarrhea in the four weeks prior to the interview. As shown in Table 3-4, the overall diarrhea prevalence rates from these surveys range from 5 to 11 percent, resulting in an estimated incidence of around 0.7 to 1.4 episodes/person/year. Diarrhea prevalence rates were consistently higher in children under five years of age.

Other CDC estimates based on the FoodNet data and other sources suggest that there are 211 million episodes of acute gastroenteritis in the United States each year that result in over 900,000 hospitalizations and 6,000 deaths (Mead et al., 1999). Mead et al. (1999) estimated the incidence of gastrointestinal illness to be 0.79 episodes/person/year. These FoodNet data are valuable for providing a measure of baseline diarrhea incidence in the U.S. population and the public health and economic burden associated with diarrheal diseases in an industrialized country. However, it is important to point out that these data offer no information on the proportion of diarrheal disease attributable to drinking water. Furthermore, these data probably underestimate the total burden of acute gastroenteritis in the population because cases with only vomiting were not included in the estimate (Imhoff, 2004), and vomiting is a common symptom for most gastroenteritis due to noroviruses and other viral agents.

TABLE 3-4 Burden of Diarrheal Disease in the U.S. based on FoodNet Telephone Survey Data

Year	No. of States	Total # respondents in analysis	Overall prevalence of acute diarrheal illness in past four weeks	Estimated incidence of episodes/person/ year	Diarrhea prevalence in children < five years old
1996–1997	5	8,624	11%	1.4	10%
1998–1999	7	12,075	6%	0.72	9%
2000–2001	8	14,046	5%	NA	9%
2002–2003	9	15,578	5%	NA	9%

NA = the authors did not report an estimate of the incidence rate.
SOURCES: Herikstatd et al. (2002); Imhoff et al. (2004) ; Hawkins et al. (2002) ; McMillan et al. (2004).

Analytical Epidemiological Studies

Determining the proportion of diarrheal disease that is attributable to water contamination is best done through analytical, experimental epidemiological studies. There have been four analytical epidemiological studies of acute gastroenteritis and drinking water systems relevant to distribution systems, all of which focused on risks from microbiological agents.

Laval Studies

Payment et al. conducted two epidemiology studies (Payment et al, 1991; Payment et al, 1997) in a suburb of Montréal known as Laval that examined the health of people who drank tap water and compared the group to people receiving water treated by reverse osmosis to determine which group had higher levels of gastrointestinal illness. In the 1991 study, reverse osmosis units were installed in 299 households (1,206 persons), and another 307 households (1,202 persons) were followed as controls with no device installed. Both groups were monitored for a 15-month period. Highly credible gastrointestinal illness (HCGI) was defined as (1) vomiting or liquid diarrhea with or without confinement to bed, consultation with a doctor, or hospitalization, or (2) nausea or soft diarrhea combined with abdominal cramps with or without absence from school or work, confinement to bed, consultation with a doctor, or hospitalization. The water source for the study area was a river that was contaminated by human sewage discharges, including combined sewer overflows. The community had a single water treatment plant with pre-disinfection, alum flocculation, rapid sand filtration, ozonation, and final disinfection with chlorine or chlorine dioxide. The quality of the finished water leaving the plant included an average of 0.6 mg/L total chlorine and approximately 0.4 mg/L free chlorine, an average turbidity of 0.26 NTU, and no detection of indicator bacteria or human enteric viruses in weekly samples (Payment et al., 1991). The overall incidence of highly credible gastroenteritis was 0.66 episodes/person/year and was highest in children five years of age and younger. The authors concluded that approximately 35 percent of the self-reported gastrointestinal illnesses was attributed to tap water consumption.

The 1997 study included groups receiving (1) regular tap water, (2) tap water from a continuously purged tap, (3) bottled plant effluent water, or (4) bottled plant effluent water purified by reverse osmosis. Differences in gastroenteritis rates between groups 1 and 2 versus group 3 was assumed to be due to changes in water quality that occurred between the time the water left the treatment plant and the time the water reached the household. The water ingested by group 1 represented tap water that had gone through the distribution system and also had residence time in the household plumbing. The water ingested by group 2 represented tap water quality in the distribution system without any significant residence time in the household plumbing. It should be noted that between the time of the first and second study, the water treatment plant was significantly upgraded with higher disinfection doses and better filtration. Estimated *Giardia* removal/inactivation exceeded 7.4 logs, and estimated virus inactivation by chlorine exceeded 10 logs. The average turbidity of the finished water was 0.1 NTU and never exceeded 0.5 NTU. However, periods of "micro-failures" in individual filters were reported (Susan Shaw, EPA, personal communication, 2006).

This second study attributed 14 percent to 40 percent of the gastrointestinal illness to the consumption of tap water (which met Canadian guidelines). Payment et al. (1997) concluded that the distribution system played a role in waterborne disease because the rates of HCGI were similar for group 3 (ingested purified bottled water) and group 4 (ingested bottled water from the treatment plant), but groups 1 and 2 (ingested water from the distribution system) had higher HCGI rates than group 4. Interestingly, there appeared to be no correlation between the relatively short residence time of the water in the distribution system (which varied from 0.3 to 34 hours) and the incidence of HCGI in a family. Furthermore, microbiological testing of the water in the distribution system did not indicate any bacterial indicators of contamination, but these water samples were not tested for viruses or protozoa. Contrary to their expectation, the investigators observed higher HCGI rates in families that ingested water from the continuously purged taps compared to families with regular tap water that may be subject to bacterial regrowth in household pipes. The investigators suggested that the shorter residence time for water from the continuously purged taps may have transported pathogens in the distribution system to the household sooner than regular tapwater and that there may have been inadequate contact time with residual chlorine in the distribution system to inactivate any introduced pathogens.

Transient pressure modeling (Kirmeyer et al., 2001) found that the distribution system studied by Payment et al. was extremely prone to negative pressures, with more than 90 percent of the nodes within the system drawing negative pressures under certain modeling scenarios (e.g., power outages). The system reported some pipe breaks, particularly during the fall and winter when temperature changes placed added stresses on the distribution system. Although the system employed state-of-the-art treatment, the distribution network suffered from low disinfectant residuals, particularly at the ends of the system. Low disinfectant residuals and a vulnerability of the distribution system to pressure transients (suggesting intrusion as a possible mechanism of contamination) could account for the observed illnesses.

Melbourne Study

A double-blinded, randomized trial was recently completed in Melbourne, Australia, to determine the contribution of drinking water to gastroenteritis (Hellard et al., 2001). Melbourne, with a population of about 3 million, draws its drinking water from protected forest catchments of the Upper Yarra and Thomson rivers. The catchments, which are approximately 1,550 square kilometers (600 square miles) in area, are closed to public access and have no permanent human habitation or activity except for logging in limited areas. Water from these catchments is stored in two major reservoirs (Silvan and Cardinia) with detention times of approximately two and 33 months, respectively. Water from both reservoirs is treated by chlorination, fluoridation (slurry or acid), and pH adjustment with lime.

Routine water quality monitoring at sampling points in the distribution system included total and fecal coliforms, HPC bacteria, and total and free chlorine. Free chlorine levels in the distribution system ranged from 0 to 0.94 mg/L, with a median of 0.05 mg/L, and 90 percent of samples had < 0.20 mg/L. Total coliform bacteria were detected in 18.9 percent of 1,167 routine 100-mL water samples, but fecal coliform bacteria were not detected. Median HPC concentrations were 37 CFU/mL with 13 percent of samples greater than 500 CFU/mL. During the study, water quality monitoring included testing a weekly composite sample from four water mains for selected pathogens: *Campylobacter sp., Aeromonas sp., Clostridium perfringens, Cryptosporidium sp.* and *Giardia sp.* These distribution system samples were positive for *Aeromonas* spp. (50 percent of 68 weekly samples), *Campylobacter* (one occasion), and *Giardia* (two positive samples by reverse transcriptase-PCR). No samples had detectable *C. perfringens* spores or *Cryptosporidium parvum* oocysts.

The study area in Melbourne is a growing area with relatively new houses and many families with young children. Six hundred (600) families (with at least two children one to 15 years of age) were recruited into the study. Approximately one third of the study households lived in areas of the distribution system with average water residence times of one to 1.5 days. Approximately two thirds of the study households lived in areas of the distribution system with average water residence times of three to four days (maximum six days).

Study households were randomly assigned to receive either a real or placebo water treatment unit installed under the kitchen sink. Functional units were designed to remove viruses, bacteria, and protozoa using microfiltration and ultraviolet light treatment. The study participants completed a weekly health diary reporting gastrointestinal symptoms during the 68-week observation period. The rates of HCGI ranged from 0.79/person/year for those with functional treatment units and 0.82/person/year with the sham devices. The study concluded that the water was not a source of measurable gastrointestinal disease (the ratio of illness rates between the group drinking treated water compared to the normal tap water was 0.99, with a 95 percent confidence interval of 0.85–1.15; p = 0.85). Analysis of 795 fecal specimens from participants with gastroenteritis did not reveal any difference in pathogen detection rates between the two groups.

This study was not designed to examine the risks from the distribution system separately from the risks associated with the entire water system. However, since there appeared to be no measurable contribution to illness due to drinking water, one may assume that the risks from degraded water quality in the distribution system were also below the detection limit of the study.

Davenport Study

The 1996 amendment to the Safe Drinking Water Act included a mandate to the CDC and the EPA to conduct studies to determine the occurrence of waterborne disease. To address this mandate, EPA scientists conducted several epidemiological studies of waterborne disease, and EPA funded several studies by external investigators, including the pilot study and full-scale study in Davenport, Iowa.

As a preliminary trial to the subsequent epidemiology study, a randomized, triple-blinded, home drinking water intervention trial of 77 households was conducted for four months in Contra Costa County, California (Colford et al., 2002). The drinking water was treated using an under-the-kitchen-sink device that incorporated ultraviolet light and microfiltration. Although the purpose of the trial was to evaluate the "blinding" of the study (e.g., could the participating households detect the active and identical-looking placebo devices), analysis of the data showed that the incidence rate ratio of highly credible gastrointestinal illness (HCGI) (incidence rate of the placebo group divided by the active device group, adjusted for clustering) was 1.32, with a 95 percent confidence interval of 0.75 to 2.33. Given the small study size, the higher rate of HCGI among the placebo group was not statistically significant. The authors concluded, however, that the relative rates of HCGI were consistent with those observed by Payment et al. (1991, 1997). This pilot study is interesting because it provides another estimate of self-reported HCGI rates in a cohort of households followed over time, and it confirmed that study subjects could successfully be blinded to the type of water treatment device they had during the intervention trial.

The full-scale Water Evaluation Trial was conducted in Davenport, Iowa to determine the incidence of gastrointestinal illness associated with consumption of drinking water meeting all federal and state treatment guidelines (LeChevallier et al., 2004; Colford et al., 2005). The municipal water system used a single source (the Mississippi River) and was treated at a single plant with conventional treatment consisting of coagulation, flocculation, sedimentation, prechlorination, filtration (dual filters with granular activated carbon and sand), and post-filtration chloramination. The average turbidity of the finished water was 0.05 NTU.

A total of 456 households with 1,296 participants were randomized into two groups. One group received a household water treatment device with a 1-micron absolute ceramic filter and UV light with 35,000–38,000 uW-second/cm^2 output. The other group received a sham device that was identical to the active device but had an empty filter chamber and a UV light that was shielded to block the transmission of radiation but still generated the same light and heat as the active unit. Each study household had an active device for six months and a sham device for six months and was blinded to the status of their device during the study. Study participants recorded the occurrence of any symptoms in daily health diaries. HCGI was defined as in the previous studies as (1) vomiting, (2) watery diarrhea, (3) soft diarrhea and abdominal cramps, or (4) nausea and abdominal cramps.

Incidence of HCGI varied by season and ranged during the study period from 1.64 to 2.80/person/years at risk (Wade et al., 2004). The overall HCGI rate for households with the sham device was 2.12 episodes/person/year and 2.20 episodes/person/year for households with the active device. The overall HCGI rate for the entire study population was 2.16 episodes/person/year. Multivariate analyses showed no effect of the household water treatment device on illness rates during the 12-month study period. As in the studies by Payment et al., the highest illness rates were in children five years of age and younger. The overall conclusion was that less than 11 percent of the gastrointestinal illness observed in this community was due to drinking water. Unlike the studies by Payment et al., this study included households without children, and it is possible that the number of young children in the study was too small to be able to detect an effect in this more vulnerable group.

United Kingdom Study

A study conducted in Wales and northwest England from 2001 to 2002 found a very strong association ($p < 0.001$) between self-reported diarrhea and reported low water pressure at the home tap based on a postal survey of 423 subjects (Hunter et al., 2005). This study was part of a larger case-control study of risk factors associated with sporadic cryptosporidiosis and was not specifically designed to study waterborne disease. Cryptosporidiosis cases and controls were identified from family physician practices in Wales and northwest England, and a postal survey asking a number of questions about potential risk factors for diarrhea was mailed to 662 cases of cryptosporidiosis and 820 controls. The survey included questions on travel outside the U.K., eating habits, food preparation habits, contact with animals, contact with young children, consumption of unboiled water, contact with other persons with diarrhea, and age. Questionnaires were returned by 427 controls, and 423 were included in the analyses. Of these, 28 (6.6 percent) reported having diarrhea in the two weeks before receiving the survey.

Four risk factors for diarrhea in the control group remained significant in the logistic regression model using a stepwise comparison strategy: feeding a child under five years old, contact with another person who had diarrhea, loss of water pressure at home, and how often the subject ate yogurt. The first three risk factors had a positive association with diarrhea (Odds Ratios of 2.5, 7.0, and 12.5, respectively, after adjusting for the effects of the other variables in the model). Yogurt consumption had a protective effect against diarrhea and showed a dose-response relationship (more frequent consumption was associated with lower risk). The investigators suggested that the strength of the association between loss of water pressure and risk of diarrhea indicates that this was not a spurious association and was not likely to be affected by recall bias because it was just one of many potential risk factors that was investigated.

The study populations were drawn from two large regions that include both heavily industrialized areas and rural areas and about 240 water treatment plants. The overall microbiological water quality for the utilities in these regions was described to be excellent with less than 0.05 percent of water samples positive for *E. coli* during this study period. The investigators hypothesized that most of the reported episodes of pressure loss were due to main breaks in which contamination entered the distribution system. However, no attempt was made to collect information on recorded main breaks in the systems where the controls lived. The investigators concluded that up to 15 percent of gastrointestinal illness may be associated with consumption of drinking water that was contaminated from main breaks or other pressure loss events, and that the associated costs of this illness should be taken into account when weighing the costs of replacing aging water supply distribution systems. Although there had previously been concern about possible health risks from pressure loss and pathogen intrusion in water distribution systems (LeChevallier et al., 2003), this was the first study to provide solid evidence of that risk, with policy implications for how to manage low pressure events in public water supplies.

***

The body of evidence from these epidemiological studies does not eliminate consumption of tap water that has been in the distribution system from causing increased risk of gastrointestinal illness. The conflicting results between the Laval and U.K. studies, which indicated risk associated with distribution system water, versus the Melbourne and Davenport studies, which showed no increased risk of gastrointestinal illness associated with tap water, may be due to a number of differences between the study designs and the individual water systems.

With respect to the latter, all four cohort studies were in cities that used surface water supplies. In Laval and Davenport, the rivers received upstream sewage discharges and were known to be contaminated. With the Davenport study in particular, it is possible that the reason they found no contribution to disease from the water supply was because the investigators chose a well-operated and maintained system. In Melbourne, the source water came from a highly protected watershed. In Laval and Davenport, the water treatment plants used conventional filtration and disinfection—indeed, Laval had both ozonation and chlorination although the average turbidity of the finished water during the first study was quite high (0.26 NTU). The water treatment plant in Melbourne did not practice filtration. There is no information on the water supplies in the U.K. study. Little to no information on the distribution systems was provided in the descriptions of the Laval or Melbourne studies except that the residence time in the Laval system was relatively short (0.3 to 34 hours), while the residence time for most of the study families in the Melbourne study was 72 to 96 hours.

Differences in study design such as population size and composition and follow-up period also played a role. As shown in Table 3-5, the size of the study population in the Davenport study is approximately half of the study population in the Laval and Melbourne studies (although the Davenport study used a cross-over design to try to compensate for the smaller sample size). The Davenport study also had the shortest follow-up period of the four studies. Unlike the Laval and Melbourne studies that only recruited households with children, households enrolled in the Davenport study were not required to have children, and the average household size was smaller in the Davenport study (2.84 persons) compared to the Laval and Melbourne studies (Laval 1988–1989: 3.97 persons; Laval 1993–1994: 3.84 persons; Melbourne: 4.69 persons). The smaller sample size, shorter follow-up period, and possibly lower proportion of children (a vulnerable sub-population), may be reasons why the Davenport study did not detect a significant risk of waterborne illness.

TABLE 3-5 Comparison of Population Parameters from the Epidemiology Studies

Study	Laval 1988-1989	Laval 1993-1994	Melbourne 1997-1999	Davenport 2000-2002
# households in tapwater group	307	346 (tap water) 330 (tap w/valve)	300	229
# of persons in tapwater group	1,202	1,296 (tap water) 1,300 (tap w/valve)	1,399	650
# households in purified water group	299	339 (purified) 354 (bottled plant)	300	227
# of people in purified water group	1,206	1,360 (purified) 1,297 (bottled plant)	1,412	646
% children in tapwater group	6.2 <6 yrs	12.8 <6 yrs (tap) 16.5 <6 yrs (tap valve)	40.2 < 10 yrs	NA
% children in purified water group	9.6 <6 yrs	15.1 <6 yrs (purified) 15.4 <6 yrs (bottled plant)	40.9 < 10 yrs	NA
Weeks of observation time	Approx 60	Approx 69	68	54

Statistical power in a cohort study is determined by the size of the study population, the follow-up time, and the frequency of the health outcome of interest (incidence of HCGI), with the number of outcomes being more relevant that the size of the study population (Hulley and Cummings, 1988). The Davenport study was designed to have the statistical power to detect an 11 percent or greater risk of HCGI due to water (Colford et al., 2005). The Melbourne study, with the larger sample size and longer follow-up period, was designed to detect a 15–20 percent reduction in the overall rate of HCGI in the group with the active point-of-use treatment devices. However, the total number of HCGI episodes measured in both study populations was very similar (tap water: Melbourne = 1,500 episodes, Davenport = 1,431 episodes; purified water: Melbourne = 1,459 episodes, Davenport = 1,476 episodes). Thus, the higher HCGI rates detected in the Davenport study and the cross-over design appear to have mitigated the effects of the smaller sample size and shorter follow-up period on the statistical power of the study. As shown in Table 3-5, all of these studies had relatively large study populations and measured thousands of illness episodes, and thus had similar statistical power.

There was limited assessment of exposure among the studies. All of the studies monitored water quality at the treatment plant, but there was a wide range in the amount of sampling and analyses of water in the distribution system. For example, monitoring in the Davenport study was extensive, with tap water samples and treatment device samples collected from about one-fourth of the study households at three times during the study. They documented higher coliform and HPC levels in water from the treatment devices compared to tap water (LeChevallier et al., 2002). None of the studies reported pathogen detection in the tap water, except for three occasions in the Melbourne study. It should be noted that the microbiological analyses of water differed for each study. Finally, all four studies attempted to measure the volume of tap water ingested via surveys, and these surveys indicated that subjects in the purified water groups also consumed regular tap water (reported range 14.5 to 40 percent).

All four cohort studies used similar approaches for recording symptoms of gastrointestinal illness and similar definitions of HCGI. Different rates of HCGI were observed in the four cohort studies. It is striking that the rates reported by the Davenport study and the Contra Costa County pilot study are more than twice as high as the rates reported by the Laval and Melbourne studies and about three times higher than the FoodNet rates of diarrheal disease (see Table 3-6). The reason for these higher rates is unknown because the investigators state that they used similar case definitions as the Laval and Melbourne studies. If there were several significant transmission routes of enteric pathogens in these communities that were responsible for these higher reported illness rates, then an intervention study targeted only to waterborne disease transmission may not show any effect (see Briscoe, 1984). However, the use of the cross-over design in Davenport should have been valuable in this regard because the effect of other transmission routes is better controlled for using this design.

TABLE 3-6 Rates of Highly Credible Gastrointestinal Illness from the Epidemiology Studies

Study	Estimated rate* of HCGI in tap water groups	Estimated rate* of HCGI in purified water group	Estimated rate* of HCGI in all study participants
Laval 1988-1989	0.76	0.50	0.66
Laval 1993-1994	0.66 (tap) 0.70 (tap valve)	0.58	0.60
Melbourne 1997-1999	0.82	0.79	0.80
Contra Costa County, CA 1999	3.48	2.63	3.05
Davenport 2000-2002	2.12	2.20	2.16
FoodNet	ND	ND	Approx 0.72

* rate expressed as episodes/person/year

The conflicting results of these epidemiological studies raise a number of questions. The fact that these were carefully conducted studies by research teams with considerable experience implies that there are detectable elevated risks of waterborne disease associated with some water systems and not others. However, not enough information was gathered to know what characteristics of the water systems posed increased risk, whether it be the source water, the treatment plant, or the distribution system.

For the studies that showed no detectable association between gastrointestinal symptoms and consumption of tap water (Melbourne and Davenport), it is not clear if they suffered from an inadequate design and sample size in order to detect an association, or if there simply was no association. The randomized clinical trial design used in Laval, Melbourne, and Davenport is one of the most rigorous analytical study designs and is less likely to be affected by error and confounding. However, it is possible that selection bias in the recruitment of the study population, misclassification of drinking water exposure, or inaccurate reporting of health outcomes may have affected the results of these studies. It must be kept in mind that epidemiological studies are not able to prove that there is zero risk associated with a specific exposure; they can only report that the risk is below the level that the study had the power to detect, which was 15 to 20 percent (Melbourne) or 11 percent (Davenport).

For the studies that did show an association between gastrointestinal symptoms and consumption of tap water (Laval study), or an association between gastrointestinal symptoms and a water pressure drop (UK study), it is not clear what portion of the observed risk was due to water contamination in the distribution system as opposed to water contamination at the source and/or inadequate water treatment. The second Laval study examined the risks associated with the distribution system by including a study group that received bottled plant effluent as well as groups that ingested tap water and continuous-flow tap water ("tap valve" group). Tap water drinkers had elevated risk of HCGI compared to those who ingested bottled water from the treatment plant or purified bottled water, suggesting that water in the distribution system posed an increased health risk (although routine water quality monitoring of the distribution system did not provide evidence of compromised quality). However, there was also an indication of some increased risk of illness from water with *reduced* residence time in the distribution system (tap valve group) compared to water with average residence times (from 0.3 to 34 hours in this system). This suggests that additional contact time with disfinectants in the distribution system may be helpful in reducing risks. The UK study suggests that pressure drops in the distribution system was associated with increased gastrointestinal illness, but this association needs to be tested more systematically and rigorously in further studies.

One of the major challenges for designing an epidemiology study of health risks associated with water quality in the distribution system is separating the effect of source water quality and treatment from the effect of distribution system water quality. Knowledge of how water distribution systems become contaminated from anecdotal evidence and outbreak data (main breaks, sudden changes in pressure and intrusion, backpressure or backsiphonage, etc.) suggests that the exposure to contamination in the distribution system is likely to be intermittent and may be very difficult to capture in an epidemiological study. Nonetheless, new approaches to deal with this challenge were tested in a pilot study in the southeastern U.S. and a third approach is being tested in a study in the Midwestern U.S. These studies were designed by multidisciplinary teams of university and research foundation scientists with input from outside experts including EPA and CDC staff. Support for these studies came from the EPA STAR Grant Program, and they are part of a series of studies funded by or conducted by the EPA to develop a national estimate of waterborne disease risks. These three approaches are described in Box 3-3 as examples. Other study designs may also be useful for addressing the question of endemic disease risks associated with water quality in the distribution system.

RISKS FROM *LEGIONELLA*

The role of biofilms and microbial risk can best be illustrated by the example of the bacterium *Legionella pneumophila* in water systems, for which occurrence data, outbreak data, and epidemiological data are available. *Legionella* are widely distributed in the aqueous environment and have been found in drinking water (Stout et al., 1985; Rogers et al., 1994) and biofilms (Rogers et al., 1994; Pryor et al., 2004; Thomas et al., 2006). Although the bacteria have been isolated from biofilms in water distribution systems, there is evidence that the

BOX 3-3
Three Approaches to Designing an Ideal Epidemiology Study that would Determine the Distribution System Component to Waterborne Disease

Method 1

This method relies on conducting a vulnerability assessment of the water distribution system and identifying areas in the distribution system that are more vulnerable and less vulnerable to contamination—based on pipe age and composition, history of main breaks, history of coliform detections, estimates of residence time, and chlorine residual. The study population (families with one or more children < six years old) should be recruited in the most vulnerable and the least vulnerable geographic areas of the distribution system. It is important to randomize the study population in each geographic area into two groups. The researchers would provide purified bottled water to half of the study households, and ask the other half of the study population to drink tap water. All study households would be asked to record health symptoms in a health diary. The difference in the rates of reported gastrointestinal symptoms (GI) for families drinking tap water to the rates for families drinking purified bottled water would then be compared. This difference (GI_{tap}-GI_{bottle}) represents the risk of GI symptoms due to source water and distribution system water. Part of the analysis would be to compare this difference (GI_{tap}-GI_{bottle}) for the study populations in the most vulnerable areas (where the degradation of distribution system water quality would be the greatest) to the difference (GI_{tap}-GI_{bottle}) for the study populations in the least vulnerable areas (where there should be little or no impact from degradation of water quality in the distribution system). This difference between the study groups in different parts of the distribution system should represent the impact of the distribution system on risk of GI illness (see Figure 3-5). Although the study is not blinded, the technique of "comparing the difference of the difference" controls for lack of blinding. This "double-difference methodology" is commonly used in economics studies and program evaluation to assess the impact of a specific intervention by comparing the differences between intervention and control groups at baseline and at a follow-up time point (Maluccio and Flores, 2005).

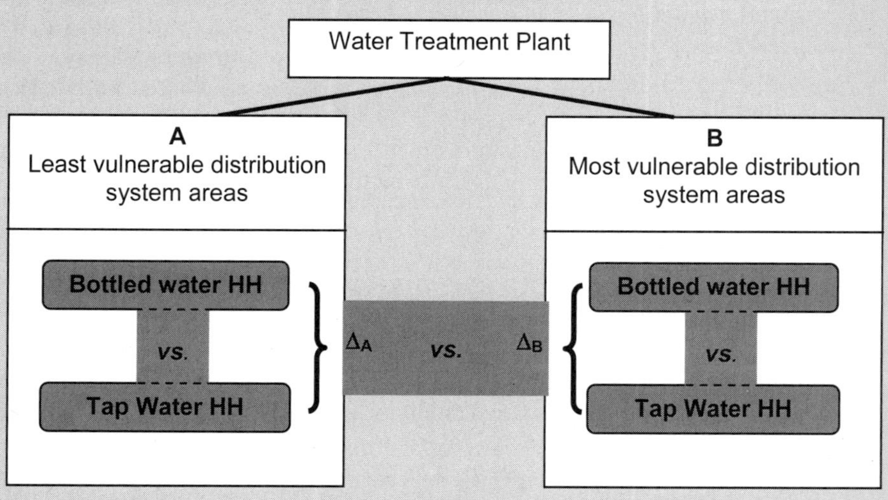

FIGURE 3-5 Study Design to Examine Risks from Water Quality in the Distribution System: Method 1.

Method 2

This approach is identical to the first, except that the study population in each geographic area is randomized into *three* groups. The researchers would provide purified bottled water to one-third of the study households, bottled finished water directly from the treatment plant to the other third of the study households, and bottled water from the most vulnerable part of the distribution system to the final third of the study population. As before, study households would be asked to record health symptoms in a health diary. This study, which has a cross-over design, is shown in Figure 3-6. The advantage of this approach over the first approach is that the study is blinded because everyone receives bottled water. Furthermore, one can recruit study subjects in any geographic location because drinking water is delivered to their home. This design is similar to a human challenge study because the investigators control exposure to the study water. The disadvantages are that bottled distribution system water will not capture temporal changes in water quality. Also, possible changes in water quality during bottling and storage may not reflect quality of distribution system water. However, these disadvantages could be mitigated by detailed microbiological studies of distribution system water quality in the study site prior to starting the epidemiologic study, bottling the distribution system water more frequently, bottling composite samples of the distribution system water over time and geographic area, and characterizing changes in distribution system water quality during bottling and storage.

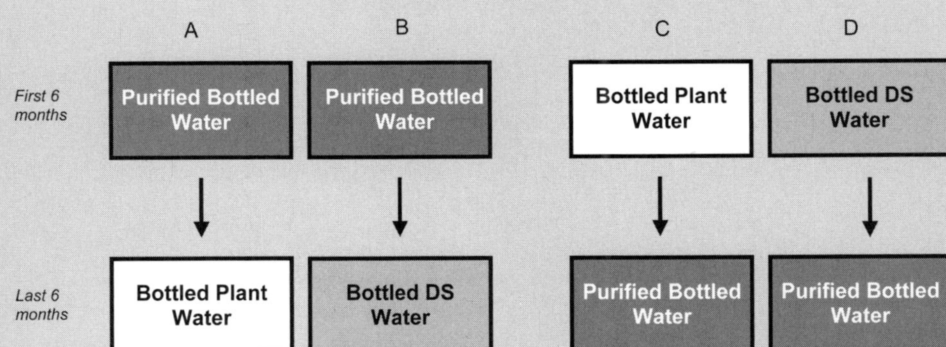

176 households per group (average); Total 704

<u>Power estimates</u>:
GI risk due to source water quality and treatment efficacy: 91%
GI risk due to distribution system: 86%

Assumptions:
20% attrition
20% variance inflation due to clustering

FIGURE 3-6 Study Design to Examine Risks from Water Quality in the Distribution System: Method 2 Cross-over Study.

continues

BOX 3-3 Continued

Method 3

A third approach is being attempted in the Wisconsin groundwater study (WAHTER) in several communities that use untreated groundwater. This study uses a community level intervention where UV disinfection is added at the wellhead, and community gastrointestinal symptom rates are compared before and after the UV intervention. The risk from the distribution system will be estimated using a risk assessment approach. Enteric virus concentrations are being measured in water samples from well heads (representing contamination in the groundwater) and compared to virus concentration measurements in water samples from study households (representing contamination from both the groundwater and the distribution system). The difference in virus concentration will be attributed to the distribution system. In those study communities with UV disinfection installed at the wellheads, viruses measured at the households could only have originated from intrusions into the distribution system. Note that the feasibility of this approach depends on studying a water supply where pathogens are detected with some frequency. For a water supply where a high proportion of water samples do not have detectable pathogens, the application of this study design is uncertain.

The study also measures the incidence of gastrointestinal symptoms in a cohort of children in the study communities using a health diary. The researchers intend to model the illness rate in the study population as a function of household pathogen concentration using dose-response models where incidence of acute gastrointestinal illness in the study population is a function of the pathogen dose in the household water (calculated as concentration of virus in the volume of water ingested over a defined period of time). The investigators will then use quantitative risk assessment to estimate the community illness rates if the population drank water directly from the wellhead. The difference between the measured illness rates in the study population and the estimated illness rates associated with source water will represent the risk from pathogens in distribution system.

One of the challenges of this approach is that there are different dose-response relationships for different waterborne viruses. Thus, information on the etiology of the predominant viral infections in the community will be used to guide the modeling analyses.

SOURCE: Available online at *http://cfpub.epa.gov/ncer_abstracts/index.cfm/fuseaction/display.abstractDetail/abstract/7430/report/0*. Accessed August 10, 2006.

organism must be taken up by protozoa to proliferate (Nahapetian et al., 1991; Barbaree et al., 1986; Barbaree, 1991; Murga et al., 2001). Some studies have reported that the presence of amoebae is a predictor of *Legionella* colonization in plumbing systems (Moore et al., 2006).

Levels of legionellae in potable water systems are typically low, but amplification can occur in cooling towers, recirculating hot water systems, and hot tubs (EPA, 1999). *Legionella* species have been shown to proliferate in biofilms in institutional and premise plumbing (Pryor et al., 2004; Thomas et al., 2006) and can be found in water heaters, shower heads, and cooling towers (Wadowsky and Yee, 1983, 1985; Stout et al., 1985; Rogers et al., 1994). Indeed, in a study of legionellosis in the United Kingdom, 528 of the examined 604 cases were attributed to contaminated cooling towers, 70 (or 12 percent) were caused by contaminated drinking water, and six were caused by contaminated whirlpools (VROM, 2005).

Legionella is an example of an organism that is an efficient pulmonary pathogen when inhaled as large aggregates or biofilm fragments. Inhalation of large numbers of the bacteria overwhelms the pulmonary defenses, and Pontiac fever results. Aspiration of smaller numbers of organisms as biofilm fragments may cause Legionnaire's disease. Epidemiological studies have linked water contaminated with both *Legionella* and protozoa to outbreaks of legionellosis (Fields et al., 1989; Breiman et al., 1990). A review paper by Lin et al. (1998) suggests that hospitals take routine samples for the organism in their distribution systems and determine the efficacy of any disinfection processes by measuring a reduction in *Legionella* counts.

Legionella are specifically mentioned in the EPA's Surface Water Treatment Rule, with the MCLG set at zero. For this reason, the bacterium was not included on the Contaminant Candidate List for methods development and potential future regulation. However, there is little evidence that filtration and disinfection of surface water prevents the growth of *Legionella* species in distribution system plumbing. In fact, since *Legionella* was incorporated into the waterborne disease outbreak surveillance system starting in 2001, several outbreaks have been attributed to the microorganism. During 2001–2002, the six drinking water outbreaks attributed to *Legionella* species (19.4 percent of the total) caused illness in 80 persons and resulted in 41 hospitalizations and four deaths. All of these outbreaks occurred in large buildings or institutional settings and were related to multiplication of *Legionella* species in the respective distribution systems. As mentioned previously, *Legionella* is now the single most common cause of outbreaks involving drinking water (Liang et al., 2006). These outbreaks underscore the importance of remaining vigilant about the possibility of growth of *Legionella* species in building complexes and the need to take measures to reduce this threat (see Chapter 8).

In an epidemiological study, Kool and colleagues (1999) examined 32 nosocomial outbreaks of Legionnaires' disease from 1979 to 1997 where drinking-water was implicated and tabulated the characteristics of the hospital (size, transplant program) and the primary disinfectant treatment, disinfectant residual, water source, community size, and pH of the water. The researchers found that the odds of a nosocomial *Legionella* outbreak was 10.2 (95 percent confidence interval of 1.4–460) times higher in systems that maintained free chlorine than in those using a chloramine residual. They estimated that 90 percent of waterborne *Legionella* outbreaks could be prevented if chloramine was universally used. Heffelfinger et al. (2003) reported that 25 percent (38) of 152 hospitals surveyed had definite reported cases or outbreaks of hospital-acquired Legionnaires' disease during the period 1989 to 1998. However, hospitals supplied with drinking water disinfected with monochloramine were less likely (odds ratio 0.20; 95 percent confidence interval, 0.07 to 0.56) to have hospital-acquired Legionnaires' disease than other hospitals. Cunliffe (1990) reported that suspensions of *Legionella pneumophila* were more sensitive to monochloramine disinfection, with a 99 percent level of inactivation when exposed to 1.0 mg monochloramine/L for 15 minutes, compared with the 37-

minute contact time required for *Escherichia coli* inactivation under similar conditions. Donlan et al. (2002) reported that monochloramine was significantly more effective than free chlorine at eradicating laboratory-grown biofilms of *L. pneumophila*.

Legionella has also been the subject of pathogen occurrence measurements. Researchers at the CDC conducted a study of *Legionella* occurrence in 53 public buildings before and after the conversion of the San Francisco water supply from free chlorine to chloramine (Fields, 2005; Flannery et al., 2006). They showed that the concentration of legionellae was reduced more than 20-fold by the conversion from free chlorine to chloramine. Interestingly, the incidence rate of *Legionella* infections was low (only one laboratory-confirmed case in the two years prior to the switch to chloramine) despite the fact that the major serotype detected included the clinically significant *Legionella pneumophila* serogroup 1. The results illustrate the difficulty in relating the detection of microbes in drinking water to a documented risk of waterborne disease.

Another recent study examined the impact of switching from chlorine to monochloramine disinfection on *Legionella* occurrence in Pinellas County, Florida (Moore et al., 2006). In this study, water samples were collected from 96 buildings (public buildings and individual homes) for a four-month period when chlorine was the primary disinfectant and from the same sampling sites for a four-month period after monochloramine was introduced into the municipal water system. In the first period, 20 percent of the buildings were colonized with *Legionella* in at least one sampling site. *Legionella* colonization was reduced by 69 percent within a month after monochloramine introduction. Monochloramine appeared to be more effective in reducing *Legionella* in hotels and single-family homes than in county government buildings, perhaps because of more consistent water usage. As in the San Francisco study, the reported incidence of legionellosis in the study area during this time was too low (nine cases) to determine if the change to monochloramine had an impact on human disease.

Given that 20 percent of reported outbreaks involving drinking water are attributed to *Legionella*, additional attention should be given to the control of this potential pathogen, especially in institutional and premise plumbing (see Chapter 8).

CONCLUSIONS AND RECOMMENDATIONS

Accurate estimates are not yet available for the prevalence of adverse health effects attributable to deficiencies in distribution systems from pathogen occurrence measurements, waterborne disease outbreak surveillance, or epidemiological studies. Pathogen occurrence measurements are rare due to limitations in detection methods and cost issues. Models to quantitatively predict pathogen occurrence in distribution systems (e.g., by cross-connections, main breaks, or intrusion) have not yet been developed. Despite under-reporting and limited data on risk factors, the voluntary waterborne disease outbreak surveillance system provides the best available evidence of public health risks associated with distribution systems in the United States. These data suggest that about one-third to one-half of reported waterborne disease outbreaks are associated with distribution system problems. To date, only one epidemiological study (the second Laval study) has been specifically designed to examine the contribution of the distribution system to endemic disease occurrence. Until better data are available from these three approaches, it will not be possible to accurately assess the magnitude of the health impacts resulting from distribution system deficiencies. The following conclusions and recommendations are made.

The distribution system is the remaining component of public water supplies yet to be adequately addressed in national efforts to eradicate waterborne disease. This is evident from data indicating that although the number of waterborne disease outbreaks including those attributable to distribution systems is decreasing, the *proportion* of outbreaks attributable to distribution systems is increasing. Most of the reported outbreaks associated with distribution systems have involved contamination from cross-connections and backsiphonage. Furthermore, *Legionella* appears to be a continuing risk and is the single most common etiologic agent associated with outbreaks involving drinking water. Initial studies suggest that the use of chloramine as a residual disinfectant may reduce the occurrence of *Legionella*, but additional research is necessary to determine the relationship between disinfectant usage and the risks of *Legionella* and other pathogenic microorganisms.

Distribution system ecology is poorly understood. There is very little information available about the types, activities, and distribution of microorganisms in distribution systems. Limited HPC data are available for some systems, but these data are not routinely collected, they underestimate the numbers of organisms present, and they include many organisms that do not necessarily present a health risk. To more adequately assess risk, more information on the microbial ecology of distribution systems, including premise plumbing, is needed.

There is inadequate investigation of waterborne disease outbreaks associated with distribution systems, especially in premise plumbing. *Legionella* has only recently been added to the outbreak surveillance system. Existing data on outbreaks due to other etiologic agents would rarely implicate premise plumbing because backflow and regrowth events likely would not be recognized and reported unless an institutional building with large numbers of people was affected. The Centers for Disease Control and Prevention are commended for revising the format used to report waterborne disease outbreaks to the surveillance system such that outbreaks arising from events in premise plumbing are now more clearly identified.

Epidemiology studies that specifically target the distribution system component of waterborne disease are needed. Recently completed epidemiological studies have either not focused on the specific contribution of distribution system contamination to gastrointestinal illness, or they have been unable to detect any link between illness and drinking water. Epidemiological studies of the risk of endemic disease associated with drinking water distribution systems need to be performed and must be designed with sufficient power and resources to adequately address the deficiencies of previous studies.

This chapter highlights the lack of information available to assess the public health risk of contaminated distribution systems. One of the consequences of this fact is that the committee was forced to rely heavily on its best professional judgment to prioritize contamination events into high, medium, and low priority (see Appendix A). Better public health data, including data on waterborne outbreaks, from epidemiological studies, and on distribution system water quality, could help refine distribution system risks and provide additional justification for the rankings.

The following three chapters consider the roles of physical, hydraulic, and water quality integrity. Protection of public health requires that water professionals incorporate approaches that combine all three into a comprehensive program of best practices to maintain the highest level of water quality.

REFERENCES

Abbaszadegan, M., M. LeChevallier, and C. Gerba. 2003. Occurrence of viruses in U.S. groundwaters. J Amer. Water Works Assoc. 95(9):107–120.

Alavandi, S. V., M. S. Subashini, and S. Ananthan. 1999. Occurrence of haemolytic and cytotoxic *Aeromonas* species in domestic water supplies in Chennai. Indian J. Med. Res. 110:50–55.

Amann, R. I., B. J. Binder, R. J. Olson, S. W. Chisholm, R. Devereux, and D. A. Stahl. 1990. Combination of 16S rRNA-targeted oligonucleotide probes with flow cytometry for analyzing mixed microbial populations. Appl. Environ. Microbiol. 56:1919–1925.

Amann, R. I., W. Ludwig, and K.-H. Schleifer. 1995. Phylogenetic identification and in situ detection of individual microbial cells without cultivation. Microbiol. Rev. 59:143–169.

American Water Works Association (AWWA). 1999. Waterborne pathogens: manual of water supply practices—M48, First Edition. Denver, CO: AWWA.

Anand, S., and K. Hanson. 1997. Disability-adjusted life years: a critical review. J. Health Econ. 16(6):685–702.

Arbuckle, T. E., S. E. Hrudey, S. W. Krasner, J. R. Nuckols, S. D. Richardson, P. Singer, P. Mendola, L. Dodds, C. Weisel, D. L. Ashley, K. L. Froese, R. A. Pegram, I. R. Schultz, J. Reif, A. M. Bachand, F. M. Benoit, M. Lynberg, C. Poole, and K. Waller. 2002. Assessing exposure in epidemiologic studies to disinfection by-products in drinking water: Report from an international workshop. Environ. Health Perspect. 110 (Suppl 1):53–60.

Balbus, J., R. Parkin, and M. Embrey. 2000. Susceptibility in microbial risk assessment: definitions and research needs. Environmental Health Perspectives 108(9):901–905.

Barbaree, J. M. 1991. Controlling *Legionella* in cooling towers. ASHRAE Journal 33:38–42.

Barbaree, J. M., B. S. Fields, J. C. Feeley, G. W. Forman, and W. T. Martin. 1986. Isolation of protozoa from water associated with a legionellosis outbreak and demonstration of intracellular multiplication of *Legionella pneumophila*. Appl. Environ. Microbiol. 51:422–424.

Bartram, J., J. Cotruvo, M. Exner, C. Fricker, and A. Glasmacher (eds). 2003. Heterotrophic Plate Counts and Drinking-water Safety. London: IWA Publishing.

Barwick, R. S., D. A. Levy, G. F. Craun, M. J. Beach, and R. L. Calderon. 2000. Surveillance for waterborne-disease outbreaks—United States, 1997–1998. MMWR 49(SS-4):1–34.

Bays, L. R., N. P. Burman, and W. M. Lavis. 1970. Taste and odour in water supplies in Great Britain: a survey of the present position and problems for the future. Journal Society Water Treatment Examination 19:136–160.

Bellen, G. E., S. H. Abrishami, P. M. Colucci, and C. Tremel. 1993. Methods for assessing the biological growth support potential of water contact materials. Denver, CO: AwwaRF.

Bennett, J. V., S. D. Holmberg, M. F. Rogers, and S. L. Soloman. 1987. Infectious and parasitic diseases. *In*: Closing the Gap: The Burden of Unnecessary Illness. R. W. Amler and H. B. Dull (eds.). New York: Oxford University Press.

Blackburn, B. G., G. F. Craun, J. S. Yoder, V. Hill, R. L. Calderon, N. Chen, S. H. Lee, D. A. Levy, and M. J. Beach. 2004. Surveillance for waterborne-disease outbreaks associated with drinking water—United States, 2001–2002. MMWR 53(SS-8):23–45.

Borchardt, M. A., N. L. Haas, and R. J. Hunt. 2004. Vulnerability of drinking-water wells in La Crosse, Wisconsin, to enteric-virus contamination from surface water contributions. Appl Environ Microbiol. 70(10):5937–5946.

Boorman, G. A., V. Dellarco, J. K. Dunnick, R. E. Chapin, S. Hunter, F. Hauchman, H. Gardner, M. Cox, and R.E. Sills. 1999. Drinking water disinfection by-products: review and approach to toxicity evaluation. Environ. Health Perspect. 107:207–17.

Breiman, R. F., B. S. Fields, G. N. Sauden, L. Volmer, A. Meier, and J. S. Spilka. 1990. Association of shower use with Legionnaires' disease. J. Amer. Med. Assoc. 263:2924–2926.

Briganti, L. A., and S. C. Wacker. 1995. Fatty acid profiling and the identification of environmental bacteria for drinking water utilities. Denver, CO:AWWA.

Briscoe, J. 1984. Intervention studies and the definition of dominant transmission routes. Am. J. Epidemiol. 120:449–55.

Camper, A. K. 1996. Factors Influencing Microbial Growth in the Distribution System: Laboratory and Pilot Experiments. Denver, CO: AwwaRF.

Cantor, K. P., C. F. Lynch, M. E. Hildesheim, M. Dosemeci, J. Lubin, M. Alavanja, and G. Craun. 1998. Drinking water source and chlorination byproducts—I. Risk of bladder cancer. Epidemiology 9:21–28.

Carnahan, A. M., and M. Altwegg. 1996. Taxonomy. *In:* The genus *Aeromonas*. S. Joseph (ed.). New York: John Wiley & Sons.

Carson, L. A., L. A. Bland, L. B. Cusick, M. S. Favero, G. A. Bolan, A. L. Reingold, and R. A. Good. 1988. Prevalence of nontuberculous mycobacteria in water supplies of hemodialysis centers. Appl. Environ. Microbiol. 54:3122–3125.

Chaidez, C., and C. Gerba. 2004. Comparison of the microbiological quality of point-of-use (POU)-treated water and tap water. Int. J. Environ. Health Res. 14:253–261.

Characklis, W. G. 1988. Bacterial Regrowth in Distribution Systems. Denver, CO: AWWA and AwwaRF.

Characklis, W. G. and K. C. Marshal. 1990. Biofilms. New York: John Wiley & Sons, Inc.

Chauret, C., C. Volk, R. Creason, J. Jarosh, J. Robinson, and C. Warnes. 2001. Detection of *Aeromonas hydrophila* in a drinking-water distribution system: a field and pilot study. Can. J. Microbiol. 47:782–786.

Clark, R. M., E. E. Geldreich, K. R. Fox, E. W. Rice, C. H. Johnson, J. A. Goodrich, J. A. Barnick, and F. Abdesaken. 1996. Tracking a Salmonella serovar typhimurium outbreak in Gideon, Missouri: role of contaminant propagation modeling. Journal of Water Supply Research and Technology - Aqua. 45(4):171–183.

Colford, J. M., Jr., T. J. Wade, S. K. Sandhu1, C. C. Wright, S. Lee, S. Shaw, K. Fox, S. Burns, A. Benker, M. A. Brookhart, M. van der Laan, and D. A. Levy. 2005. A randomized, controlled trial of in-home drinking water intervention to reduce gastrointestinal Illness. Am. J. Epidemiol. 161(5):472–482.

Colford, J. M., J. R. Rees, T. J. Wade, A. Khalakdina, J. F. Hilton, I. J. Ergas, S. Burns, A, Benker, C. Ma, C. Bowen, D. C. Mills, D. J. Vugia, D. D. Juranek, and D. A. Levy. 2002. Participant blinding and gastrointestinal illness in a randomized, control trial of an in-home drinking water intervention. Emerging Infectious Diseases 8(1):29–36.

Colbourne, J. S., D. J. Pratt, M. G. Smith, S. P. Fisher-Hoch, and D. Harper. 1984. Water fittings as sources of *Legionella pneumophila* in hospital plumbing system. Lancet 1(8370):210–213

Collins, C. H., J. M. Grange, and M. D.Yates. 1984. Mycobacteria in water. J. Appl. Bacteriol. 57:193–211.

Covert, T. C., M. R. Rodgers, A. L. Reyes, and G. N. Stelma. 1999. Occurrence of non-tuberculous mycobacteria in environmental samples. Appl. Environ. Microbiol. 65(6):2492–2496.

Craun, G. F., and R. L. Calderon. 2001. Waterborne disease outbreaks caused by distribution system deficiencies. J. Amer. Water Works Assoc. 93(9):64–75.

Craun, G. F., ed. 1986. Waterborne Diseases in the United States. Boca Raton, FL: CRC Press, Inc.

Cunliffe, D. A. 1990. Inactivation of *Legionella pneumophila* by monochloramine. J. Appl. Bacteriol. 68(5):453–459.

De Victoria, J., and M. Galvan. 2001. *Pseudomonas aeruginosa* as an indicator of health risk in water for human consumption. Water Sci. Technol. 43:49–52.

Dodds, L., W. King, C. Woolcott, and J. Pole. 1999. Trihalomethanes in public water supplies and adverse birth outcomes. Epidemiology 10:233–237.

Donlan, R., R. Murga, J. Carpenter, E. Brown, R. Besser, and B. Fields. 2002. Monochloramine disinfection of biofilm-associated *Legionella pneumophila* in a potable water model system. Pp. 406–410 *In: Legionella.* R. Marre, Y. Abu Kwaik, C. Bartlett, N. P. Cianciotto, B. S. Fields, M. Frosch, J. Hacker, and P. C. Luck (eds.). Washington, DC: American Society for Microbiology.

Dott, W., and D. Waschko-Dransmann. 1981. Occurrence and significance of actinomycetes in drinking water. Zbl. Bakt. Hyg. I., Abt. Orig. B, 173:217–232.

Doyle, T. J., W. Zheng, J. R. Cerhan, C.-P. Hong, T. A. Sellers, L. H. Kushi, and A. R. Folsom. 1997. The association of drinking water source and chlorination by-products with cancer incidence among postmenopausal women in Iowa: a prospective cohort study. Am. J. Public Health 87:1168–1176.

duMoulin, G. C., and K. D. Stottmeir. 1986. Waterborne mycobacteria: an increasing threat to health. Am Soc. Microbiol. News 52:525–529.

duMoulin, G. C., K. D. Stottmeier, P. A. Pelletier, T. A. Tsang, and J. Hedley-Whyte. 1988. Concentration of *Mycobacterium avium* by hospital hot water systems. J. Amer. Medical Assoc. 260:1599–1601.

Environmental Protection Agency (EPA). 1989. Drinking water; national primary drinking water regulations; total coliforms (including fecal coliforms and *E. coli*); final rule. Federal Register 54:27544–68.

EPA 1998. Announcement of the drinking water contaminant candidate list; notice. Federal Register 63:10274–87.

EPA. 1999. Legionella: human health criteria document. EPA-822-R-99-001. Washington, DC: EPA Office of Science and Technology.

EPA. 2002. Potential contamination due to cross-connections and backflow and the associated health risks: an issues paper. Available on-line at: *http://www.epa.gov/safewater/tcr/pdf/ccrwhite.pdf.* Accessed May 11, 2006.

EPA. 2004. Drinking water contaminant candidate list 2; notice. Federal Register 69(64):17406–17415.

Falkinhan, J. O., III, C. D. Norton, and M. W. LeChevallier. 2001. Factors influencing numbers of *Mycobacterium avium*, *Mycobacterium intracellulare*, and other mycobacteria in drinking water distribution systems. Appl. Environ. Microbiol. 67(3):1225–1231.

Ferroni, A., L. Nguyen, B. Pron, G. Quesne, M. C. Brusset, and P. Berche. 1998. Outbreak of nosocomial urinary tract infection due to *Pseudomonas aeruginosa* in a pediatric surgical unit associated with tap-water contamination. J. Hosp. Infect. 39:301–307.

Fewtrell, L. and J. M. Colford, Jr. 2004. Water, Sanitation and Hygiene: Interventions and Diarrhoea - A Systematic Review and Meta-analysis. 2004. Washington, DC: World Bank.

Fields, B. S., G. N. Sanden, J. M. Barbaree, W. E. Morril, R. M. Wadowsky, E. H. White, and J. C. Feeley. 1989. Intracellular multiplication of *Legionella pneumophila* in amoebae isolated from hospital hot water tanks. Current Microbiol. 18:131–137.

Fields, B. 2005. *Legionella* in distribution system biofilms. January 13, 2005. Presented to the NCR Committee on Public Water Supply Distribution Systems. Irvine, CA.

Fischeder, R. R., R. Schulze-Robbecke, and A. Weber. 1991. Occurrence of mycobacteria in drinking water samples. Zbl. Hygiene 192:154–158.

Flannery, B., L. B. Gelling, D. J. Vugia, J. M. Weintraub, J. J. Salerno, M. J. Conroy, V. A. Stevens, C. E. Rose, M. R. Moore, B. S. Fields, and R. E. Besser. 2006. Reducing *Legionella* colonization of water systems with monochloramine. Emerg. Infect. Dis. [serial on the Internet]. Available on-line at *http://www.cdc.gov/ncidod/EID/vol12no04/05-1101.htm.* Accessed April 13, 2006.

Fout, G. S., B. C. Martinson, M. W. N. Moyer, and D. R. Dahling. 2003. A multiplex reverse transcriptase-PCR method for detection of human enteric viruses in groundwater. Appl. Environ. Microbiol. 69: 3158–3164.

Frost, F. J., M. Roberts, T. R. Kunde, G. Craun, K. Tollestrup, L. Harter, and T. Muller. 2005. How clean must our drinking water be: the importance of protective immunity. Journal of Infectious Diseases 191:809–814.

Geldreich, E. E., K. R. Fox, J. A. Goodrich, E. W. Rice, R. M. Clark, and D. L. Swerdlow. 1992. Searching for a water supply connection in the Cabool, Missouri disease outbreak of *Escherichia coli* 0157:H7. Water Research 26(8):1127–1137.

Geldreich, E. E. 1996. Microbial Quality of Water Supply in Distribution Systems. Boca Raton, FL: CRC Press, Inc.

Geldreich, E. E., and M. W. LeChevallier. 1999. Microbial water quality in distribution systems. Pp. 18.1–18.49 *In:* Water Quality and Treatment. 5th edition. R. D. Letterman (ed.). New York: McGraw-Hill.

Gerba, C. P., J. B. Rose, and C. N. Haas. 1996. Sensitive populations: who is at the greatest risk? International Journal of Food Microbiology 30(1–2):113–123.

Glover, N. A., N. Holtzman, T. Aronson, S. Froman, O. G. W. Berlin, P. Dominguez, K. A. Kunkel, G. Overturf, G. Stelma, Jr., C. Smith, and M. Yakrus. 1994. The isolation and identification of *Mycobacterium avium* complex (MAC) recovered from Los Angeles potable water, a possible source of infection in AIDS patients. International J. Environ. Health Res. 4:63–72.

Graves, C. G., G. M. Matanoski, and R. G. Tordiff. 2001. Weight of evidence of an association between adverse reproductive and developmental effects and exposure to disinfection byproducts: a critical review. Regul. Toxicol. Pharmacol. 34:103–24.

Haas, C. N., M. A. Meyer, and M. E. Paller. 1983. The ecology of acid-fast organisms in water supply, treatment, and distribution systems. J. Amer. Water Works Assoc. 75:39–144.

Haudidier, K., J. L. Paquin, T. Francais, P. Hartemann, G. Grapin, F. Colin, M. J. Jourdain, J. C. Block, J. Cheron, O. Pascal, Y. Levi, and J. Miazga. 1988. Biofilm growth in drinking water networks: a preliminary industrial pilot plant experiment. Water Sci. Technol. 20:109–115.

Havelaar, A. H., A. E. Hollander, P. F. Teunis, E. G. Evers, H. J. Van Kranen, J. F. Versteegh, J. E. Van Koten, and W. Slob. 2000. Balancing the risks and benefits of drinking water disinfection: disability adjusted life-years on the scale. Environmental Health Perspective 108(4):315–321.

Hawkins, M. A., S. M. DeLong, R. Marcus, T. Jones, S. Shallow, S. M. Zansky, K. G. McCombs, A. K. Courtney, C. Medus, B. C. Imhoff, and the EIP FoodNet Working Group. 2002. The Burden of Diarrheal Illness in FoodNet, 2000–2001 (Draft report). Atlanta, GA: CDC.

Heffelfinger, J. D., J. L. Kool, S. K. Fridkin, V. J. Fraser, J. C. Carpenter, J. Hageman, J. Carpenter, and C. G. Whitney. 2003. Risk of hospital-acquired Legionnaires ' disease in cities using monochloramine versus other water disinfectants. Infection Control and Hospital Epidemiol. 24(8):569–574.

Hellard, M. E., M. I. Sinclair, A. B. Forbes, and C. K. Fairley. 2001. A randomized, blinded, controlled trial investigating the gastrointestinal health effects of drinking water quality. Environ. Health Perspect. 109:773–778.

Herikstad, H., S. Yang, T. J. Van Gilder, D. Vugia, J. Hadler, P. Blake, V. Deneen, B. Shiferaw, F. J. Angulo, and The FoodNet Working Group. 2002. A population-based estimate of the burden of diarrhoeal illness in the United States: FoodNet, 1996–97. Epidemiol. Infect. 129:9–17.

Herwaldt, B. L., G. F. Craun, S. L. Stokes, and D. D. Juranek. 1991. Waterborne-disease outbreaks, 1989–1990. MMWR 40(No. SS-3):1–21.

Horsburgh, C. R. 1991. *Mycobacterium avium* complex infection in the acquired immunodeficiency syndrome. New England Journal of Medicine 324:1332–1338.

Hulley, S. B., and S. R. Cummings. 1988. Designing Clinical Research: An Epidemiologic Approach. Baltimore, MD: Williams and Wilkins.

Hunter, P. R., R. M. Chalmers, S. Hughes, and Q. Syed. 2005. Self-reported diarrhea in a control group: a strong association with reporting of low-pressure events in tap water. Clinical Infectious Diseases 40:e32–34.

Imhoff, B., D. Morse, B. Shiferaw, M. Hawkins, D. Vugia, S. Lance-Parker, J. Hadler, C. Medus, M. Kennedy, M. R. Moore, and T. Van Gilder. 2004. Burden of self-reported acute diarrheal illness in FoodNet surveillance areas, 1998–1999. For the Emerging Infections Program (EIP) FoodNet Working Group. Clinical Infectious Diseases 38 Suppl 3:S219–26.

Jacquemin, J. L., A. M. Simitzia, and N. Chaneoau. 1981. Free-living amoebae in fresh water: a study of the water supply of the town of Poiters. Bull. Soc. Pathol. Exot. 74:521–534.

Jo, W. K., C. P. Weisel, and P. J. Lioy. 1990. Routes of chloroform exposure and body burden from showering with chlorinated tap water. Risk Analysis 10(4):575–580.

Keinanen, M. M., P. J. Martikainen, and M. H. Kontro. 2004. Microbial community structure and biomass in developing drinking water biofilms. Can. J. Microbiol. 50(3):183–191.

Kelley, J., G. Kinsey, R. Paterson, D. Brayford, R. Pitchers, and H. Rossmoore. 2003. Identification and control of fungi in distribution systems. Denver CO: AWWA and AwwaRF.

King, W. D., and L. D. Marrett. 1996. Case-control study of bladder cancer and chlorination by-products in treated water (Ontario, Canada). Cancer Causes Control 7:596–604.

King, W. D., L. Dodds, and A. C. Allen. 2000. Relation between stillbirth and specific chlorination by-products in public water supplies. Environ. Health Perspect. 108:883–886.

Kirmeyer, G. J., M. Friedman, K. Martel, D. Howie, M. LeChevallier, M. Abbaszadegan, M. Karim, J. Funk, and J. Harbour. 2001. Pathogen intrusion into the distribution system. #90835. Denver, CO: AWWA and AwwaRF.

Klotz, J. B., and L. A. Pyrch. 1999. Neural tube defects and drinking water disinfection by-products. Epidemiology 10:383–390.

Koivusalo, M., E. Pukkala, T. Vartiainen, J. J. K. Jaakkola, and T. Hakulinen. 1997. Drinking water chlorination and cancer – a historical cohort study in Finland. Cancer Causes Control 8:192–200.

Kool, J. L., J. C. Carpenter, and B. S. Fields. 1999. Effect of monochloramine disinfection of municipal drinking water on risk of nosocomial Legionnaires' disease. Lancet 353(9149):272–277.

Kramer, M. H., and T. E. Ford. 1994. Legionellosis: ecological factors of an environmentally 'new' disease. Zentralbl Hyg Umweltmed. 195(5–6):470–482.

Kramer, M. H., B. L. Herwaldt, G. F. Craun, R. L. Calderon, and D. D. Juranek. 1996. Surveillance for waterborne-disease outbreaks—United States, 1993–1994. MMWR 45(No. SS-1):1–33.

LeChevallier, M. W. 2004. Control, treatment, and disinfection of *Mycobacterium avium* complex in drinking water. Pp. 143–168 *In:* Pathogenic Mycobacteria in Water. S. Pedley, J. Bartram, G. Rees, A. Dufour, J. Cotruvo (eds.). ISBN: 1843390590. Geneva: World Health Organization.

LeChevallier, M. W., R. W. Gullick, M. R. Karim, M. Friedman, and J. E. Funk. 2003. The potential for health risks from intrusion of contaminants into distribution systems from pressure transients. J. Water and Health 1(1):3–14.

LeChevallier, M. W., and R. J. Seidler. 1980. *Staphylococcus aureus* in rural drinking water. Appl. Environ. Microbiol. 39:739–742.

LeChevallier, M. W., T. J. Wade, S. E. Shaw, D. Levy, R. L. Calderon, J. M. Colford, Jr. 2004. Results of the big WET: an epidemiology study of the microbiological quality of drinking water in Davenport, Iowa. AWWA Water Quality and Technology Conference, San Antonio, Texas, November 14–17, 2004.

Lee, S. H., D. A. Levy, G. F. Craun, M. J. Beach, and R. L. Calderon. 2002a. Surveillance for waterborne disease outbreaks—United States, 1999–2000. MMWR 51(SS-8):1–47.

Lee, S., D. Levy, A. Hightower, B. Imhoff, and the EIP FoodNet Working Group. 2002b. Drinking water exposures and perceptions among 1998–1999 FoodNet survey respondents. International Conference on emerging Infectious Diseases. Atlanta, GA, March 2002.

Levy, D. A., M. S. Bens, G. F. Craun, R. L. Calderon, and B. L. Herwaldt. 1998. Surveillance for waterborne-disease outbreaks—United States, 1995–1996. MMWR 47(SS-5):1–34.

Liang, J. L., E. J. Dziuban, G. F. Craun, V. Hill, M. R. Moore, R. J. Gelting, R. L. Calderon, M. J. Beach, and S. L. Roy. 2006. Surveillance for Waterborne Disease and Outbreaks Associated with Drinking Water and Water not Intended for Drinking—United States, 2003–2004. MMWR (in print).

Lin, Y. E., R. D. Vidic, J. E. Stout, and V. L. Yu. 1998. *Legionella* in water distribution systems. J. Amer. Water Works Assoc. 90:112–121.

Maluccio, J. A., and R. Flores. 2005. Impact evaluation of a conditional cash transfer program. The Nicaraguan *Red de Protección Social.* Research Report No. 141. Washington, DC: International Food Policy Research Institute.

Martiny, A. C., T. M. Jorgensen, H. J. Albrechtsen, E. Arvin, and S. Molin. 2003. Long-term succession of structure and diversity of a biofilm formed in a model drinking water distribution system. Appl. Environ. Microbiol. 69(11):6899–6907.

McMillan, M., T. F. Jones, A. Banerjee, D. Vugia, A. Cronquist, S. Segler, P. Ryan, C. Medus, P. Smith, B. Shiferaw, and F. Angulo. 2004. The Burden of Diarrheal Illness in FoodNet, 2002–2003. *In:* Proceedings of the International Conference on Emerging Infectious Diseases, Atlanta, GA.

Mead P.S., L. Slutsker L, V. Dietz, L. F. McCaig, J. S. Bresee, C. Shapiro, J. M. Griffin, and R. V. Tauxe. 1999. Food-related illness and death in the United States. Emerg. Infect. Dis. 5(5):607–25.

Moore, A. C., B. L. Herwaldt, G. F. Craun, R. L. Calderon, A. K. Highsmith, and D. D. Juranek. 1993. Surveillance for waterborne disease outbreaks—United States, 1991–1992. MMWR 42(SS-5):1–22.

Moore, M. R., M. Pryor, B. Fields, C. Lucas, M. Phelan, and R. E. Besser. 2006. Introduction of monochloramine into a municipal water system: Impact on colonization of buildings by *Legionella* spp. Appl Environ Microbiol 72(1):378–383.

Morris, R. D., and R. Levine. 1995. Estimating the incidence of waterborne infectious disease related to drinking water in the United States. Assessing and Managing Health Risk from Drinking Water Contamination: Approaches and Application. Proceedings of the Rome Symposium, September 1994. IAHS 233: 75–88.

Murga, R., T. S. Forster, E. Brown, J. M. Pruckler, B. S. Fields, and R. M. donlan. 2001. Role of biofilms in the survival of *Legionella pneumophila* in a model potable water system. Microbiology 147:3121–3126.

Nagy, L. A., and B. H. Olson. 1982. The occurrence of filamentous fungi in drinking water distribution systems. Canadian Journal Microbiol. 28:667–671.

Nahapetian, K., O. Challemel, D. Beurtin, S. Dubrou, P. Gounon and F. Squinazi. 1991. The intracellular multiplication of *Legionella pneumophila* in protozoa from hospital plumbing systems. Res. Microbiol. 142:677–685.

National Research Council (NRC). 1983. Risk Assessment in the Federal Government: Managing the Process. Washington, DC: National Academy Press.

NRC. 2004. Review of the Army's Technical Guides on Assessing and Managing Chemical Hazards to Deployed Personnel. Washington, DC: National Academy Press.

Niemi, R. M., S. Knuth, and K. Lundstrom. 1982. Actinomycetes and fungi in surface waters and in potable water. Appl. Environ. Microbiol. 43:378–388.

Nightingale, S. D., L. T. Byrd, P. M. Southern, J. D. Jockusch, S. X. Cal, and B. A. Wynne. 1992. *Mycobacterium avium*-intracellulare complex bacteremia in human immunodeficiency virus positive patients. Journal of Infectious Disease 165:1082–1085.

Nieuwenhuijsen, M. J., M. B. Toledano, N. E. Eaton, J. Fawell, and P. Elliott. 2000. Chlorination disinfection byproducts in water and their association with adverse reproductive outcomes: a review. Occup. Environ. Med. 57:73–85.

Norton, C. D., and M. W. LeChevallier. 2000. A pilot study of bacteriological population changes through potable treatment and distribution. Appl. Environ. Microbiol. 66(1):268–276.

Payment, P., J. Siemiatycki, L. Richardson, G. Renaud, E. Franco, and M. Prevost. 1997. A prospective epidemiological study of gastrointestinal health effects due to the consumption of drinking water. Intern. J. Environ. Health Res. 7:5–31.

Payment, P., L. Richardson, J. Siemiatycki, R. Dewar, M. Edwardes, and E. Franco. 1991. Randomized trial to evaluate the risk of gastrointestinal disease due to consumption of drinking water meeting microbiological standards. Am. J. Pub. Health 81(6):703–708.

Piriou, P., K. Helmi, M. Jousset, N. Castel, E. Guillot, and L. Kiene. 2000. Impact of biofilm on *C. parvum* persistence in distribution systems. *In:* Proceedings of an International Distribution System Research Symposium, June 10–11. Denver, CO: AWWA.

Pryor, M., S. Springthorpe, S. Riffard, T. Brooks, Y. Huo, G. Davis, and S. A. Satter. 2004. Investigation of opportunistic pathogens in municipal drinking water under different supply and treatment regimes. Water Sci. Technol. 50(1):83–90.

Quignon, F., M. Sardin, L. Kiene, and L. Schwartzbrod. 1997. Poliovirus-1 inactivation and interaction with biofilm: a pilot-scale study. Appl. Environ. Microbiol. 63(3):978–982.

Raina, P. S., F. L. Pollari, G. F. Teare, M. J. Goss, D. A. Barry, and J. Wilson. 1999. The relationship between *E. coli* indicator bacteria in well-water and gastrointestinal illnesses in rural families. Can. J. Public Health 90(3):172–175.

Regli, S., J. B. Rose, C. N. Haas, and C. P. Gerba. 1991. Modeling risk from *Giardia* and viruses in drinking water. J. Amer. Water Works Assoc. 83(11):76–84.

Rogers, J., A. B. Dowsett, P. J. Dennis, J. V. Lee, and C. W. Keevil. 1994. Influence of materials on biofilm formation and growth of *Legionella pneumophila* in potable water systems. Appl. Environ. Microbiol. 60:1842–1851.

Savitz, D., and C. L. Moe. 1997. Water: chlorinated hydrocarbons and infectious agents. Pp. 89–118 *In*: Topics in Environmental Epidemiology. N. K. Steenland and D. A. Savitz (eds.). Oxford: Oxford University Press.

Schoenen, D. 1986. Microbial growth due to materials used in drinking water systems. *In:* Biotechnology, Vol. 8. H. J. Rehm and G. Reed (eds.). Weinheim: VCH Verlagsgesellschaft.

Schulze-Robbecke, R., and R. Fischeder. 1989. *Mycobacteria* in biofilms. Zbl. Hyg. 88:385–390.

Schwartz J., R. Levin, and K. Hodge. 1997. Drinking water turbidity and pediatric hospital use for gastrointestinal illness in Philadelphia. Epidemiology 8:615–620.

Shepherd, J. L.; R. L. Corsi, and J. Kemp. 1996. Chloroform in indoor air and wastewater: the role of residential washing machines. Journal of the Air & Waste Management Association 46(7):631–642.

Singh, N., and V. Yu. 1994. Potable water and *Mycobacterium avium* complex in HIV patients: is prevention possible? Lancet 343:1110–1111.

Smith, D. B., A. F. Hess, and S. A. Hubbs. 1990. Survey of distribution system coliform occurrences in the United States. *In*: Proceedings of the Water Quality Technology Conference. Denver, CO: AWWA.

Smith, C. A., C. Phiefer, S. J. Macnaughton, A. Peacock, R. S. Burkhalter, R. Kirkegaard, and D. C. White. 2000. Quantitative lipid biomarker detection of unculturable microbes and chlorine exposure in water distribution system biofilms. Water Research 34(10):2683–2688.

Stanwell-Smith, R., Y. Anderson, and D. Levy. 2003. National surveillance systems. Pp. 25–40 *In:* Drinking Water and Infectious Disease: Establishing the Links. P. R. Hunter, M. Waite, and E. Ronchi (eds.). Boca Raton, FL: CRC Press

Steenland, K., and C. Moe. 2003. Epidemiology and drinking water—are we running dry? Epidemiology 14(6):635–636.

Steinmaus, C., Y. Yuan, M. N. Bates, and A. H. Smith. 2003. Case-control study of bladder cancer and drinking water arsenic in the western United States. American Journal of Epidemiology 158:1193–1201.

Stout, J. E., V. L. Yu, and M. G. Best. 1985. Ecology of *Legionella pneumophila* within water distribution systems. Appl. Environ. Microbiol. 49:221–228.

Strauss, B., W. King, A. Ley, and J. R. Hoey. 2001. A prospective study of rural drinking water quality and acute gastrointestinal illness. BMC Public Health 1:8.

Taylor, R. H., J. O. Falkinham, III, C. D. Norton, and M. W. LeChevallier. 2000. Chlorine, chloramine, chlorine dioxide, and ozone susceptibility of *Mycobacterium avium*. Appl. Environ. Microbiol. 66(4):1702–1705.

Thofern, E., D. Schoenen, and G. J. Tuschewitzki. 1987. Microbial surface colonization and disinfection problems. Off Gesundh.-wes., 49:Suppl:14.

Thomas, V., K. Herrera-Rimann, D. S. Blanc, and G. Greub. 2006. Biodiversity of amoebae and amoeba-resisting bacteria in a hospital water network. Appl. Environ. Microbiol. 72(4):2428–2438.

Trautman, M., T. Michalsky, H. Wiedeck, V. Radlosavljevic, and M. Ruhnke. 2001. Tap wate colonization with *Pseudomonas aeruginosa* in a surgical intensive care unit (ICU) and relation to *Pseudomonas* infections of ICU patients. Infect. Control. Hosp. Epidemiol. 22:49–52.

Vaerewijck, M. J. M., G. Huys, J. P. Palomino, J. Swings, and F. Portaels. 2005. Mycobacteria in drinking water distribution systems: ecology and significance for human health. FEMS Microbiology Review 29:911–934.

von Reyn, C. F., R. D. Waddell, T. Eaton, R. D. Arbeit, J. N. Maslow, T. W. Barber, R. J. Brindle, C. F. Gilks, J. Lumio, J. Lahdevirta, A. Ranki, D. Dawson, and J. O. Falkinham, III. 1993. Isolation of *Mycobacterium avium* complex from water in the

United States, Finland, Zaire, and Kenya. Journal of Clinical Microbiology 31:3227–3230.

von Reyn, C. F., J. N. Maslow, T. S. Barber, J. O. Falkinham, III, and R. D. Arbeit. 1994. Persistent colonisation of potable water as a source of *Mycobacterium avium* infection in AIDS. Lancet 343:1137–1141.

VROM. 2005. *Legionella* in Europe: Problems and Prevention: Summary by the Chairman. International Congress September 28–29, Amsterdam, The Netherlands. Available on-line at *http://international.vrom.nl/docs/internationaal/congres%20verslag%20Engels%2025-11-04.pdf.* Accessed May 11, 2006.

Wade, T. J., S. K. Sandhu, D. Levy, S. Lee, M. W. LeChevallier, L. Katz, and J. M. Colford, Jr. 2004. Did a severe flood in the Midwest cause an increase in the incidence of gastrointestinal symptoms? Amer. J. Epidemiol. 159(4):398–405.

Wadowsky, R. M., and R. B. Yee. 1983. Satellite growth of *Legionella pneumophila* with an environmental isolate of *Flavobacterium breve*. Appl. Environ. Microbiol. 46:1447–1449.

Wadowsky, R. M., and R. B. Yee. 1985. Effect of non-legionellaceae bacteria on the multiplication of *Legionella pneumophila* in potable water. Appl. Environ. Microbiol. 49:1206–1210.

Waller, K., S. H. Swan, G. DeLorenze, and B. Hopkins. 1998. Trihalomethanes in drinking water and spontaneous abortion. Epidemiology 9:134-140.

Wolinsky, E. 1979. Nontuberculous *Mycobacteria* and associated diseases. Am. Rev. Respir. Dis. 119:107–159.

World Health Organization (WHO). 2002. Heterotrophic Plate Count Measurements in Drinking Water Safety Management. WHO/SDE/WSH/02.10. Geneva, Switzerland: WHO.

Yang, C. Y., H. F. Chiu, M. F. Cheng, and S. S. Tsai. 1998. Chlorination of drinking water and cancer mortality in Taiwan. Environmental Research 78:1–6.

4
Physical Integrity

This chapter focuses on physical integrity—the ability of the distribution system to act as a physical barrier that prevents external contamination from affecting the quality of the internal, drinking water supply. Water distribution system engineers have defined the physical integrity of the distribution system to be its ability to handle external and internal stresses such that the physical material of the system does not fail (Male and Walski, 1991). Here failure is interpreted more broadly to encompass the absence of a critical component, the improper installation of a component, or the installation of an already contaminated component.

The physical integrity of the distribution system is always in a state of change, and the aging of the nation's distribution systems and eventual need for replacement are growing concerns. Maintaining such a vast physical infrastructure is a challenge because of the complexity of individual distribution systems, each of which is comprised of a network of mains, fire hydrants, valves, auxiliary pumping or booster disinfection substations, storage reservoirs, standpipes, and service lines along with the plumbing systems in residences, large housing projects, high-rise buildings, hospitals, and public buildings. This is further complicated by factors that vary from system to system such as the size of the distribution network for the population served, the predominant pipe material and age of pipelines, water pressure, the number of line breaks each year, water storage capacity, and water supply retention time in the system. When considering the replacement of a given component of the distribution system, decision makers must weigh its potential remaining life versus the potential that the component will fail, which could result in costly consequences and compromise the water utility's service.

The physical integrity of the distribution system, from the entry point to the customer's tap, is a primary barrier against the entry of external contaminants and the loss in quality of the treated drinking water. This barrier includes such materials as the pipe wall and reservoir cover as well as physical connections to nonpotable water sources. The barrier must be non-permeable since contaminants can enter through breaks or failures in materials as well as through the materials themselves. Table 4-1 gives examples of the infrastructure components that constitute this physical barrier, what they protect against, and the materials of which they are commonly constructed.

A variety of components and materials make up this physical barrier. Four major component types are delineated and referred to repeatedly in this chapter: (1) pipes including mains, services lines, and premise plumbing; (2) fittings and appurtenances such as crosses, tees, ells, hydrants, valves, and meters;

TABLE 4-1 Infrastructure Components, What They Protect Against, and Common Materials

Component	External Contamination the Barrier Protects Against	Materials Used
Pipe	Soil, groundwater, sewer exfiltration, surface runoff, human activity, animals, insects, and other life forms	Asbestos cement, reinforced concrete, steel, lined and unlined cast iron, lined and unlined ductile iron, PVC, polyethylene and HDPE, galvanized iron, copper, polybutylene
Pipe wrap and coatings	Supporting role in that it preserves the pipe integrity	Polyethylene, bitumastic, cement-mortar
Pipe linings	Supporting role in that it preserves the pipe integrity	Epoxy, urethanes, asphalt, coal tar, cement-mortar, plastic inserts
Service lines	Soil, groundwater, sewer exfiltration, surface runoff, human activity, animals, insects, and other life forms	Galvanized steel or iron, lead, copper, chlorinated PVC, cross-linked polyethylene, polyethylene, polybutylene, PVC, brass, cast iron
Premise plumbing	Air contamination, human activity, sewage and industrial nonpotable water.	Copper, lead, galvanized steel or iron, iron, steel, chlorinated PVC, PVC, cross-linked polyethylene, polyethylene, polybutylene
Fittings and appurtenances (meters, valves, hydrants, ferrules)	Soil, groundwater, sewer exfiltration, surface runoff, human activity, animals, insects, and other life forms	Brass, rubber, plastic
Storage facility walls, roof, cover, vent hatch	Air contamination, rain, algae, surface runoff, human activity, animals, birds, and insects	Concrete, steel, asphaltic, epoxy, plastics
Backflow prevention devices	Nonpotable water	Brass, plastic
Liquids	Not applicable	Oils, greases, lubricants
Gaskets and joints	Soil, groundwater, sewer exfiltration, surface runoff, human activity, animals, insects, and other life forms	Rubber, leadite, asphaltic, plastic

(3) storage facilities including reservoirs (underground, open, and covered), elevated storage tanks, ground level storage tanks, and standpipes; and (4) backflow prevention devices. The materials used by the water industry for these components, particularly pipes, have changed significantly over time (AWWA, 1986; Von Huben, 1999). For example, cast iron pipe (lined or unlined) has been largely phased out due to its susceptibility to both internal and external corrosion and associated structural failures. Ductile-iron pipe (with or without a cement lining) has taken its place because it is durable and strong, has high flexural strength, and has good resistance to external corrosion from soils. It is, however, quite heavy, it might need corrosion protection in certain soils, and it requires multiple types of joints. Concrete, asbestos cement, and polyvinyl chloride (PVC) plastic pipe have been used to replace metal pipe because of their relatively good resistance to corrosion. Polyethylene pipe is growing in use, especially for trenchless applications like slip lining, pipe bursting, and directional drilling (Morrison, 2004). High-density polyethylene pipe is the second most commonly used pipe. It is tough, corrosion resistant both internally and externally, and flexible. The manufacturer estimates its service life to be 50 to 100 years (AWWA, 2005a). Chapter 1 discusses the rate of pipe replacement in the United States and notes that much of the current infrastructure is nearing the end of its usable lifetime.

FACTORS CAUSING LOSS OF PHYSICAL INTEGRITY

Losses in physical integrity are caused by an abrupt or gradual alteration in the structure of the material barrier between the external environment and the drinking water, by the absence of a barrier, or by the improper installation or use of a barrier. These mechanisms are summarized in Table 4-2.

Infrastructure components break down or fail over time due to chemical interactions between the materials and the surrounding environment, eventually leading to holes, leaks, and other breaches in the barrier. These processes can occur over time scales of days to decades, depending on the materials and conditions present. For example, plastic pipes can be very rapidly compromised by nearby hydrophobic compounds (e.g., solvents in the vadose zone that result from surface or subsurface contamination), with the resulting permeation of those compounds into the distribution system through the pipe materials. Both internal and external corrosion can lead to structural failure of pipes and joints, thereby allowing contaminants to infiltrate into the distribution system via leaks or subsequent main breaks. Materials failure can be hastened if the distribution system water pressure is too high, from overburden stresses on pipes, and during natural disasters. Indeed, hurricanes and earthquakes have caused extensive sudden damage to distribution systems, including broken service lines and fire hydrants, pipes disconnected or broken by the uprooting of trees, cracks in cement water storage basins, and seam separations in steel water storage tanks (Geldreich, 1996).

TABLE 4-2 What Causes a Loss in Physical Integrity?

Component	Mechanism of Integrity Loss		
	Alteration in material structure leading to failure	*Absence of the barrier or material*	*Improper application or installation of the barrier*
Pipe	• Corrosion • Permeation • Too high internal water pressure or surges • Shifting earth • Exposure to UV light • Stress from overburden • Temperature fluctuations, freezing	• Absence of external or internal linings, wraps, coatings to protect the pipe	• Unsanitary activity during construction, replacement, or repair • Unintentional creation of cracks and breaks • Use of faulty materials
Fitting and appurtenance	• Corrosion • Permeation	• Appurtenance in a flooded meter or valve pit (absence of appropriate structures)	• Unsanitary activity during construction, replacement, or repair • Unintentional creation of cracks and breaks • Use of faulty materials • Contact between dissimilar metals
Storage facility wall, roof, cover, vent, hatch	• Corrosion • Permeation • Natural disasters • Failure due to aging and weathering	• Missing cover, roof, hatch, vent, can lead to unprotected access to the storage facility. Could be unintentional or intentional (vandalism)	• Unsanitary activity during construction, replacement, or repair • Unintentional or intentional creation of cracks and breaks • Poor drainage for runoff • Use of faulty materials
Backflow prevention device	• Corrosion	• Missing device will allow a backflow event via a cross connection	• Use of faulty materials • Improper installation • Inadequate drainage of meter pit • Operational failure

A second major contributor to the loss of physical integrity is when certain critical components are absent, either by oversight or due to vandalism. For example, the absence of backflow prevention devices and covers for storage facilities can allow external contaminants to enter distribution systems. For the purposes of this discussion, pipes are assumed to always be present.

Finally, human activity involving distribution system materials can allow contamination to occur such as through unsanitary repair and replacement practices, unprotected access to materials, or the improper handling of materials leading to unintentional damage. One must even consider the installation of flawed materials, which might, for example, be brought about because of a lack of protection of materials during storage and handling.

Structural Failure of Distribution System Components

Metallic pipe failures are divided generally into two categories: corrosion failures and mechanical failures. Common types of failures for iron mains include (Male and Walski, 1991; Makar, 2002):

- Bell splits or cracks that require cutting out the joint and replacing it with a mechanical fitting; these are typical for leadite joints
- Splits at tees and offsets and other fittings that require replacement
- Circumferential cracks or round cracks and holes, more typical in smaller diameter pipe (< 10 in.). These can result from a lack of soil support, causing the pipe to be called upon to act as a beam
- Splits or longitudinal cracks or spiral cracks that will blow out. Longitudinal cracks are more common for larger pipe (> 12 in.) and can result from crushing under external loads or from excessive internal pressure
- Spiral failures in medium diameter pipe
- Shearing failures in large diameter pipe
- Pinholes (corrosion hole) caused by internal corrosion
- Tap or joint blowout
- Crushed pipe

A simpler categorization can be found in Romer et al. (2004), who summarized three types of pipe failures as weeping failures, pipe breaks, and sudden failures. A weeping failure is where a leak allows an unnoticeable exchange of water to and from the surrounding soil. A pipe break includes a hole in the pipe or a disengagement of a bell-and-spigot joint. A sudden failure is the bursting of a pipe wall or shear of the pipe cross section, as would occur for a concrete pipeline, or a blow out, which refers to a complete break in a pipe.

Pipe breaks can occur for a myriad of reasons such as normal materials deterioration, joint problems, movement of earth around the pipe, freezing and thawing, internal and external corrosion, stray DC currents, seasonal changes in internal water temperature, heavy traffic overhead including accidents that damage fire hydrants, changes in system pressure, air entrapment, excessive overhead loading, insufficient surge control (such as with water hammer and pressure transients), and errors in construction practices (Male and Walski, 1991). This last factor is especially troubling since it should be entirely preventable. Nonetheless, there is evidence that poor quality workmanship during initial pipe installation can lead to early structural failure of pipes (Clark and Goodrich, 1989). Burlingame et al. (2002) reported on premature (within one year of installation) failures in service lines that resulted from the combination of using hard copper tubing and poor workmanship during cutting and flaring of the ends. AwwaRF (1985) has also reported that failures with copper tubing can be due to poor workmanship. One of the goals of proper installation of water mains is to account for and circumvent these issues; unfortunately, failure to do so translates into a substantial number of unnecessary main breaks.

One overriding factor in determining the potential for pipe failure is the force exerted on the water main. Contributors to this force include changes in temperature, which cause contraction and expansion of the metal and the surrounding soil, the weight of the soil over the buried main, and vibrations on the main caused by nearby activities such as traffic. An important consideration in this regard is the erosion potential of the supporting soil beneath the buried main. In the construction of a main, special sand and soil can be laid beneath it to help it bear external forces. But the movement of water in the ground beneath the main can wash away the finer material and create small or large caverns under the pipe. The force now bearing down on top of the pipe must be taken by the pipe itself, without the help of supporting material underneath. If these forces exceed the strength of the pipe, the main breaks. Most often these breaks occur at the weakest part of the main, i.e., the joint.

The factors that cause pipe failures can compound one another, hastening the process. For example, if a main develops small leaks because of corrosion, water within the distribution system can exfiltrate into the area surrounding the pipe, eroding away the supporting soil. Leakage that undermines the foundation of a water main can also occur from nearby sewer lines, go on essentially unnoticed, and eventually lead to water main collapse (Morrison, 2004).

Table 4-3 summarizes common problems that lead to pipe failures for pipes of differing materials. These are some of the principal factors, but they are not the only factors that act individually or in combination to lead to a main break. Other factors could include a street excavation that accidentally disturbs a water main and the misuse of fire hydrants. At most utilities, overall pipe break rates have been relatively low and stable (Damodaran et al., 2005) even though the infrastructure is aging.

Other components of distribution system also experience structural failure, although they have not historically received the attention afforded to pipes. For storage facilities, structural failure is less of a problem than external contamination due to the absence or failure of an essential component such as a cover or vent. Fittings and appurtenances can suffer from the effects of corrosion and permeation.

TABLE 4-3 Most Common Problems that Lead to Pipe Failure for Various Pipe Materials

Pipe Material (common sizes)	Problems
PVC and Polyethylene (4–36 in.)	Excessive deflection, joint misalignment and/or leakage, leaking connections, longitudinal breaks from stress, exposure to sunlight, too high internal water pressure or frequent surges in pressure, exposure to solvents, hard to locate when buried, damage can occur during tapping
Cast/Ductile Iron (4–64 in,) (lined and unlined)	Internal corrosion, joint misalignment and/or leakage, external corrosion, leaking connections, casting/manufacturing flaws
Steel (4–120 in.)	Internal corrosion, external corrosion, excessive deflection, joint leakage, imperfections in welded joints
Asbestos-Cement (4–35 in.)	Internal corrosion, cracks, joint misalignment and/or leakage, small pipe can be damaged during handling or tapping, pipe must be in proper soil, pipe is hard to locate when buried
Concrete (12–16 to 144–168 in.) (prestressed or reinforced)	Corrosion in contact with groundwater high in sulfates and chlorides, pipe is very heavy, alignment can be difficult, settling of the surrounding soil can cause joint leaks, manufacturing flaws

SOURCES: Morrison (2004) and AWWA (1986).

Corrosion as a Major Factor

Corrosion is the degradation of a material by reaction with the local environment. In water distribution systems, the term corrosion refers to dissolution of concrete linings and concrete pipe, as well as to the deterioration of metallic pipe and valves via redox reactions (e.g., iron pipe rusting). Degradation originating from the inside of the pipe via reactions with the potable water is termed internal corrosion. Degradation originating outside the pipe on surfaces contacting moist soil is referred to as external corrosion. Both internal and external corrosion can cause holes in the distribution system and cause loss of pipeline integrity. In some cases holes are formed directly in pipes by corrosion, as is the case with pinholes, but in many other instances corrosion weakens the pipe to the point that it will fail in the presence of forces originating from the soil environment.

The type of corrosion and mode of failure causing loss of physical integrity are highly system specific. External corrosion can be exacerbated by a low soil redox potential, low soil pH, stray currents, and dissimilar metals or galvanic corrosion (Von Huben, 1999; Szeliga and Simpson, 2002; Romer et al., 2004; Bonds et al., 2005). The life of the pipe is also influenced by the material used, thickness of the pipe wall, use of protective outer wraps or coatings, application of cathodic protection, and backfill materials and techniques. Internal corrosion is influenced by pH, alkalinity, disinfectant type and dose, type of bacteria present in biofilms, velocity, water use patterns, use of inhibitors, and many other factors.

Corrosion is not well understood, particularly at the level of the local water utility, such that insufficient attention has been given to its control (see a later section in this chapter). Some utilities have tried to avoid the issue by using plastic pipe. Even so, unprotected metal materials are regularly used at the present time, illustrating the water industry's lack of attention to the problem. According to Romer et al. (2004), "approximately 72 percent of the materials reported in use for water mains are iron pipe, approximately two-thirds of the reported corrosion is in corrosive soils, and approximately two-thirds of the corrosion is on the pipe barrel." In addition, metallic or cementitious pipe are often designed on the basis of their hydraulic capabilities first and foremost, and corrosion resistance is often a secondary consideration. The annual direct costs of corrosion are estimated to be $5 billion (Romer et al., 2004) for the main distribution system (not counting premise plumbing).

Issues with Service Lines

Recent evidence indicates that service lines (the piping between the water main and the customer's premises) and their fittings and connections (ferrules, curb stops, corporation stops, valves, and meters) can account for a significant proportion of the leaks in a distribution system (AWWA Water Loss Control Committee, 2003). However, much less is known about what causes structural failures in service lines compared to distribution mains and other system components. Possibilities include improper techniques used during installation that damage materials, improper tapping and flaring to make connections, lack of corrosion prevention or use of corrosive backfill material, damage during handling to plastic tubing, and kinks in copper tubing, and excessive velocity. The Uniform Plumbing Code and International Plumbing Code do not clearly address these issues, and local plumbing codes may not either.

Many galvanized and lead pipe service lines are being replaced with copper or plastic pipe (chlorinated polyvinyl chloride or CPVC) (Von Huben, 1999). CPVC and copper each have their benefits and weaknesses. Installation of CPVC requires less skill compared to installation of copper, although if workers are not careful installation can result in cracking and damage to CPVC pipe. CPVC is better for corrosive soils and waters, while copper is more resistant to internal biofilm growth. Buried CPVC pipe is difficult to locate compared to metal or copper pipe because it does not conduct electrical current for tracing. CPVC can impart a "plastic" flavor to water while the copper pipe can impart a "metallic" flavor. With CPVC, low levels of vinyl chloride can leach into the water. If manufacturers follow American Society for Testing and Materials (ASTM) standards and are ISO 9002 certified, and certification includes NSF

International standards 14 and 61, the adverse conditions above can be minimized.

Permeation

Permeation refers to a mechanism of pipe failure in which contaminants external to the pipe materials and non-metallic joints compromise the structural integrity of the materials and actually pass through them into the drinking water. Permeation is generally associated with plastic pipes and with chemical solvents such as benzene, toluene, ethylbenzene, and xylenes (BTEX) and other hydrocarbons associated with oil and gasoline, all of which are easily detected using volatile organic chemical gas chromatography analyses. These chemicals can readily diffuse through the plastic pipe matrix, alter the plastic material, and migrate into the water within the pipe. Such compounds are common in soils surrounding gasoline spills (leaking storage tanks), at abandoned industrial sites, and near bulk chemical storage, electroplaters, and dry cleaners (Glaza and Park, 1992; Geldreich, 1996). Permeation incidents have occurred at high-risk sites, such as industrial sites and near underground chemical storage tanks, as well as at lower risk residential sites (Holsen et al., 1991). In some cases the integrity of the pipe has been irreversibly compromised, requiring the complete replacement of the contaminated section.

Common pipe materials such as PVC, polybutylene, and polyethylene differ in their chemical and physical structure, and thereby differ in their susceptibility to being altered upon exposure to solvents and in permeation rates. In studying BTEX and 1,3-dichlorobenzene, PVC pipe was found to be more permeable than polyethylene pipe unless the polyethylene pipe was altered by the solvents in contact, after which it can become more permeable to the pollutants (Burlingame and Anselme, 1995).

Human Activities that Lead to Contamination

A second major cause of physical integrity loss is human activity surrounding construction, repair, and replacement that can introduce contamination into the distribution system. Any point where the water distribution system is opened to the atmosphere is a potential source of contamination. This is particularly relevant when laying new pipes, engaging in pipe repairs, and rehabilitating sites. For example, a Midwestern water utility experienced a noticeable increase in the heterotrophic bacterial population of water from a newly installed pipe and identified *Pseudomonas fluorescens, Ps. Maltophilia*, and *Ps. putida* as the bacteria responsible for the increase (Geldreich, 1996). The same strains of *Pseudomonas* were recovered from the sand used as an aggregate in making the concrete lining for the new ductile iron pipe, implicating contamination during construction and installation. More recently, workers in Camden, New Jersey, were cleaning and lining a 30-inch water main when a parallel sewer line from the post-Civil War era broke. Because of the proximity of the sewer line and the possibility of contamination, officials decided to issue a boil-water alert until water quality testing could show that no external contamination had entered the main. Between 1997 and 1999, the Philadelphia water supply measured elevated turbidity (>1 NTU) in about 12 to 14 percent of the samples that were collected from newly installed water mains. This turbidity, or the particulate debris captured on filters, was found to be largely iron oxides and rust (from the existing water mains still in service), vegetable material such as plant roots, and backfill sand.

Incidents like these are not uncommon, as revealed in a survey by Pierson et al. (2002), who point out that pipe repair and installation have not been accomplished using the best available sanitary practices. This is captured generally in Table 4-4, which summarizes the survey of distribution system workers at three different utilities (eastern U.S., western U.S., and western Canada) on the potential for external contamination to occur during water main repair and replacement activities. Given that the average number of main repairs a year for a single utility ranges from 66 to 901 (which corresponds to 7.9–35.6 repairs per 100 miles of pipe per year) (Clark and Goodrich, 1989), it is clear that exposure of the distribution system to contamination during repair is an inescapable reality.

Unsanitary activity during construction, replacement, or repair can also lead to the contamination of fittings and appurtenances. The use of inappropriate or inferior materials, and the contact between dissimilar metals within fittings, can also cause failures where they should not occur. Appurtenances can be improperly installed in a flooded meter or valve pit which can allow contaminants to enter under intrusion or can create corrosive conditions.

Backfill sand contaminating a new pipe at a water main construction site. Photo courtesy of Bureau of Laboratory Service, Philadelphia Water Department.

TABLE 4-4 Potential for Contaminant Entry during Water Main Activities

Activity	Percent of Responses from Workers at 3 Different Utilities (A, B, C)					
	Occurs Often			*Occurs Sometimes*		
	A	B	C	A	B	C
Broken service line fills trench during installation	46	75	56	39	25	33
Pipe gets dirty during storage before installation	53	75	22	43	25	33
Trench dirt gets into pipe during installation	24	100	39	37	0	44
Rainwater fills trench during installation	20	25	5	60	75	83
Street runoff gets into pipe before installation	30	0	11	61	38	67
Pipe is delivered dirty	4	25	17	33	63	22
Trash gets into pipe before installation	24	0	0	56	50	11
Vandalism occurs at the site	15	0	0	35	0	5
Animals get into pipe before installation	0	0	0	11	0	11

SOURCE: Reprinted, with permission, from Pierson et al. (2002). © 2002 by American Water Works Association.

New pipe materials are not sterile, whether they have been kept well protected or not. Indeed, according to a survey (Geldreich, 1996) about 18 percent of new pipe, irrespective of pipe material and size, failed upon testing the water to approve it for release. In one case, Geldreich reported the finding of a piece of wood construction material embedded in a new main that contributed to coliform contamination. Thus, new materials need inspection and some form of disinfection before they are exposed to drinking water. The physical cleanliness of new pipe is important to guarantee that post-installation disinfection will be successful (Geldreich, 1996). The installation or rehabilitation of facilities such as storage reservoirs with floating covers must include water quality checks for health and aesthetic considerations and not assume that new materials and their installation will be free of contaminants (Krasner and Means, 1986).

The installation process for buried pipe is not the only place where contamination can occur. The storage of pipe, pipe fittings, and valves along roadways or in pipe yards prior to installation can expose them to contamination from soil, stormwater runoff, and pets and wildlife. Damage to pipes prior to their installation is also possible, such as during pipe storage and handling or actual manufacturing defects such as surface impurities or nicks.

Regardless of where and how materials become contaminated, the hope is that post-installation disinfection will be sufficient to kill any introduced bacteria. This is not always the case, however, as evidenced by a coliform event in Florissant, Missouri in 1984 (Geldreich, 1996). The coliforms detected in a storage tank were thought to be the result of inadequate disinfection following new pipe installation or repair. Unfortunately, contaminated water subsequently passed into the distribution network. No direct public health outcome was reported; however, the "repeated reissuance of boil-water orders caused a loss of confidence" in the water utility by the public (Geldreich, 1996).

It is unclear how often faulty materials are installed or good materials are improperly installed because most utilities do not keep records that would facilitate the evaluation of this problem. Sufficient standards exist for materials quality and for the certified testing of materials quality. Water utilities can incorporate existing standards into contracts and specifications for materials and materials installations, and most if not all water utilities already do this. Water utilities can also certify and decertify manufacturers and contractors.

Absence of a Barrier

Points in a plumbing system where nonpotable water comes into contact with the potable water supply are called cross connections, and a backflow event occurs when nonpotable water flows into the drinking water supply through a cross connection. The use of backflow prevention devices can be extremely effective in eliminating this type of contamination event. The absence of such devices, which is widespread given the highly variable nature of cross-connection control programs across the country, constitutes a potential threat to the physical integrity of distribution systems. Backflow protection devices are seldom installed on domestic service lines and even on many small business service lines. Operational failure of devices that are in place is akin to having the device not be present.

Similar issues surface for storage facilities that do not have adequate protection to prevent their contamination. There are 154,000 treated water storage facilities in the United States (AWWA, 2003) encompassing a variety of types including elevated tanks, standpipes, open and covered reservoirs, underground basins, and hydropneumatic storage tanks. Storage facilities are susceptible to external contamination from birds, insects, other animals, wind, rain, and algae. Indeed, coliform occurrences have been associated with birds roosting in the vent ports of covered water reservoirs (Geldreich, 1996). This is most problematic for uncovered storage facilities, although storage facilities with floating covers are also susceptible to bacterial contamination due to rips in the cover from ice, vandalism, or normal operation. Even with covered storage facilities, contaminants can gain access through improperly sealed access openings and hatches or faulty screening of vents and overflows. Four reported waterborne disease outbreaks have been associated with covered storage tanks, in particular, a *Salmonella typhimurium* outbreak due to a bird contamination of a covered municipal water storage tank (Clark et al., 1996). Such events can be aggravated by the loss of disinfectant residual that storage tanks typically experience with increasing water age.

Preparing to sample stored drinking water from the access hatch of a floating cover on a distribution system reservoir. Photo courtesy of Bureau of Laboratory Service, Philadelphia Water Department.

CONSEQUENCES OF A LOSS IN PHYSICAL INTEGRITY

A loss of physical integrity implies a breakdown in the barrier that prevents contact between the external, unsanitary environment and the internal, drinking water environment. The water quality effects that can result include the introduction into the distribution system of microbial and chemical contaminants, debris, and particulate matter, sometimes accompanied by changes in water color, turbidity, taste, and odor. Whether a breach in physical integrity results in exposure of the public to contaminants at levels posing an unacceptable risk is dependent on site-specific conditions. As revealed in Chapter 3 and Appendix A, **most documented cases of waterborne disease outbreaks that can be attributed to distribution systems have been caused by breaches in physical integrity.** For example, a review of 619 reported waterborne disease outbreaks in the U.S. between 1971 and 1998 found that over one-half of the outbreaks in distribution systems were due to cross connections and backflow (Craun and Calderon, 2001). Of the 12 largest outbreaks, seven were associated with cross connections, three with contaminated storage tanks, and two with water main contamination during installation or repair. Overall, in community water systems, cross connections were the number one cause of distribution system-related outbreaks, contaminated mains were number two, and contaminated storage facilities were number three. In non-community water systems, contaminated storage facilities were the second leading cause. The contaminants involved have ranged from pathogens such as *Giardia,* Norwalk virus-like agents, hepatitis A virus, *Campylobacter*, *Salmonella*, *Shigella*, and *E. coli* 0157:H7 to chemical contaminants such as copper (the most commonly reported chemical), chlordane, nitrite, ethylene glycol, and oil (Craun and Calderon, 2001).

Not all of what can enter a distribution system from a failure in a physical barrier will have a known or direct health impact. Particulate matter and other debris can gain entry during main breaks; reservoir cover, hatch, or vent failures; and during repair, installation, and maintenance activities. Utilities have reported particulates in distribution system water that included such things as sand, patina, pipe joint materials, rubber gasket chunks, insect pieces, plant fibers, and glass chips, many of which are likely to have no direct health impacts (Booth and Brazos, 2005).

Changes in taste and odor, turbidity, and color typically provoke customers to complain (Burlingame, 1999a,b; McGuire et al., 2004), but may present little direct public health risk. This is because aesthetic problems often occur at contaminant concentrations far below the known health effects levels. For example, color problems derived from iron or manganese introduced into drinking water during a backflow event from a fire service connection or a heating system are unlikely to pose a health risk. On the other hand, color problems can also indicate backflow events that have health risks associated with them such as with ethylene glycol or corrosion inhibitors from HVAC and fire service connections. The sections below discuss the typical consequences of the loss of physical integrity in pipes, fittings and appurtenances; storage facilities; and backflow prevention devices.

Contamination of Mains, Fittings, and Appurtenances

Pipe interior, appurtenances, and related materials can be exposed to microbial and chemical contaminants in the external environment (1) during water main failures and breaks and (2) due to human activities to install new, rehabilitate old, or repair broken mains and appurtenances. When a pipe break or failure occurs, there is immediate potential for external contamination from soil, groundwater, or surface runoff to enter the distribution system or come into contact with the pipe interior in the area of the failure. Other less dramatic types of structural failure, such as the development of cracks or leaks in pipe, pipe joints, or appurtenances, can also provide avenues for distribution system contamination during periods of low pressure or a pressure transient—an event known as intrusion. Intrusion refers to the flow of nonpotable water into drinking water mains through leaks, cracks, submerged air valves, faulty seals, and other openings resulting from low or negative pressures. Discussed in greater detail in the next chapter, intrusion can exist undetected for long periods of time. A prominent example of a waterborne disease outbreak being caused by a main break and intrusion is presented in Box 4-1.

BOX 4-1
Waterborne Disease Outbreak Associated with Main Breaks and Intrusion: Cabool, Missouri

In the winter of 1989–1990, Cabool, Missouri, a town of approximately 2,100 people, experienced a large outbreak of *E. coli* O157:H7. A total of 243 cases were reported, with 32 hospitalizations and four deaths. This was the first documented waterborne outbreak of *E. coli* O157:H7 and the largest waterborne outbreak of *E. coli* O157:H7 before the 2000 outbreak in Walkerton, Canada.

The town's water system (untreated groundwater) was implicated in the outbreak. Two of the town's four wells were operating at the time of the outbreak: one was 305 meters deep and the other was 396 meters deep. Both wells had protected wellheads, and the monitoring data from the ten years before the outbreak indicated that no coliforms had been detected in either well. Investigation of the outbreak indicated that the distribution system was not well maintained and was vulnerable to sewage contamination at several points. Approximately 35 percent of the total flow was lost in the system—suggesting leaks, inaccurate meters, or unmetered connections. The town sewer system was also in poor condition and operating beyond capacity, resulting in regular sewage back-ups and overflows.

As with most waterborne disease outbreaks, a constellation of risk factors contributed to this outbreak. In mid-December 1989, unusually cold weather caused two large water mains and 45 in-ground water meters to fail (Figure 4-1). Ten cases of bloody diarrhea were reported to the local health department on January 4, 1990. A boil-water order was issued on January 5, and water chlorination was initiated on January 12. Analyses of the temporal distribution of the cases indicated that the first cases occurred seven days before the first water main break (December 23), and the last case occurred three days after the implementation of water chlorination (Figure 4-2). The early cases may have been due to leaks and holes that developed prior to the main break. There was a small increase in the incidence of diarrhea after the first main break and a large increase in diarrhea cases about four days after the second main break on December 26.

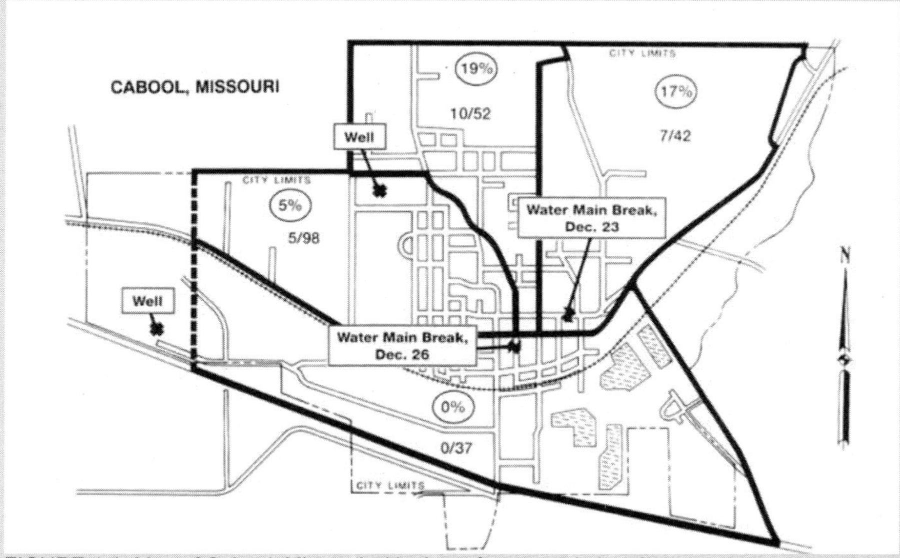

FIGURE 4-1 Map of Cabool, Missouri with sites of water main breaks.
SOURCE: Reprinted, with permission, from Swerdlow et al. (1992). © 1992 by American College of Physicians.

In addition, replacement of the failed water meters may have further contributed to contamination of the distribution system. During the replacement of the meters and main break repairs, the lines were subjected to "limited flushing" but were not disinfected, and no water samples were tested for microbial indicators to examine the water quality before bringing the lines back into service. Although sewage overflow into the distribution system via the main breaks and intrusion was believed to be responsible for the outbreak, microbial contamination of the distribution system could not be confirmed. Only two water samples from the distribution system were collected (on December 18 and January 3) and analyzed, but neither sample was collected from the areas with the highest concentration of cases. Hydraulic modeling of the system by Geldreich et al. (1992) reinforced the evidence that the second main break had the potential to contaminate a greater portion of the distribution system, including the northern part of the town where 36 percent of the cases occurred.

This outbreak illustrates how, despite a clean groundwater source, lack of disinfection combined with poorly maintained water and sewer lines, unusually cold weather, and casual line replacement practices led to a large drinking water outbreak with fatalities in a small town in an industrialized country.

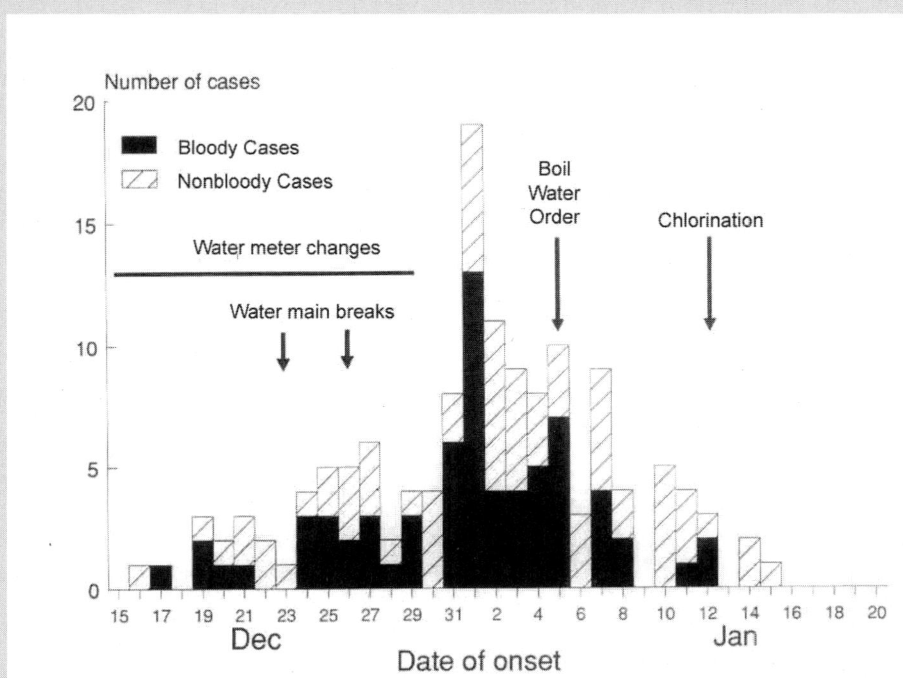

FIGURE 4-2 Cases of diarrheal disease among city residents. SOURCE: Swerdlow et al. (1992). SOURCE: Reprinted, with permission, from Swerdlow et al. (1992). © by American College of Physicians.

SOURCES: Swerdlow et al. (1992), Geldreich et al. (1992), Hrudey and Hrudey (2004).

The storage, installation, rehabilitation, and repair of water mains, fittings, and appurtenances provide another opportunity for microbial and chemical contamination of materials that come into direct contact with drinking water. Pierson et al. (2001) noted that this was particularly prevalent during the handling and storage of distribution system materials and during their installation in the trench. Indeed, a survey of water utilities found that about 14 percent experienced positive coliform samples from 1 to 10 percent of the time that new mains are checked before they are released (Haas et al., 1998).

Studies have demonstrated that the soil surrounding buried pipe can be contaminated with fecal indicator microorganisms and pathogens (Kirmeyer et al., 2001). Besides contaminated soil, runoff from streets and agricultural land can be highly concentrated with microbiological and chemical contaminants (Makepeace et al., 1995), and this runoff can contaminate pipes during a main break, during the unprotected storage of pipe materials, and even during pipe installation in the trench. One of the culprits in this regard are sewer lines that run in close proximity to distribution system mains. Leaking sewers can contaminate the soil and groundwater in the area of a water main or a trench where main activity will take place. The general rule is that there should be a horizontal separation of at least 10 ft (3 m) between water and sewer lines, and that the water line should be at least 1 ft (0.3 m) above the sewer (although variations to this general rule may occur from state to state). This rule, however, is fairly recent in comparison to the average age of the nation's buried infrastructure.

A second major mechanism of pipe failure is permeation, where contaminants external to the pipe materials and non-metallic joints compromise the structural integrity of the materials and actually pass through them into the drinking water. Taste and odor events are common consequences of permeation of plastic pipe given the types of contaminants involved. For example, in one case solvents trapped beneath a polyethylene wrap and soil migrated through plastic pipe and pipe connections to contaminate the drinking water in the service lines (Burlingame and Anselme, 1995). Because the solvents were derived from a hot-butyl rubber coating applied to the external surface and ferrules of a ductile iron main, they included toluene, indan, indene, naphthalene, xylene, and benzofuran. The event was initially detected first by customer complaints about off odors. In addition to the taste and odor issues, continued exposure to solvents can change a pipe's integrity and eventually lead to pipe failure.

Although there is the potential for water quality degradation as a result of the permeation of plastic pipe, the health impacts associated with such permeation are not well documented nor are they expected to be significant. In some permeation incidents, the concentrations of certain chemicals have been shown to reach levels in the low parts per million, which are well above their respective maximum contaminant levels (MCLs) (AWWA and EES, Inc., 2002). However, these MCLs are based on long-term exposure, and the short-term risk levels for these chemicals are generally much higher. In the case of permeation by gasoline components, the taste or odor thresholds of the majority of these chemicals are below the levels that would pose a short-term risk (EPA,

2002a,b,c,d), such that customers would notice an objectionable taste or odor in the water before significant exposure. In addition, these high concentrations would be expected to occur during worst case situations where water has been in contact with the affected pipe for a considerable length of time. During periods of normal water use, these concentrations would be expected to be much lower. It should be noted that the taste and odor thresholds for some contaminants may be above the MCL, in which case permeation of these chemicals could result in undetected long-term exposure if monitoring of these chemicals in the distribution system is not conducted.

Contamination of Storage Facilities

Although they may suffer from structural failures, storage facilities are most susceptible to external contamination due to the absence or failure of an essential component, such as a cover, vent, hatch, etc. The complete absence of a cover or vent on a storage facility can allow birds access to the tank and subsequently introduce microbial pathogens such as bacteria and parasites to the water within. For example, in the winter of 1993 a waterborne disease outbreak of salmonellosis in Gideon, Missouri, was traced back to the contamination by birds of the public water supply's distribution system storage tank (see Box 4-2; Clark et al., 1996). Indeed, one water storage tank connected to the distribution system was found to have holes in the top and bird feathers floating in the water. Two other storage tanks were found to be in similar need of maintenance, and pigeons were found roosting on the tanks. Birds, and consequently bird excrement, are probably the biggest concern for storage tanks and reservoirs with floating covers. Sea gulls, for example, can be found roosting at storage facilities. Open reservoirs also offer the opportunity for detrimental changes in water quality because of exposure to the atmosphere or sunlight, such as changes in pH, dissolved oxygen, and algal growth.

Even when covered, storage facilities can suffer from algal growth on the tops of floating covers that can gain entry into the tank through rips and tears or missing hatches. Algae can also be airborne or carried by birds and gain entry into storage tanks through open hatches and vents. Algae increase the chlorine demand of the stored water, reduce its oxygen content upon their degradation, affect taste and odor, and in some cases release byproducts.

Chemical contaminants gain access to storage facilities via air pollution and surface-water runoff into open storage reservoirs. For example, accidental spills of chemicals during truck transport on highways adjacent to reservoirs are a potential threat, and can be very serious if the chemicals are present in a concentrated form and highly toxic (Geldreich, 1996). Surface-water runoff into open reservoirs can also introduce pesticides, herbicides, fertilizers, silt, and humic materials from nearby land. The potential for chemical contamination of storage facilities continues to be overlooked in regulations in comparison to microbial

BOX 4-2
Waterborne Disease Outbreak Associated with Contamination of Water Storage Tanks: Gideon, Missouri

In 1993, the town of Gideon, Missouri (pop. 1,100) suffered from an outbreak of salmonellosis that affected more than 650 people and caused seven deaths. The Gideon water system consisted of two deep wells (396 meters) with no treatment or disinfection and a water distribution system that dated to the 1930s. In early November 1993, a cold snap caused a thermal inversion in the water storage tanks that resulted in taste and odor problems. In response, the water system was systematically flushed on November 10. The first cases of acute gastroenteritis were reported on November 29 and diagnosed as *Salmonella typhimurium*. However, the outbreak investigation later revealed that diarrhea cases in Gideon started around November 12 with a peak incidence around November 20. By early December, there was a 250 percent increase in absenteeism in the Gideon schools and a 600 percent increase in anti-diarrheal medication sales. Over 40 percent of nursing home residents suffered from diarrhea and seven people died. The outbreak was not linked to the water system until December 15 when the water system samples were reviewed and investigative water sampling was initiated. A boil-water advisory was issued on December 18. On December 22nd, emergency chlorination was added to the production well and the two municipal storage tanks were superchlorinated. The last reported cases occurred on December 28.

Water samples collected from a hydrant in the distribution system on December 16, 17, 20, and 21 were positive for total coliforms, and the samples from December 20 and 21 were also positive for fecal coliforms. Inspection of the two municipal water storage tanks suggested that the outbreak was probably caused by bird feces in one or more of the tanks. The larger of the tanks was in disrepair and had birds roosting on the roof. A third private storage tank had an unscreened overflow pipe and a hole at the top of the tank that was large enough for birds to enter. This private tank had been drained on December 30, but tJhe outbreak strain of *S. typhimurium* was detected in samples of sediment collected on January 5, 1994. The remaining water on the bottom of the tank was described as black

contamination. For example, the Long Term 2 Enhanced Surface Water Treatment Rule requires that water systems with uncovered finished water storage reservoirs cover the reservoir *or* treat the reservoir discharge to the distribution system to achieve a 4-log virus, 3-log *Giardia*, and 2-log *Cryptosporidium* inactivation, the latter of which would not protect against chemical contamination. However, it should be noted that EPA has published a Guidance Manual on Uncovered Finished Water Reservoirs (EPA, 1999) that addresses chemical contamination. Although the actions contained in the manual are not mandated, some states (such as California) are requiring water systems to implement them.

Contamination Due to the Absence or Operational Failure of Backflow Prevention Devices

Backflow events via unprotected domestic, commercial, industrial, and fire connection services can introduce contaminants into the potable water supply, with potentially profound health implications. A recent survey (USC, 2002) found that more than 95 percent of sampled homes had direct or indirect cross and very turbid, with rust, suspended particles, and bird feathers floating on the top. Initially attention was focused on the private tank as the source of the outbreak as reported by Skala (1994). However, an in-depth hydraulic analysis of the Gideon system, conducted as part of the outbreak investigation, raised questions about the possibility of the private tank being the source of the outbreak. A subsequent review of as-built drawings of the Gideon system by Missouri Department of Natural Resources personnel revealed that the private tank was separated from the municipal system by a functioning backflow prevention valve. In a subsequent hydraulic analysis the private tank was eliminated as a contamination source for the outbreak, which led to results that were consistent with the behavior of the system as observed during the outbreak scenario. This analysis also pointed to the largest municipal tank as the most likely source of the outbreak. A visual inspection of the large municipal tank revealed broken and rusted hatches and bird parts and feathers on the top of the tank and floating on the surface of the tank water. Both Clark et al. (1996) and Angulo et al. (1997) concluded that the large municipal tank was the source of the outbreak.

In the end, the outbreak investigation concluded that the cold weather in early November caused a thermal inversion in the water storage tanks that mixed the contaminated upper layers of stored water with the water entering the distribution system. The widespread flushing program on November 10 served to draw more contaminated storage tank water into the distribution system than under normal operation. The large discharge of the stored water over a short period of time may also have stirred up sediments in the tank and introduced them into the distribution system. Hydraulic modeling indicated that the part of the distribution system that served the school and the contaminated fire hydrant would have received water from the problem municipal tank within the first six hours of flushing. Other contributing factors included late recognition of the outbreak by the public health authorities, late recognition that the outbreak was linked to the public water supply, and a low rate of compliance with the boil-water order.

SOURCES: Skala (1994), Clark et al. (1996), Angulo et al. (1997), Hrudey and Hrudey (2004).

connections ("direct" meaning a cross connection subject to both backpressure and backsiphonage while an "indirect" cross connection is subject to backsiphonage only). Because of the enormous range of contaminant sources involved, as well as the number of unprotected cross connections, backflow events collectively constitute the greatest potential health risk from distribution system contamination. Whether an individual backflow event poses a risk depends on the type of the contamination, the length of an individual's exposure to the contaminated water, and other factors. A survey of water utilities in North America found that 28.8 percent of cross connections resulted in bacteriological contamination whereas 26.1 percent resulted in chemical contamination, and 29.8 percent resulted in both bacteriological and chemical contamination (Lee et al., 2003).

Although their potential to occur is high in all systems, backflow events are a particular concern in dual distribution systems where one line carries a nonpotable water source that may become connected with a potable source in the other line. Generally, the nonpotable line is a substantial health risk because it carries reclaimed water containing chemicals and microbial pathogens at levels exceeding water quality standards.

One of the most well known backflow events affected two Chicago hotels in 1893 during the World's Fair (Columbian Exhibition). A loss in water pressure in the distribution system caused backsiphonage through cross connections which contaminated the hotels' drinking water. An amoebic dysentery outbreak resulted in over 1,400 illnesses and at least 98 deaths (Von Huben, 1999). It is likely that the frequency and magnitude of contamination events due to cross-connections is underreported, especially where premise plumbing is involved. Box 4-3 describes a waterborne disease outbreak associated with an unprotected cross connection.

DETECTING LOSS OF PHYSICAL INTEGRITY

In some cases, a loss in physical integrity might actually be observed, such as a hole or tear in a reservoir cover, a missing vent or hatch on a storage facility, or a flooded meter or valve pit. Other structural failures, such as pipe leaks, tend to be much less obvious. The ability to predict and detect a failure in a material barrier is a desired capability for any water supplier. Structural failure is predictable for all major infrastructure components given information about materials composition and age and the surrounding environment. Structural integrity and operational performance should be confirmed on a regular basis via testing and inspections, particularly for backflow devices and storage facilities. The lack of standards and proper training can be predictive of a loss of physical integrity due to the improper installation, repair, or replacement of infrastructure components.

Predictions of structural failure can often be made based on historical information. For example, much is known about iron pipe based on years of actual experience. Cast iron pipe has been shown, under the right conditions, to last 100 years and more. When first introduced, cast iron pipe had no internal lining or external coatings to protect it from corrosion. After 1860, most pipes were lined with a molten tar pitch, and after 1922 some pipes were lined with a cement-mortar lining which in turn is sometimes protected by an asphaltic seal-coat. By the mid 1950s, ductile iron pipe came into use that has about twice the strength as cast iron but with a reduced wall thickness. This thinner wall pipe is more forgiving during installation and is more resistant to damage (AWWA, 2005b). It can be protected from external corrosion by a polyethylene wrap. AWWA/ANSI standards exist for pipes, joints, wraps and epoxy coatings, and fittings, and they provide information on the lifetimes of these materials.

The drive to predict and prevent failures varies depending on the consequence of the failure (Makar and Kleiner, 2002). For example, a branched distribution system has greater consequences associated with failure compared to a grid/looped system. There has been much attention given to predicting pipe failures, and more attention is needed in order to better predict overall system reliability (Grigg, 2005). Most useful would be a user-friendly guidance manual for utilities regarding the failure mechanisms of different types of pipes and how to use the various types of information on the current condition of the pipe to determine its expected lifetime.

Table 4-5 summarizes common methods used to detect a failure in a material barrier based on the types of failures that can occur. Inspection, direct testing, and consumer complaints play a significant role. Water quality testing may be the least effective means for detecting a loss in physical integrity, and thus is not discussed extensively.

The sections below discuss the role of inspections, condition assessment of infrastructure, leak detection, main break monitoring, and water quality monitoring for both prediction and detection of physical integrity. It is hoped that water utilities will embrace these activities and keep appropriate records in order to identify those factors that lead to failures, recognize early warning conditions, and improve their overall prediction capabilities. Integrating all of these data streams in order to plan how and where to rehabilitate, repair, or replace infrastructure is a significant challenge for water utilities and yet essential to being proactive in deterring contamination events that would pose a risk to public health (Martel et al., 2005). This chapter does not focus on failure analysis, which is the systematic investigation into the causes of pipe failure by visual

TABLE 4-5 Examples of Ways to Detect a Loss in Physical Integrity

Component	Mechanism of Integrity Loss	Detection by
Pipe	Permeation	VOC testing, investigate customer complaints about taste/odor
	Structural failure (leak)	Leak detection, investigate customer complaints
	Structural failure (break)	Investigate customer complaints, pressure monitoring
	Improper installation	Inspection
	Unsanitary activity	Inspection, water quality testing
Fitting and appurtenance	Structural failure	Inspection, pressure monitoring, investigation of customer complaints, leak detection, detection of operational failures
	Improper installation	Inspection
	Unsanitary activity	Inspection
Storage facility wall, roof, cover, vent, hatch	Structural failure (crack, hole)	Inspection, water quality testing
	Absence of	Inspection, water quality testing
	Improper installation	Inspection
	Unsanitary activity	Inspection, water quality testing
Backflow prevention devices	Absence of	Inspection, investigate customer complaints
	Improper installation	Inspection, investigate customer complaints
	Operational failure	Inspection, investigate customer complaints

BOX 4-3
Waterborne Disease Outbreak Associated with a Cross-Connection in the Water Distribution System: The Netherlands

A new housing development in the central part of the Netherlands was built with a dual distribution system. One set of pipes carried drinking water and a second set of pipes carried water from the same source that received partial treatment and was designed to be used for toilet flushing, laundry, and garden irrigation ("economy water"). Both the drinking water and the economy water originated from a surface water source and were treated by coagulation, flocculation, sedimentation, and rapid sand filtration. The drinking water was further treated by dune filtration. Approximately 30,000 households were served by this dual distribution system.

On December 3, 2001, the water utility received complaints from two people living in one neighborhood of the development that the drinking water had an unusual taste and odor. Drinking water samples collected on December 4 indicated unusually high coliform levels. On December 5 and 6, the water utility issued a boil-water advisory. On December 6, a local physician informed the public health service that he had seen an unusually number of cases of gastroenteritis with nausea, vomiting, and diarrhea in his clinic over the past few days. Further investigation of the water system revealed that when maintenance work was done on November 29, the drinking water system had been connected to the economy water system in order to flush and clean it, and that the workers had failed to remove the cross-connection when the economy water system was put on-line again. In addition, the economy water supply lines were under higher pressure than the drinking water lines, which forced the economy water to enter the drinking water distribution system. This cross-connection was removed on December 6, *E. coli* concentrations in the drinking water system dropped to below detection limits on December 12, and the boil-water advisory was lifted on December 17. A 1,000-liter sample of the economy water collected on December 20 was found to contain approximately 1.6×10^3 PCR-detectable units of norovirus. Southern blot hybridization identified the norovirus isolate as a genogroup I virus.

Two retrospective studies were conducted to determine the effect of this cross-connection on the rates of gastroenteritis in the housing development. The first was a retrospective cohort study that compared the incidence of gastrointestinal symptoms and other health symptoms during the period of November 29 through December 9 among 412 households in the area exposed to the cross-connection to the incidence of symptoms among 486 households in an adjacent control area that also had the dual distribution system but was not affected by the cross-connection. Data on symptoms and normal daily water consumption were collected by a one-time questionnaire that was mailed out to over 900 households in the exposed area and over 1,600 households in the control area. In addition, over 400 stool collection kits were mailed to randomly selected households equally divided between the exposed neighborhood and the adjacent control neighborhood, and the households were asked to provide a stool specimen from one member of the household who had recently experienced gastroenteritis.

The results of this study indicated that during the period of November 29 through December 9, households in the exposed area experienced significantly higher illness rates than households in the adjacent control area. In the exposed area, the rates of diarrhea, vomiting, nausea, abdominal pain, and chills were twice as high (38–54 percent of exposed households) as the rates reported from the control area (19–28 percent). The reports of blood in stool were about four times higher, and the reports of itching were over six times higher in the exposed neighborhood compared to the control neighborhood.

However, these symptoms were rare (1–3 percent blood in stool, 1–5 percent itching). Households in the exposed area were 1.5 times more likely to seek medical care during the period of November 29 to December 9 than households in the control area. There was no significant difference in the reported rates of coughing and sneezing in both neighborhoods or in the reports of symptoms that occurred after December 9. Interestingly, the distribution of symptom reports over time indicated that a peak in gastroenteritis symptoms occurred in both communities around December 3–5, but the peak was lower and shorter in the control community. Although the adjacent control community experienced a lower incidence of gastrointestinal symptoms, the rates during this time were still higher than normal. Also, there was clear evidence of a dose-response effect in both communities with significantly higher rates of households with diarrhea in those who reported higher water consumption (chi square for trend of 51.26 for the exposed area and 23.47 for the adjacent control area, $p<0.01$ for both communities).

Analyses of the 53 stool samples that were sent from 31 exposed households and 22 households in the adjacent control area yielded one norovirus genogroup I strain and one *Giardia lamblia* isolate from the exposed neighborhood and one norovirus genogroup II strain from the control neighborhood. The second norovirus strain came from a household that reported gastroenteritis symptoms after December 9.

The second retrospective investigation was a survey of two health care facilities for cases of gastroenteritis during the period of November 26 – December 12, 2001. One health facility served both the exposed (pop. 1,866) and adjacent control (pop. 2,875) areas in the cohort study. The other health facility was farther away and served a different part of the housing development (pop. 5,788) that was not exposed to the cross-connection incident in the distribution system but was still served by the dual distribution system. Based on the computer database of date of visit, patient address, and diagnosis codes, the incidence of gastroenteritis cases seeking medical care was compared between the three communities. Residents in the exposed area had a rate of 19.8 cases per 1,000 inhabitants compared with 7.0 cases per 1,000 in the adjacent control area and 3.3 cases per 1,000 in the more distant control area. The rate of gastroenteritis cases seeking medical care increased markedly in the exposed area and moderately in the adjacent control area on December 3 –5 and again on December 10–11 (weekdays). There was no change in the rate of diagnosed cases of gastroenteritis in the distant control area during this time period.

Taken together, the results of these two retrospective studies suggest that an outbreak of gastroenteritis, probably due to noroviruses, occurred shortly after the cross connection between the economy water distribution system and the drinking water distribution system was created. It is notable that there appeared to be an increased risk of gastroenteritis in the adjacent control community that reportedly was not affected by the cross-contamination incident. This may have been due to secondary transmission from the exposed community to the adjacent control community by other routes (food, person-to-person) because these two communities shared several facilities located in the control community (schools, health center, supermarket) or consumption of contaminated water by visitors from the control community to the exposed community. Studies of the surface water source in the spring of 2001 indicated high concentrations of noroviruses of up to 1.4×10^4 PCR detectable units per liter, and it is unlikely that these viruses would have been completely eliminated by the treatment processes used for the economy water. Noroviruses (1.6×10^3 PCR detectable units per liter) were also detected in the economy water on December 20. It is possible that exposure to the economy water in this system through aerosols from toilets, laundry, or garden irrigation may have posed some risk to the inhabitants in this development even without the cross-connection incident.

SOURCE: Cooperative Research Centre for Water Quality and Treatment (2003).

means and other inspection tools in order to determine the point where the failure started and the specific type of failure. Makar et al. (2005) discusses the actors that can cause failure such as flaws in the material (inherent to its manufacture or produced afterwards), forces that exceeded the design strength, design that did not account for normal operating loads, or some combination of the above.

Inspections

Regular inspections of the distribution system, either visual examination of the various structures or via acoustic leak detection and pressure monitoring (discussed below), provide the most direct way to detect a failure in the material barrier. Storage facilities need to be inspected on a routine basis for vandalism, settling, cracking and spalling, seepage, leakage at seams and joints and in the roof, missing hatches and vents, rust and corrosion, cathodic protection, and failing structures (AWWA, 1986). A second critical type of inspection is to check both the material integrity and the cleanliness of pipe prior to installation. Even though it is often assumed that pipe is inspected before it leaves the factory, damage to the spigot end of the pipe, the exterior, and the internal lining can occur during pipe storage and handling. Another reason to inspect pipe prior to installation is to detect manufacturing defects such as surface impurities or nicks, which are likely to induce corrosion and pitting once installed (Von Huben, 1999). Finally, pipe should be examined before installation for oil, dirt, grease, animals, and foreign matter; if found, the pipe should be cleaned out with a strong hypochlorite solution.

An important opportunity in this regard is the sanitary survey, which is a broad review and inspection program for a water utility that occurs once every three to five years. The survey might reveal an absence of (1) training and certification, (2) use of standards, and (3) routine inspections, all of which could be predictive of a loss of physical integrity. This is because a lack of training, certification, inspection, and standards often lead to the improper installation and application of materials (for example, using the wrong backflow prevention device or installing plastic pipe in contaminated soils).

Monitoring the Condition of Buried Infrastructure

The various tools available for locating buried pipe include ground-probing radar, metal detectors, magnetic locators, and radio transmission units for metallic pipe (Von Huben, 1999). Similar methods can be used to detect non-metallic pipe if metallic tapes or tracer wire was installed with the pipe. Locating pipe is the first step of *condition assessment*. A condition assessment is based on the assumption that materials or infrastructure components deteriorate, with the goal of gathering information to predict the need for repair, rehabilitation, or replacement (Grigg, 2005). The three steps of condition assessment are (Morrison, 2004):

1. Develop an up-to-date inventory of assets. With pipes, a Geographic Information System (GIS) can be used to collect the following data: diameter, material, classification/grade, wall thickness, joint type, installation date, lining and coating types, corrosion protection system, depth of burial, soil conditions, groundwater level, bedding classification, and history of problems (Shamsi, 2005).
2. Inspect the internal pipe condition, pipe wall condition, pipe environment condition, and leakage (which can be difficult and costly to accomplish for buried, in-use pipe).
3. Rate the condition of the asset.

There are five categories of pipe rating used during condition assessment (Morrison, 2004):

Rank 5. In danger of immediate failure, requires emergency repair or replacement as soon as possible to avoid jeopardizing public health and safety.
Rank 4. Severely deteriorated and in need or repair, renewal, or replacement. Should be addressed immediately.
Rank 3. Mildly deteriorated, short-term performance just adequate; however, will require renewal or replacement soon. Capital improvement plans are needed with more frequent inspections.
Rank 2. Minor deterioration, performance adequate. An inspection or assessment plan should exist.
Rank 1. Little to no deterioration, performance more than adequate.

Condition assessment requires information from existing pipe to help predict the lifetime of pipe still in use. To make the exercise more economically feasible, it might be done for selected pipes that represent a cross section of installed pipe materials and installation dates. Within some utilities, condition assessment is conducted whenever pipe repairs are made or new pipe is installed, because existing pipe is exposed, which facilitates the assessment. Other utilities carry out a regularly scheduled assessment program independent of other activities. Whatever the final outcome, how and when condition assessments will be conducted should be determined and standardized at each utility.

The technologies available to carry out condition assessment are varied and mature. Destructive testing includes the use of coupons or cuts from actual sections of pipe and spot condition assessment. Nondestructive testing include magnetic, electromagnetic, sonic, acoustic, infrared thermography, and ground-penetrating radar equipment for locating pipe; global positioning system (GPS)/GIS databases for managing information; and ultrasonics, acoustic emission, magnetic flux leakage, and remote field eddy current for assessing pipe. Finally, closed-circuit television has been used in some situations. For the most part there are sufficient tools available to allow utilities to conduct condition assessments and to utilize the data that are collected to guide decisions. What is needed is for these tools to be utilized more uniformly as well as fine-tuned for use in small systems that have more limited capabilities. It would be extremely useful for the results of condition assessments to be shared among utilities, and even benchmarked, so that utilities can build upon shared experience and knowledge.

Future tools for assessing the condition of buried pipe include real-time tools that travel through pipe and collect information; small chips set in pipe; sensors to record sounds of breaks; fiber optics to record breaks in light; and improved metering to identify leaks (Grigg, 2005). These tools are in development and likely show promise for specific situations rather than globally for all materials in all circumstances. For example, different sensors are needed for plastic pipe than for iron or concrete pipe.

Leak Detection

The early detection of leaks and their remediation is a goal for water utilities. Leak detectors include listening devices, such as an aquaphone or a more complicated amplified detection kit that detects sound caused by flowing or escaping water (Von Huben, 1999). Another way of detecting leaks is to conduct a water audit which uses flow meters around smaller districts of a system at night when water use should be low. Acoustic methods are easy to use and widely applied on metallic pipes, with improvements being made for use on plastic pipes (Lange, 2002). Morgan et al. (2005) recently used a fixed-based acoustic monitor system called MLOG to scan the distribution system at night for leaks. The system was highly effective, detecting 17 previously unknown leaks within the first three months of use.

In addition to improvement in leak detection, water meters have been developed to detect and record backwards flow through the meter in order to determine the magnitude and frequency of backflow events (Neptune Technology Group, 2005). Although the majority of this flow may simply be service line water, the use of advance meter reading can detect these backflows in real time. The ability to detect and track backflow events will allow more focused monitoring to determine their impact on drinking water quality.

Main Break Monitoring

Main break monitoring consists of utilities recording responses to water main breaks such as time and date of response, location of break, valves operated to shut down the main, properties affected by the shut down, repaired or replaced portion of main, and shut-down time. Transmission mains are given a higher priority for main break monitoring and the prevention of failure than smaller size distribution mains, given their potential seriousness (for example, destruction to local structures such as roadways, bridges, and buildings or interruption of automobile traffic and the evacuation of residences).

If main break monitoring data are maintained in a computerized database, then quarterly, annual, or five-year historical evaluations can be done, trends can be predicted, and both can be compared to changes in related practices (such as replacing cast iron with ductile iron pipe). If the utility can also collect data on leaks and repairs leading up to a break, as well as failure analysis results during the break, it becomes possible to develop better predictive models for the distribution system's pipe infrastructure. Water utilities have successfully trended water main break rates and have adjusted their practices to minimize the occurrence of failures for various types of pipe in their systems. The trending of water main breaks and leaks along with condition assessment provide an important tool to minimize public health risk. Not only will a water utility reduce its risk of serious consequences from an unexpected failure, but in reducing the seriousness of such failures the water utility will gain control in minimizing the potential for water quality contamination. This will happen in two ways, because the severity of any single failure is reduced, and because the frequency of failure is reduced.

Water Quality Testing

Much of the monitoring needed to assess the physical integrity of a distribution system is accomplished by other means than water quality monitoring, such as by leak detection, customer complaint response, inspections, or the exercising of valves and hydrants. However, water quality testing can play a role. Typically, water quality analyses are limited to common chemical parameters (total chlorine residual, pH), physical parameters (turbidity, color), and biological parameters (heterotrophic plate count, total coliform count) (see Chapter 6 for a more thorough discussion). For those parameters that are routinely monitored under the Safe Drinking Water Act, a detection of a change in a parameter would not in itself identify the occurrence of a loss in physical integrity that resulted in contamination, since water quality changes could be from internal conditions in the system or from a treatment breakthrough or failure. However, a thorough follow-up response to a change in water quality (such as high turbidity, colored water, or non-detectable chlorine residual) could include valve checks, hydrant flushing, and other techniques that might identify the cause of the loss in water quality as being an external contamination event.

Water quality data can also be useful in identifying problems with physical integrity when integrated with others sets of data, such as customer complaints, water main break occurrences, timing of newly installed water mains, cleaning of storage facilities, or backflow events (see Chapter 7 for more discussion of data integration). Water quality testing is particularly useful if it can be correlated with customer complaints. For example, consumer complaints of chemical-type odors along with the utility's detection of volatile organic chemicals (VOCs) could signal that permeation of plastic pipe has occurred, which could be further studied through groundwater and soil testing for the same VOCs. Once remediation is put in place, the same water quality parameters could be used to gauge the success of the remediation. Backflow events are another example of where customer complaints used in conjunction with water quality monitoring may be informative. Depending on the contaminants present, backflow events can affect the color of water, can introduce debris and particles, and can cause an off-odor or taste.

As with any environmental sampling, increasing the frequency of water quality monitoring, for example going to on-line monitoring of storage facility effluent as opposed to daily or weekly grab sampling, will make it more likely that a contamination event will be detected. For example, when doing water quality sampling on a new water main prior to its release for use, a typical number of samples would be four to five. If this new addition or replacement involved 100 feet of 6-inch pipe, the total volume would be around 150 gallons (568 liters), such that four water quality samples of about 250 mL would only test $1/568^{th}$ of the potentially contaminated water. Clearly, the approval of a new water main should not rely solely on the final water quality check, but also on inspections at every stage of the process guaranteeing that materials were handled in a sanitary manner and protected from exposure to contamination. Thus, in isolation water quality data are not sufficient to identify failures in physical integrity. But combined with other data, they may be useful for detecting external contamination events.

MAINTAINING PHYSICAL INTEGRITY

Every water supplier's goal is to develop the means by which to better maintain the physical integrity of its distribution system so that a failure or loss rarely occurs, or when it does occur its impact is minimized. Table 4-6 summarizes some common measures used to prevent a loss of physical integrity in the distribution system.

The maintenance issues for pipes, fittings and appurtenances, storage facilities, and backflow prevention devices are similar in a general sense. Materials selection must meet standards and best practices. Installations of all components must be followed up with routine inspections. A regular program of valve operation and maintenance must be in place so that shut downs can be effective when needed. Many valves and hydrants are unused for a number of years, and debris within the distribution system may cause hydrants to become heavily encrusted leading to a significant reduction in discharge flow and fire protection. Furthermore, valves and hydrants should be carefully manipulated to maintain positive pressures and mitigate pressure transients that could result in pipe breakage. Good construction practices, conducted by those with training and certification and that follow standards and specifications, are essential. Standard

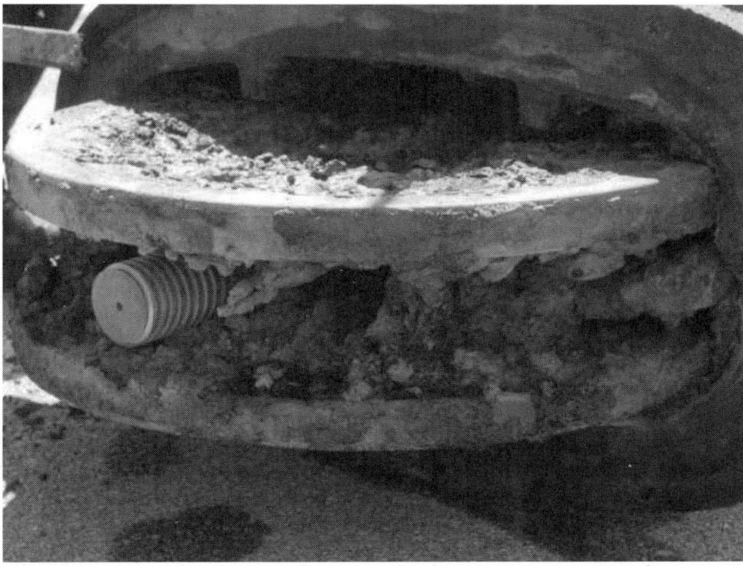

Valves should be inspected and operated on a regular basis to prevent rust and encrustation from interfering with their performance. Photo courtesy of Bureau of Laboratory Services, Philadelphia Water Department.

parts should be used to ensure consistency in repairs, and they should be stored in a sanitary fashion. Designing the distribution system to minimize sections of pipe and appurtenances that cannot be adequately tested, flushed, and disinfected would be beneficial (Pierson et al., 2002). Finally, funding and staffing must support all of these activities. These and other preventive measures are discussed below. It should be noted that maintaining the appropriate operating pressure to prevent main breaks and intrusion is discussed more thoroughly in Chapter 5.

Materials Quality

Materials that make up drinking water infrastructure range in type and value. Pumps have various components from pipe to valves to impellers, all made of differing materials. System piping includes valves and fittings, ferrules, and hydrants. Storage facilities range in their composition from concrete to steel with linings of cement, asphaltic, and epoxy. Customer premise plumbing includes meters, backflow prevention devices, valves, fittings, tubing, and faucets made of a plethora of materials. Rubber gaskets and plastic seats can be found

TABLE 4-6 Examples of Ways to Maintain Physical Integrity

Component	Mechanism of Integrity Loss	Prevention by
Pipe	Permeation	Standards on pipe applications, local assessments of soil and groundwater for contamination
	Structural failure (leak or break)	Better design and installation, early leak detection with rehabilitation and repair, optimized scheduling of pipe renewals, optimized placement of valves for effective shut-offs and isolations
	Improper installation	Standards and certification for installation, followed by inspection
	Unsanitary activity	Strict requirements and inspection during repair, rehab, installation
Fitting and appurtenance	Structural failure	Improved materials quality as well as quality in the operating components of valves and hydrants, periodic valve exercising followed by maintenance or replacement as needed
	Improper installation	Strict requirements on installation and design
	Unsanitary activity	Strict requirements during repair, rehab, installation and inspection
Storage facility wall, roof, cover, vent, hatch	Structural failure (crack, hole)	Better design and installation, early leak detection with rehabilitation and repair
	Absence of	Inspection and better design with inspection
	Improper installation	Strict requirements on installation and design
	Unsanitary activity	Strict requirements during repair, rehab, installation and inspection
Backflow prevention device	Absence of	Inspection and certification
	Improper installation	Strict requirements on installation and design
	Operational failure	Annual testing and maintenance

in valves and meters and in the joints of mains. In addition to these solid materials there are greases, lubricants, fluxes, and coatings. The diversity, complexity, and value of materials used in drinking water infrastructure are important to distribution system management, especially given the increasing emphasis on system reliability and more stringent water quality demands. The following factors should be considered when choosing distribution system materials:

- health effects of the material when in contact with drinking water;
- hazards and safety in working with the materials;
- structural capabilities of the material;
- water quality impacts of the material;
- cost and availability of the material;
- compatibility of the material with other materials in the system and with the conveyed water and surrounding soils;

- environmental effects of the material;
- whether the manufacturers of the material are ISO certified and meet NSF and ASTM standards; and
- future changes that could impact on the above.

Not all of these factors have been given equal weight. Materials selection is typically based on tensile strength, flexural strength, durability, corrosion resistance, roughness coefficient (Hazen Williams C value), and economy (e.g., the cost of materials and installation lifetime value) (AWWA, 1986). Indeed, economic considerations and the availability of the material can weigh in heavily and may dominate the choice of material.

As shown in Table 2-3, 30 of 34 responding states have some basic requirements for the types of pipe materials allowed in distribution systems. Indeed, a variety of standards and guidelines are available to help utilities choose the correct materials for their infrastructure, including the ASTM Annual Book of Standards, standards from NSF International, AWWA standards, and other publications (Nayyar, 1992). In practice, the larger water utilities tend to apply material standards and test whether they are being met, but small water utilities likely have no way to test for compliance. Furthermore, water suppliers in the United States have underutilized the services that materials engineering can provide such as manufacturer and supplier certification, development of materials specifications for procurement, and evaluation of materials (chemical and physical) according to specifications after procurement (Burlingame et al., 2002). Testing of materials to ensure they meet the standards used for procurement should be a broader practice within the water industry, and not limited to only the largest water utilities.

In addition to making an informed choice of materials, water utilities should strive to protect the quality of the materials after initial purchase. This includes inspections during materials manufacture; proper storage, handling, and transport of the material to the utility; inspection and testing of the material upon delivery; protection during onsite storage; inspection during and after installation; failure analysis to detect early failures; and finally replacement of the material when its lifetime is exceeded (see section below on asset management). Failure analysis involves using a standard approach to record events around material failures; take soil and pipe samples and collect background records; conduct a preliminary investigation to determine the type of failure; and conduct structural analyses, visual examinations, metallographic and mechanical testing, and inspections for graphitization and manufacturing flaws on pipe surfaces (Makar, 2002). Because failure analysis has not been widely embraced by the water community, there is limited information on many of the materials in common use today. Thus, additional support is needed for technology transfer about materials, funding of materials testing programs, better materials development and information management, better training, and better cross-industry networking. For example, there have been no studies to date on the consequences of material failures due to workmanship/installation errors or manufacturer variability.

Another consideration is that many materials still in use today were not originally designed to meet the system reliability and water quality standards expected today and for the near future. Existing materials standards may not be complete and up-to-date for all applications. Furthermore, manufacturers are not always responsive to customer or end-user needs, especially as these needs change due to water quality regulations. Although improvement is needed in many areas, a substantial first step would be to improve installation workmanship. This could be accomplished by requiring that all trades people who work with materials being installed or repaired that come in contact with potable water be trained and certified for the level of sanitary and materials quality that their work demands.

Corrosion Control

The historical use of metallic pipes and the many environmental conditions they come in contact with have made both external and internal corrosion an issue for the water industry for some time. Although most utilities use some form of internal corrosion control to minimize color and turbidity problems and to meet the Lead and Copper Rule requirements, not all utilities practice external corrosion control, even though it is important for maintaining the physical integrity of their distribution systems, as acknowledged by 14 of 34 responding states (see Table 2-3). There is no regulatory motivation for external corrosion control in the water utility industry as there is in the oil and gas pipeline industry where corrosion control such as cathodic protection of its pipelines is mandated (Romer et al., 2004). Nonetheless, understanding the conditions that lead to corrosion and implementing a consistent corrosion control methodology can result in significant operation and maintenance savings because of the longer pipeline life.

As mentioned previously, the extent of external corrosion depends on soil conditions such as resistivity, pH, and water content; the occurrence of stray currents; contact between dissimilar metals; and bacterial activity in the environment surrounding the pipe. The testing and GIS mapping of soil conditions can help water utilities predict and plan for corrosive problems and design corrosion control (Romer et al., 2004). Unfortunately, the tools for analyzing soils prior to making water main construction decisions require further development. In addition, there is no standardized corrosivity testing method used by all water utilities. The Ductile Iron Pipe Research Association has promoted a qualitative corrosion evaluation system based on soil conditions of resistivity, pH, redox potential, the presence of sulfides, and site drainage conditions, which has been found to be dependable and accurate for determining when external corrosion control should be applied for buried iron pipe (Bonds et al., 2005). The American Concrete Pressure Pipe Association provides recommendations based on soil chloride and resistivity. In general, methods for the analysis of corrosion include a soil corrosivity survey, a close-interval potential survey, a cell-to-cell potential survey, ultra-sonic measurements, pit depth analysis, visual inspection, corrosion rate measurements, acoustic monitoring, and failure analysis (Szeliga and Simpson, 2002).

External corrosion control methods include determining the soil conditions and then (1) selecting the appropriate distribution system materials, such as plastic pipe for use in very corrosive soils; (2) applying external metallic corrosion prevention materials at the time of manufacture, such as concrete, mortar, or asphaltic shop coat; (3) applying barrier coatings and polyurethane encasements in the field; (4) using galvanic cathodic protection or impressed cathodic protection; and (5) mitigating stray currents (Szeliga and Simpson, 2002; Romer et al., 2004). For example, Edmonton, Alberta proactively reduced the impacts of external corrosion using cathodic protection and nondestructive testing of their cast iron mains (Seargeant, 2002). Proactive measures are also important, since a variety of design options (such as using rubber-gasket bell-and-spigot joints) can affect the extent of external corrosion (Romer et al., 2004). Transmission mains are more frequently engineered for external corrosion control than distribution mains because of the greater need to prevent catastrophic failures in the larger diameter water mains.

Internal corrosion of pipe is caused by distribution system water that is corrosive to the materials with which it comes into contact. Internal corrosion is common in unlined cast-iron and steel mains and also occurs inside steel water tanks, metal service lines, and premise plumbing and appliances. Concrete pipe and cement mortar are also vulnerable to corrosion from low alkalinity, low hardness waters. Internal corrosion is generally controlled by feeding corrosion inhibitors, such as phosphates, to the water in combination with pH adjustment and alkalinity control. The mechanism of action is generally one of forming a stable scale on the pipe surface from corrosion products and water constituents that both inhibits corrosion and reduces the release of metals from scale dissolution. Inhibitors and water quality control procedures need to be tested at each site of use because of differences in source water quality, pipe materials, and pipe condition. Ductile-iron and steel pipe are generally lined with a cement mortar lining to prevent internal corrosion or contact with water. Linings can reduce the frequency of small leaks in pipes and pipe connections as a result of the high resistance of cement mortar to pressure, enhance the hydraulic characteristics of the mains, and prevent further internal corrosion damage. Finally, steel water tanks are protected by internal coatings and cathodic protection.

External and internal corrosion control practices need to be used more consistently, universally and uniformly. A manual of practice for the industry should be developed as an aid to implementing best practices. At present the best defense against corrosion relies on site-specific testing and practical experience gained at individual utilities, given the variation in materials, soils, and water quality from utility to utility. There is also a need for research to develop new materials and corrosion science to better understand how to more effectively control both external and internal corrosion, and to better match distribution system materials with the soil environment and the quality of water with which they are in contact.

Permeation Prevention

Appropriate measures can be taken to minimize the occurrence of permeation, such as issuing regulations or guidelines that define the conditions under which plastic pipe should be used. The proper selection and use of PVC, polybutylene, and polyethylene plastic pipe, such as according to the soil or potential soil conditions in which the pipe will be buried, limits the potential for permeation. For example, California precludes the use of plastic pipe in areas subject to contamination by petroleum distillates (California Code of Regulations, Title 22, Division 4, Chapter 16, Article 5, Section 64624f). In addition to the pipe material, the environmental conditions around the buried pipe are also important. Utilities that install plastic water mains need to maintain an up-to-date knowledge base of the locations of underground storage tank sites, industrial spills, other developments that could discharge solvents, and their associated solvent plumes so as to avoid the contact of such contaminants with the pipe. In general, if this information can be gathered prior to laying new pipe, most if not all permeation incidents can be avoided.

Maintaining Storage Facilities

Storage facility issues are similar to other distribution system components in that materials selection, system design following standards and specifications, installation inspection, and good construction practices by those with training and certification all play a role. Many states do have some standards for storage tank design and construction, the use of vents, screens, hatches, and overflows, and they even encourage tank inspection and maintenance (see Table 2-4). However, perhaps because of their perceived peripheral role in water supply, storage tanks have not historically received the attention afforded to pipe maintenance.

Storage facilities have many purposes (see Chapter 1), such that a disciplined storage facility management program is critical to water utilities. Such a program includes developing an inventory and background profile on all tanks, developing an evaluation and rehabilitation schedule, developing a detailed tank evaluation process, performing tank evaluations, making rehabilitations and replacements when needed, and performing a one-year warranty inspection for all tanks (Wallick and Zubair, 2002). More specifically, storage tanks should be inspected for needed repairs, barrier screen replacements, and painting. Depending on the nature of the water supply chemistry, such detailed inspections should be made every three to five years, and consist of tanks needing to be drained, sediment removed, and appropriate rust-proofing applied to the metal surfaces (such as where the water level rises and falls more frequently) (Kirmeyer et al., 1999). These inspections are in addition to daily or weekly inspections for vandalism, security, and water quality purposes (such as identifying missing vents, open hatches, leaks, and so forth).

In one of the rare documents that addresses storage facilities, Von Huben (1999) summarizes the use of air vents to allow air to enter and exit as the water level rises and falls. These vents must be screened to keep out birds and insects. In general, preventing access to the tank interior by wildlife and sediment removal are important deterrents to possible pathogen contamination and coliform colonization that should be undertaken for every tank.

Asset Management

Asset management refers to a strategy of operating, maintaining, rehabilitating, and replacing infrastructure in order to sustain a cost-effective level of service to customers. For a water utility, asset management requires collecting and analyzing data and information about all functions of the utility (customer service and support, financial, engineering, operations, maintenance) in order to make strategic decisions about the infrastructure (Paralez and Muto, 2002; Schwarzwalder, 2002; Allbee, 2004; Lockridge, 2004; Cagle, 2005). When thought of with respect to maintaining physical integrity, it refers to developing an inventory of distribution system components and determining when repair should give way to rehabilitation or replacement (EPA, 2004). Table 4-7 gives some of the typical life expectancies for pipe, storage, valves, hydrants and service lines, although it is expected that properly installed and well maintained pipes should have a service life much longer than their design lives (Morrison, 2004).

TABLE 4-7 Material Life Expectancies

Distribution System Component	Typical Life Expectancies
Concrete and metal storage tanks	30 years
Transmission pipes	35 years
Valves	35 years
Mechanical valves	15 years
Hydrants	40 years
Service lines	30 years

SOURCE: EPA (2004). EPA's Note: These expected useful lives are drawn from a variety of sources. The estimates assume that assets have been properly maintained.
The adjusted useful life will be equal to or less than the typical useful life.

In order to do asset management, the water supplier needs to have condition assessment data (see earlier discussion) and management tools (such as funding, planning, and modeling tools) (Grigg, 2004). The goal of asset management is to determine the time to failure and vulnerability of individual components (like pipes) under varying scenarios. As mentioned previously, determining the condition of in-use buried pipe is currently difficult and costly to accomplish because the pipe is usually still in use, the inside needs to be assessed, and the assessment can only look at one small area of one pipe out of many associated pipes. Thus, a water utility typically lumps pipes into classes and assigns to them average failure information, and, using statistics about the system, then predicts investment needs to maintain the assets.

Beyond maintaining physical integrity, there are many important reasons for utilities to engage in asset management, including (Morrison, 2004):

- To maintain assets at a predetermined level of service, which requires inspection and assessment in order to ascertain whether the assets are capable of providing this level of service;
- To uncover performance issues that might hinder a utility's ability to meet customer service expectations, or potentially lead to a catastrophic failure endangering public health and safety;
- To control costs of rectifying or mitigating a problem, which are always much less just after inspection than after a rupture or other emergency event;
- To tailor maintenance practices to the actual condition of the asset, and not merely base them on habit, resulting in an overall reduction in expenditure;
- To properly plan for the retirement and/or replacement of the asset, which, if done over a period of time, will avoid any unexpected surprises.

Westerhoff et al. (2004) found that most utilities engage in asset management, although it ranges from simple maintenance programs to complicated business planning processes. Indeed, terminology, data collection, reporting, mapping, inventory control, records, and operational parameters are largely defined on a utility-by-utility basis (Grigg, 2005).

Cross-Connection Control

Proven technologies and procedures are available to mitigate the impact of cross connections on potable water quality. Well-known backflow control devices include air gaps, reduced-pressure-zone backflow preventors, double check valves, vacuum breakers, and complete isolation. Lists of approved backflow prevention assemblies can be found with the University of Southern California (USC), the American Society of Sanitary Engineering (ASSE), Underwriters Laboratories, the International Association of Plumbing and Mechanical

Officials, Factory Mutual, and the Canadian Standards Association, while the three most commonly used guidance manuals are the USC Manual of Cross-Connection Control (USC, 1993), AWWA's Manual M14 (AWWA, 2004), and EPA's Cross-Connection Control Manual (EPA, 2003).

The application of backflow prevention devices is based largely on the degree of hazard thought to be present. A potential threat from chemical and biological contaminants that pose a human health risk would constitute a high hazard. A low hazard would include incidents that alter the water's aesthetic properties but do not constitute a health threat. Higher hazards are also related to the type of facility from which the threat emanates, such as hospitals, funeral homes, chemical manufacturing plants, laboratories, film processing facilities, commercial laundromats, among many others. Low hazard facilities include apartment complexes, warehouses, office buildings, and public buildings. Table 4-8 gives the recommended applications of various backflow protection devices according to the degree of hazard and whether those hazards are due to either back-siphonage (negative pressure or suction on the supply side of the device) or back-pressure (high pressure on the service side of the device).

There are generally two types of cross-connection control programs: one is a service-protection program and the other is an internal protection program (AWWA, 2004). The service-protection program is the most common one for water utilities to undertake, given their typical enforcement capabilities. This program is one of "containment," in that any backflow incident would be contained within the customer's facility and prevented from entering the public distribution system. This is accomplished by installing a backflow prevention device at the water meter. Water utilities are typically effective with this type of program because they readily have enforcement capability in the shut-off of the water service at the curb stop. The internal protection program is based on "elimination" or getting rid of the cross connection where it exists within a customer's plumbing. Because water utilities typically have no authority within the premises of their customers, it is more likely that other agencies such as the local health department or plumbing code agency would maintain such a program.

Lee et al. (2003) found that more than 80 percent of responding water utilities require approved backflow protection devices and field testing of their proper operations. However, little if any information exists on whether these devices are present in customers' premises, where 83 percent of cross connections are known to exist (Lee et al., 2003). It is probable that they are absent for a very large percentage of cross connections nationwide or not functioning properly for a small percentage of cross connections. Clearly, their increased use and regular inspection would do much to reduce public health risks from drinking water distribution systems. Indeed, for utilities operating dual distribution systems, the need for an effective cross-connection control program is paramount.

TABLE 4-8 Use of Backflow Prevention Devices by Degree of Hazard and Mechanism

Device	Degree of Hazard			
	Low Hazard		*High Hazard*	
	Back-siphonage	Back-pressure	Back-siphonage	Back-pressure
Air Gap (AG)	X		X	
Atmospheric vacuum breaker (AVB)	X		X	
Spill-resistant pressure-type vacuum-breaker assembly (SVB)	X		X	
Double check valve assembly (DC or DCVA)	X	X		
Pressure vacuum-breaker assembly (PVB)	X		X	
Reduced-pressure principle assembly	X	X	X	X
Reduced-pressure principle detector assembly	X	X	X	X
Double check valve detector check assembly	X	X		
Dual check device (internal protection only)	X	X		
Dual check with atmospheric vent device (internal protection only)	X	X		

SOURCE: Adapted from Table 3-1 in AWWA (2004).

Basically, every state has some requirement for cross connection control (see Tables 2-3 and 2-6), and state plumbing codes define the type of cross-connection control devices that are approved for use. Unfortunately, as discussed in Chapter 2, the elements of such programs, their implementation, and oversight vary widely, partly because of the variation in available resources. In a few states, local jurisdictions are responsible for implementing a cross-connection control program. In most states, testing of cross-connection control devices is the responsibility of the customer while inspection of the devices is the responsibility of the water system or the local jurisdiction. Given this variability, Chapter 2 recommends that EPA explicitly define what an acceptable cross-connection control program should be.

RECOVERING PHYSICAL INTEGRITY

It is impossible for a distribution system of any significant size to be managed in such a way as to prevent *any* loss of physical integrity over time. Even a water utility with a good program of corrosion control and pipe replacement can experience an annual pipe break rate of around 750 to 850 breaks per year (Falarski, 2002). Damodaran et al. (2005) gave an industry average of 0.1 to 0.3 breaks per mile of pipe per year, such that a low break rate would cause 1 to 3 breaks per year per 1,000 people served. Philadelphia tracks the number of breaks experienced for each 1,000 miles of main using a five-year moving average to smooth out the effect of weather variations. Based on historical information dating back to 1930, the average for 2001 was 212 breaks for every 1,000 miles of main—the lowest total in over 45 years and better than the national average of 240 to 270 breaks per 1,000 miles. Nonetheless, even with a water main replacement program that appears to be successful compared to the national average, every year over 600 water main breaks occur. Therefore, procedures need to be in place by which to recover from a failure in a material barrier and minimize the effects on water quality.

Table 4-9 summarizes some of the common methods used today to recover from a failure in a material barrier in order to prevent or minimize contamination of the water supply. There are several categories of recovery efforts. First, compromised materials can be cleaned, repaired, rehabilitated, or replaced. For example, leaks and small breaks can be repaired by repair sleeves or by joint sealing compounds. Storage facilities might have to be drained and cleaned following potential contamination. Another form of restoration is to treat the contaminated water. Chlorine and other disinfectants have been used to protect pipes and storage facilities against external microbial contamination, prevent

TABLE 4-9 Ways to Recover from a Loss in Physical Integrity

Component	Mechanism of Integrity Loss	Recovery by
Pipe	Permeation	Reline or replace and conduct water quality testing
	Structural failure (leak)	Replace or repair or rehab
	Structural failure (break)	Replace or repair, flush or disinfect, conduct water quality testing
	Improper installation	Replace, reinstall
	Unsanitary activity	Disinfect, flush, and water quality testing
Fitting and appurtenance	Structural failure	Replace, repair, rehab and disinfect
	Improper installation	Reinstall
	Unsanitary activity	Disinfect and flush
Storage facility wall, roof, cover, vent, hatch	Structural failure (crack, hole)	Repair or rehab or replace, disinfect
	Absence of	Install
	Improper installation	Reinstall
	Unsanitary activity	Disinfect, flush, and water quality testing
Backflow prevention device	Absence of	Install
	Improper installation	Reinstall
	Operational failure	Replace or repair

regrowth of nuisance organisms in response to intruded chemicals, prevent further contamination from the installation of a dirty main, and alleviate customer complaints. Both continuous disinfectant residual maintenance throughout the distribution system and dosing a section of the system with disinfectant are common. Third, recovery is often brought about by flushing the contaminated water from the system rather than treating it, generally using hydrant flushing. Although flushing is mentioned sporadically here because it accompanies many of the other recovery techniques, it is treated more comprehensively in Chapters 5 and 6 where hydraulic and water quality integrity are the focus, respectively.

In those situations where the absence of a component was the cause for the lack of physical integrity, then simply installing the component is the recovery effort. For example, the installation of backflow prevention devices or changing covers on reservoirs (say from floating to hard covers) should restore integrity. Finally, where operational failure is the problem, devices may also need to be entirely replaced, along with instituting inspections to ensure that failure does not recur.

Repairing, Rehabilitating, and Replacing Pipe

Common types of repair activities include cutting and plugging the portion of pipe associated with a leak, installing a repair sleeve or clamp, eliminating dead end mains, replacing and repairing valves, adding ferrules, and repairing or replacing hydrants. These activities are discussed extensively in Grigg (2004) and not considered further here. Improvements are being made in locating buried failure sites, excavation, and repair. For example, trenchless methods are being developed and applied, although the technology development is slow.

Rehabilitation of pipe involves the recycling and reinforcing of the existing infrastructure in order to prolong its useful life. For example, structural lining can be used to improve the structural integrity of existing pipes and involves placing a watertight structure in immediate contact with the inner surface of a cleaned pipe (Selvakumar et al., 2002; Ellison et al., 2003). The most commonly used structural lining techniques include conventional slip lining (where new PE pipe is structurally able to replace the existing pipe), cured-in-place rehabilitation or inversion lining (which inserts a non-structural material) (Hughes and Conroy, 2002), fold-and-form pipe, and close-fit slip lining (which can use a structural or non-structural replacement material). Selvakumar et al. (2002) provide a detailed description of all these methods along with their costs, benefits, and limitations. Nonstructural rehabilitation of water mains, which does not focus on recovering the physical integrity of distribution systems, includes chemical dosing for corrosion control, cement mortar lining, epoxy resin lining, and thin-walled PE lining (Hughes and Conroy, 2002; Grigg, 2004; Damodaran et al., 2005). Such rehabilitation should be internally inspected to ensure that it is done to standards.

Pipes are candidates for replacement when the pipe is severely deteriorated (e.g., the pipe has suffered a series of breaks), or when additional hydraulic capacity is needed. Box 4-4 discusses the economic considerations that play into the decision to replace a pipe rather than rehabilitate or repair it. Historically, pipeline replacement involved the construction of a new pipeline normally parallel to the one being replaced. Once constructed, the new pipeline was connected to the pipe network and the old pipeline abandoned. This approach normally involved digging a trench, installing the new pipe, backfilling the trench, and final surface restoration. This construction can be very disruptive in built-up areas plus it may be very difficult to find a location to construct a new waterline. As a result, new trenchless technologies have developed which can result in cost savings over the conventional construction methods. Horizontal directional drilling has seen considerable growth as an alternative to open trench construction, especially at crossings of waterways, rail lines, and highways. A drilling bit bores a horizontal hole that is kept open using drilling fluid. Once a predetermined length of hole is completed, a new pipe is pulled back through the horizontal hole. This method is far less disruptive than open trench construction, and in most cases would not interfere with business or residential property access.

Another type of trenchless technology that is most useful in areas where it is difficult to install new pipe is pipe-bursting. This technology is similar to horizontal directional drilling, but with pipe-bursting a new pipe is pulled in the same location as the old pipe. A burster is pulled through the old pipe, breaking it apart and making room for the new pipe. The only openings required are at the two ends and at all active service locations. The equipment can install pipe of the same size that is being replaced or a size or two larger. Selvakumar et al. (2002) give a detailed description of pipe bursting, microtunneling, and horizontal directional drilling methods along with their costs, benefits, and limitations.

BOX 4-4
Decision-making regarding Replacement vs. Ongoing Repair

There now exist fairly good models for making decisions about ongoing repair vs. replacement of infrastructure pipe components (Damodaran et al., 2005), although they do not incorporate public health risk and water quality deterioration. The traditional economic life of a component is the point at which the cost of keeping it in use equals the cost of replacing it. The "cost", though, has been expanded beyond the utility's internal costs to include external costs, like the public's costs associated with the failure of a component (loss of water and business, traffic disruptions, etc.). Expectations for customer service are rising at the same time that repair and replacement costs are rising. Decisions based on internal costs alone often favor ongoing repair over replacement. When external costs (such as the number of households affected by a failure) are counted, replacement begins to be favored over repair. When the break rate for a 20-ft long pipe exceeds once per year then it can become more economical to replace the pipe than repair it (Damodaran et al., 2005). Utilities need guidance on including external costs along with internal costs, and the advantages and disadvantages of replacement methods, so that they can make up-to-date and sound decisions in a timely manner.

Regardless of whether the situation requires repair, rehabilitation, or replacement, there are practices that can minimize the contamination potential, such as maintaining a positive pressure until the repair site is unearthed and cleared. Trench water should be removed before work is done, and street drainage should be provided to keep water and runoff out of the trench. New materials and repaired materials can be sprayed or swabbed with chlorine or appropriate sanitizing agents, as specified in ANSI/AWWA standards C600-99 for the installation of ductile iron mains and C651-99 for the disinfection of mains. During these activities, inspectors or engineers managing the site need to be aware of all issues related to water quality including the type of pipe that can be laid in soils suspected of contamination, the means by which to protect materials during storage, the methods for working in trenches to prevent contamination of materials, and what to do if materials do become contaminated.

Prior to the release for use of a new or replaced water main or facility, a water utility will typically conduct water quality testing. Total coliform bacteria have been the most common indicator that the new material is sanitary and did not become contaminated during storage or installation. In addition to total coliform testing, the water utility can also test for turbidity, HPC bacteria, total chlorine residual, pH, and odor, as unsanitary and improper installation practices can affect these parameters.

As documented in Table 2-3, 16 of 34 responding states address the storage and handling of pipes, while 29 of 34 address the need for disinfection and water quality testing following installation. Experience has shown, unfortunately, that sanitary practices vary widely. Even well-run utilities can experience a 30 percent failure rate in the approval of new mains based on water quality testing (Burlingame and Neukrug, 1993). Pipe design and construction is usually focused on existing codes (such as depth of installation to prevent freezing) and corrosion protection (such as using plastic pipe or metallic pipe with protective wrap in corrosive soils) but not on sanitary practices and rarely on permeation concerns. Pierson et al. (2001) found that although the ANSI/AWWA standards, particularly C600-99, attempt to address installation or construction practices, there is a general lack of training and the use of requirements for sanitary practices. It is possible for trenches where pipe is being laid or repaired to fill partially with water from broken lines or from precipitation or groundwater. This water can mobilize soil-related contaminants as well as carry contamination itself. Clearly, during emergency repairs or repairs made under less than favorable conditions, it becomes even more difficult to prevent the exposure of materials to environmental contamination. This could be addressed in part by requiring foremen or managers of construction sites to be certified on a regular basis, as it is for the certification of backflow installers and testers. Such training and certification can be provided through third-party organizations (non-water utility agencies) such as the New England Water Works Association and American Society of Sanitary Engineers. Not only would foremen or managers have to know the engineering requirements, but they would also have to record and understand the issues related to protecting the sanitary condition of the materials and the water supply.

Disinfection

Haas et al. (1998) reported that interior pipe surfaces are not free of microbial contaminants even under best case conditions. Furthermore, the lack of adequate distribution system maintenance (which includes flushing, disinfecting, and coliform testing of all pipe repairs and pipe replacement activities) has been found to contribute to higher coliform occurrence rates (Clement et al., 2003). Thus, when a new main is installed or a valve is repaired, it is advisable to act as if some level of contamination has occurred to both the water and the materials and to address potential contamination before the affected portion of the water system is returned to use. When the interior of pipe has become contaminated or needs cleaning due to unsanitary activities, disinfection becomes necessary.

Pipes can have a significant chlorine demand which reduces the effectiveness of disinfection (Haas et al., 1999). Fortunately, there is a current AWWA standard (C652) governing new pipe disinfection, which sets forth two options. The first is to flush followed by filling the facility/pipe with a strong (> 25 mg/L) chlorine solution and maintaining it for 24 hours providing that a residual of 10 mg/L remains. The second option is contacting the pipe or facility with a 100 mg/L free chlorine solution for at least three hours so that the residual remaining is at least 50 mg/L. The chlorine used for these disinfection operations may be supplied either as solid calcium hypochlorite powder dissolved in water, sodium hypochlorite (liquid bleach) dissolved in water, or gaseous chlorine dissolved in water.

These guidelines basically require that a "CT" (product of disinfectant and contact time) of 14,000 (first option) or 9,000 (second option) mg-min/L be achieved. Tests on actual mains indicate that these guidelines are sufficient to yield four logs (99.99 percent) inactivation of heterotrophic plate count (HPC) bacteria (Haas et al., 1998). Where unusually high levels of contamination are suspected, the design "CT" for facility disinfection should be increased.

After disinfection, the chlorinated water must be flushed from the system and the adequacy of disinfection checked by microbiological testing. In flushing the heavily chlorinated water, attention must be paid to (1) preventing leakage into the active distribution system if the newly disinfected pipe is connected to the system, (2) the potential impacts on the sewer system if the water is discharged to a sewer, or (3) dechlorinating the water (using sulfur dioxide, sulfite, or bisulfite) if the water is discharged to a surface waterbody so as to minimize adverse impacts to aquatic life.

CONCLUSIONS AND RECOMMENDATIONS

The loss of physical integrity of the distribution system—in which the system no longer acts as a physical barrier that prevents external contamination from deteriorating the internal, drinking water supply—is brought about by physical and chemical deterioration of materials, the absence or improper installation of critical components, and the installation of already contaminated components. When physical integrity is compromised, the drinking water supply becomes exposed to sources of contamination that increase the risk of negative public health outcomes. The following primary conclusions and recommendations for maintaining and restoring physical integrity to a distribution system are made.

Storage facilities should be inspected on a regular basis. A disciplined storage facility management program is needed that includes developing an inventory and background profile on all facilities, developing an evaluation and rehabilitation schedule, developing a detailed facility inspection process, performing facility inspections, and rehabilitating and replacing storage facilities when needed. Depending on the nature of the water supply chemistry, every three to five years storage facilities need to be drained, sediments need to be removed, appropriate rust-proofing needs to be done to the metal surfaces, and repairs need to be made to structures. These inspections are in addition to daily or weekly inspections for vandalism, security, and water quality purposes (such as identifying missing vents, open hatches, and leaks).

Better sanitary practices are needed during installation, repair, replacement, and rehabilitation of distribution system infrastructure. All trades people who work with materials that are being installed or repaired and that come in contact with potable water should be trained and certified for the level of sanitary and materials quality that their work demands. Quality workmanship for infrastructure materials protection as well as sanitary protection of water and materials should go hand-in-hand considering the increasing costs of infrastructure failure and repair and the increasingly stringent water quality standards. Training and certification can be provided through third-party organizations (non-water utility agencies) such as the New England Water Works Association and American Society of Sanitary Engineers.

Although it is difficult and costly to perform, condition assessment of buried infrastructure should be a top priority for utilities. Every water utility should maintain a complete, up-to-date inventory of all infrastructure components from storage facilities to pipes to valves to hydrants, including their current condition. Because failure analysis has not generally been embraced by the water community, there is limited information on many of the materials in common use today. Most useful would be a user-friendly guidance manual for utilities regarding the failure mechanisms of different types of infrastructure materials and how to use the various types of information on the current condition of the pipe to determine its expected lifetime. Finally, as an essential part of condition assessment, every water utility should have in place a leak detection program that includes checking service lines as well as transmission mains.

External and internal corrosion should be better researched and controlled in standardized ways. There is a need for new materials and corrosion science to better understand how to more effectively control both external and internal corrosion, and to match distribution system materials with the soil environment and the quality of water with which they are in contact. At present the best defense against corrosion relies on site-specific testing of materials, soils, and water quality followed by the application of best practices, such as cathodic protection. Indeed, a manual of practice for external and internal corrosion control should be developed to aid the water industry in applying what is known. Corrosion is poorly understood and thus unpredictable in occurrence. Insufficient attention has been given to its control, considering its estimated annual direct cost of $5 billion for the main distribution system (not counting premise plumbing).

Cross-connection control should be in place for all water utilities. Every utility should have a uniform and consistent cross-connection control program along with adequate support such as regulations or codes, and staffing. The program should at the least provide for service-protection or containment (i.e., making sure that customers cannot backflow contaminants into the public distribution system), and when possible should attempt to eliminate cross connections on customer's premises. Most if not all technical and administrative information already exists upon which to institute a cross-connection control program.

REFERENCES

Allbee, S. 2004. A center of excellence—a sensible step on the pathway to excellence in water utility infrastructure management. Underground Infrastructure Management (Nov/Dec.):27–29.

American Water Works Association (AWWA). 1986. Introduction to Water Distribution Principles and Practices of Water Supply Operations. Denver, CO: AWWA.

AWWA. 1999. Water Audits and Leak Detection, Manual M36, 2nd edition. Denver, CO: AWWA.

AWWA. 2003. Water Stats 2002 Distribution Survey CD-ROM. Denver, CO: AWWA.

AWWA. 2004. Recommended Practice for Backflow Prevention and Cross-Connection Control, Manual M14, 3rd edition. Denver, CO: AWWA.

AWWA. 2005a. Flexible, lightweight PE gaining ground. Opflow (July):24–25.

AWWA. 2005b. Ductile-Iron Pipe—Iron and Icon for Durability, Reliability. Opflow (February):14–15.

AWWA and EES, Inc. 2002. Permeation and leaching. Available on-line at *http://www.epa.gov/safewater/tcr/pdf/permleach.pdf*. Accessed May 5, 2006.

American Water Works Association Research Foundation (AwwaRF). 1985. Internal Corrosion of Water Distribution Systems. Cooperative report with DVGW Forschungsstelle. Denver, CO: AWWA Research Foundation.

AWWA Water Loss Control Committee. 2003. Committee report: applying worldwide BMPs in water loss control. J. Amer. Water Works Assoc. 95(8):65–80.

Angulo F. J., S. Tippen, D. J. Sharp, B. J. Payne, C. Collier, J. E. Hill, T. J. Barrett, R. M. Clark, E. E. Geldreich, H. D. Donnell, Jr., and D. L. Swerdlow. 1997. A community waterborne outbreak of salmonellosis and the effectiveness of a boil water order. American Journal of Public Health 87(4):580–584.

Bonds, R. W., L. M. Barnard, A. M. Horton, and G. L. Oliver. 2005. Corrosion and corrosion control of iron pipe: 75 years of research. J. Amer. Water Works Assoc. 97(6):88–98.

Booth, S., and B. Brazos. 2005. Qualitative Procedures for Identifying Particles in Drinking Water. Denver, CO: AwwaRF.

Burlingame, G. A. 1999a. Solving customers' taste and odor complaints—part 1: the importance of the first response. Opflow 25(10):10–11.

Burlingame, G. A. 1999b. Solving customers' taste and odor complaints—part 2: tracking odors to their source. Opflow 25(11):6–7.

Burlingame, G. A., and C. Anselme. 1995. Distribution system tastes and odor. Pp. 281–319 *In:* Advances in Taste-and-Odor Treatment and Control. Denver, CO: AwwaRF.

Burlingame, G. A., and H. M. Neukrug. 1993. Developing proper sanitation requirements and procedures for water main disinfection. Pp. 137–146 *In:* Proceedings of AWWA Annual Conference. Denver, CO: AWWA.

Burlingame, G. A. 2001. A balancing act: distribution water quality and operations. Opflow 27(7):14–15.

Burlingame, G. A., J. Rahman, E. Navera, and J. E. Durrant. 2002. Pp. 83–101 *In:* Assessing the Future: Water Utility Infrastructure Management. D. M. Hughes (ed.). Denver, CO: AWWA.

Cagle, R. F. 2005. Daddy, are we there yet? Underground infrastructure management. Jan/Feb:43–46.

Clark, R. M., and J. A. Goodrich. 1989. Developing a database on infrastructure needs. J. Amer. Water Works Assoc. 81(7):81–87.

Clark, R. M., Geldreich, E. E., Fox, K. R., Rice, E. W., Johnson, C. H., Goodrich, J. A., Barnick, J. A., and Abdesaken, F. 1996. Tracking a *Salmonella* serovar *typhimurium* outbreak in Gideon, Missouri: Role of contamination propagation modeling. Journal of Water Supply Research and Technology—Aqua 45(4):171–183.

Clement, J., C. Spencer, A. J. Capuzzi, A. Camper, K. V. Andel and A. Sandvig. 2003. Influence of Distribution System Infrastructure on Bacterial Regrowth. Denver, CO: AwwaRF.

Cooperative Research Centre for Water Quality and Treatment. 2003. Setback for Netherlands Dual Supplies. Health Stream 30:5.

Craun, G. F., and R. L. Calderon. 2001. Waterborne disease outbreaks caused by distribution system deficiencies. J. Amer. Water Works Assoc. 93:9:64–75.

Damodaran, N., J. Pratt, J. Cromwell, J. Lazo, E. David, R. Raucher, C. Herrick, E. Rambo, A. Deb, and J. Snyder. 2005. Customer acceptance of water main structural reliability. Denver, CO: AwwaRF.

Donahue, E. J., III. 2002. GASB 34 and water utilities: deferred maintenance and contributed capital. *In:* Assessing the Future: Water Utility Infrastructure Management. D. M. Hughes (ed.). Denver, CO: AWWA.

Ellison, D., S. J. Duranceau, S. Ancel, G. Deagle, and R. McCoy. 2003. Investigation of pipe cleaning methods. Denver, CO: AwwaRF.

Environmental Protection Agency (EPA). 1999. Uncovered Finished Water Reservoirs Guidance Manual. EPA 815-R-99-011. Washington, DC: EPA Office of Water. Available on-line at *http://www.epa.gov/safewater/mdbp/pdf/uncover/ufw8p.pdf.*

EPA. 2002a. Technical fact sheet on: Benzene. Available on-line at *http://www.epa.gov/OGWDW/dwh/t-voc/benzene.html.* Accessed on May 8, 2006.

EPA. 2002a. Technical fact sheet on: Xylenes. Available on-line at *http://www.epa.gov/OGWDW/dwh/t-voc/xylenes.html.* Accessed on May 8, 2006.

EPA. 2002c. Technical fact sheet on: Toluene. Available on-line *at http://www.epa.gov/OGWDW/dwh/t-voc/toluene.html.* Accessed on May 8, 2006.

EPA. 2002d. Technical fact sheet on: Ethylbenzene. Available on-line *at http://www.epa.gov/ OGWDW/dwh/t-voc/ethylben.html.* Accessed on May 8, 2006.

EPA. 2003. Cross-Connection Control Manual. Washington, DC: EPA. Available on-line at *http://www.epa.gov/safewater/crossconnection.html.* Accessed on May 8, 2006.

EPA. 2004. Taking stock of your water system—a simple asset inventory for very small drinking water systems. EPA 816-K-03-002. Washington, D.C.: EPA.

Falarski, M. R. 2002. East Bay Municipal Utility District's Pipeline Replacement Program. *In:* Assessing the Future: Water Utility Infrastructure Management. D. M. Hughes (ed). Denver, CO: AWWA.

Geldreich, E. E. 1996. Microbial Quality of Water Supply in Distribution Systems. Boca Raton, FL: CRC Press, Inc.

Glaza, E. C., and J. K. Park. 1992. Permeation of organic contaminants through gasketed pipe joints. J. Amer. Water Works Assoc. 84(7):92–100.

Government Accounting Standards Board. 1991. GASB Statement 34: basic financial statements and management's discussion and analysis for state and local governments issued in 1991.

Grigg, N. S. 2004. Assessment and Renewal of Water Distribution Systems. Denver, CO: AwwaRF.

Grigg, N. S. 2005. Assessment and Renewal of Water Distribution Systems. J. Amer. Water Works Assoc. 97:2:58–68.

Haas, C. N., M. Gupta, G. A. Burlingame, R. B. Chitluru, and W. O.Pipes. 1999. Bacterial levels of new mains. J. Amer. Water Works Assoc. 91(5):78–84.

Haas, C. N., R. B. Chitluru, M. Gupta, W. O. Pipes, and G. A. Burlingame. 1998. Development of disinfection guidelines for the installation and replacement of water mains. Denver, CO: AwwaRF.

Holsen, T. M., Park, J. K., Bontoux, L., Jenkins, D. and Selleck, R. E. 1991. The effect of soils on the permeation of plastic pipes by organic chemicals. J. Amer. Water Works Assoc. 83(11):85–91.

Hrudey, S. E., and E. J. Hrudey. 2004. Safe Drinking Water: Lessons from Recent Outbreaks in Affluent Nations. London: IWA Publishing.

Hughes, D. M., and P. J. Conroy. 2002. Matching deteriorating main conditions to replacement/rehabilitation options. *In:* Assessing the Future: Water Utility Infrastructure Management. D. M. Hughes (ed.). Denver, CO: AWWA.

Kirmeyer, G. J., L. Kirby, B. M. Murphy, P. F. Noran, K. D. Martel, T. W. Lund, J. L. Anderson, and R. Medhurst. 1999. Maintaining and operating finished water storage facilities. Denver, CO: AwwaRF.

Kirmeyer, G. K., M. Freidman, K. Martel, D. Howie, M. LeChevallier, M. Abbaszadegan, M. Karim, J. Funk, and J. Harbour. 2001. Pathogen Intrusion into the Distribution System. Denver, CO: AwwaRF.

Krasner, S. W., and E. G. Means, III. 1986. Returning newly covered reservoirs to service: health and aesthetic considerations. J. Amer. Water Works Assoc. 78(3):94–100.

Lange, G. 2002. Locating leaks in plastic water pipes. Opflow 28(7):1,4,5,10.

Lee, J. J., P. Schwartz, P. Sylvester, L. Crane, J. Haw, H. Chang, and H. J. Kwon. 2003. Impacts of Cross-Connections in North American Water Supplies. Denver, CO: AwwaRF.

Lockridge, R. 2004. The four C's of asset management. Underground Infrastructure Management Nov/Dec.:39–41.

Makar, J. 2002. Investigating large gray cast-iron pipe failures: a step-by-step approach. *In:* Assessing the Future: Water Utility Infrastructure Management. D. M. Hughes (ed.). Denver, CO: AWWA.

Makar, J., and Y. Kleiner. 2002. Maintaining water pipeline integrity. *In:* Assessing the Future: Water Utility Infrastructure Management. D. M. Hughes (ed.). Denver, CO: AWWA.

Makar, J., R. Rogge, S. McDonald, and S. Tesfamariam. 2005. The effect of corrosion pitting on circumferential failures in grey cast iron pipes. Denver, CO: AwwaRF.

Makepeace, D. K., D. W. Smith, and S. J. Stanley. 1995. Urban stormwater quality: summary of contaminant data. Critical Reviews in Environmental Science and Technology 25(2):93–127.

Male, J. W., and T. M. Walski. 1991. Water Distribution Systems: A Troubleshooting Manual. Chelsea, MI: Lewis Publishers, Inc.

Martel, K., A. Hanson, G. J. Kirmeyer, M. Besner, A. Carrier, M. Prevost, A. Lynggaard-Jensen and N. Bazzurro. 2005. Data Integration for Water Quality Management. Denver, CO: AwwaRF.

McGuire, M., N. Graziano, L. Sullivan, R. Hund, and G. Burlingame. 2004. Water Utility Self-Assessment for the Management of Aesthetic Issues. Denver, CO: AwwaRF.

Morgan, W., R. Titus, and D. M. Hughes. 2005. Simultaneous Communication of Acoustic Data and Meter Readings Automatically. *In:* Proceedings from the International Water Association Water Loss Task Force Conference, Halifax, Nova Scotia.

Morrison, R. 2004. Condition assessment—back to the basics. Underground Infrastructure Management Nov/Dec.:51–55.

Nayyar, M. L. 1992. Piping Handbook, Sixth Edition. New York: McGraw-Hill, Inc.

Neptune Technology Group. 2005. E-Coder. Available on-line at *http://www.neptunetg.com/uploadedFiles/E-Coder%20Press%20Release.pdf.* Accessed May 8, 2006.

Paralez, L. L., and D. Muto. 2002. Creating an asset management strategy: an asset management template. *In:* Assessing the Future: Water Utility Infrastructure Management. D. M. Hughes (ed.). Denver, CO: AWWA.

Pierson, G. L., G. Burlingame, and K. Martin. 2002. Establishing a tradition of contamination prevention. Opflow 28(7):6,7,11.

Pierson, G., K. Martel, A. Hill, G. Burlingame, and A. Godfree. 2001. Methods to prevent microbiological contamination associated with main rehabilitation and replacement. Denver, CO: AwwaRF.

Propato, M., and J. G. Uber. 2004. Vulnerability of water distribution systems to pathogen intrusion: how effective is a disinfectant residual? Environ. Sci. Technol. 38(13):3713–3722.

Romer, A. E., G. E. C. Bell, S. J. Duranceau, and S. Foreman. 2004. External Corrosion and Corrosion Control of Buried Water Mains. Denver, CO: AwwaRF.

Schwarzwalder, R. 2002. Asset management for the new millennium: strategic approaches for water utilities. *In:* Assessing the Future: Water Utility Infrastructure Management. D. M. Hughes (ed.). Denver, CO: AWWA.

Seargeant, D. 2002. Using new technology to optimize management of cast iron pipe assets. *In:* Assessing the Future: Water Utility Infrastructure Management. D. M. Hughes (ed.). Denver, CO: AWWA.

Selvakumar, A., R. M. Clark, and M. Sivaganesan. 2002. Costs for water supply distribution system rehabilitation. Jour. Water Resources Planning and Management ASCE 128(4):303–306.

Shamsi, U. M. 2005. GIS Applications for Water, Wastewater and Stormwater Systems. Boca Raton, FL: CRC Press.

Skala, M. F. 1994. Waterborne salmonella outbreak in southeastern Missouri. Missouri Epidemiologist 17(2):1–2.

Swerdlow, D. L., B. L. Woodruff, R. C. Brady, P. M. Griffin, S. Tippen, H. D. Donnell, Jr., E. Geldreich, B. J. Payne, A. Meyer Jr., J. G. Wells, K. D. Greene, M. Bright, N. H. Bean, and P. A. Blake. 1992. Waterborne outbreak in Missouri of *Escherichia coli* O157:H7 associated with bloody diarrhea and death. Annals of Internal Medicine 117(10):812–819.

Szeliga, M. J., and D. M. Simpson. 2002. Evaluating the conditions of existing water mains. *In:* Assessing the Future: Water Utility Infrastructure Management. D. M. Hughes (ed.). Denver, CO: AWWA.

University of Southern California (USC). 2002. Prevalence of cross connections in household plumbing systems. Available on-line at *http://www.usc.edu/dept/fcchr/epa/hhcc.report.pdf*. Los Angeles, CA: USC Foundation for Cross-Connection Control and Hydraulic Research.

USC. 1993. Manual of Cross-Connection Control, 9th Edition. Los Angeles, CA: USC Foundation for Cross-Connection Control and Hydraulic Research.

Von Huben, H. (Tech. Ed). 1999. Water Distribution Operator Training Handbook, 2nd edition. Denver, CO: AWWA.

Wallick, P. C., and M. Zubair. 2002. Tank evaluation, rehabilitation, and replacement decisions for water storage tanks. *In:* Assessing the Future: Water Utility Infrastructure Management. D. M. Hughes (ed.). Denver, CO: AWWA.

Westerhoff, G., P. Fahy, and S. Robinson. 2004. On the pathway to improved asset management. Underground Infrastructure Management Nov/Dec:35–37.

5
Hydraulic Integrity

The hydraulic integrity of a water distribution system is defined as its ability to provide a reliable water supply at an acceptable level of service—that is, meeting all demands placed upon the system with provisions for adequate pressure, fire protection, and reliability of uninterrupted supply (Cesario, 1995; AWWA, 2005). Water demand is the driving force for the operation of municipal water systems. Because water demands are stochastic in nature, water system operation requires an understanding of the amount of water being used, where it is being used, and how this usage varies with time. For most water systems the ratio of the maximum day water demand to the average day water demand ranges from 1.2 to 3.0, and the ratio of the peak hour to the average day is typically between 3.0 and 6.0. Of course, these values are system specific, and seasonal variations may make these ratios even more extreme (Walski et al., 2003). Demands may be classified as follows (Clark et al., 2004):

- Baseline demands, which usually correspond to consumer demands and unaccounted-for-water associated with average day conditions.
- Seasonal variations in demand because water use typically varies over the course of the year with higher demands occurring in the warmer months.
- Fire demands, which may be the most important consideration for water system design.
- Diurnal variations due to the continuously varying demands which are inherent in water systems.

There is a need for research that relates distribution system design to demand in a stochastic framework. Pioneering work by Buchberger and Wu (1995), Buchberger and Wells (1996), and Buchberger at al. (2003) has found that residential water use follows a Poisson arrival process with a time dependent rate parameter. Variations in demand have an important influence on water distribution system operation and in the determination of water age which in turn influences water quality, as discussed later in the chapter.

From an infrastructure perspective, a water distribution system is an elaborate conveyance structure in which pumps move water through the system, control valves allow water pressure and flow direction to be regulated, and reservoirs smooth out the effects of fluctuating demands (flow equalization) and provide reserve capacity for fire suppression and other emergencies. All these distribution system components and their operations and complex interactions can produce significant variations in critical hydraulic parameters, such that many opportunities exist for the loss of hydraulic integrity and degradation of service. This, in turn, may lead to serious water quality problems, some of which may threaten public health.

One of the most critical components of hydraulic integrity is the maintenance of *adequate pressure*, defined in terms of the minimum and maximum design pressure supplied to customers under specific demand conditions. Low pressures, caused for example by failure of a pump or valve, may lead to inadequate supply and reduced fire suppression capability or, in the extreme, intrusion of potentially contaminated water. High pressures will intensify wear on valves and fittings and will increase leakage and may cause additional leaks or breaks with subsequent repercussions on water quality. High pressures will also increase external load on water heaters and other fixtures. Pipes and pumps must be sized to overcome the head loss caused by friction at the pipe walls and thus to provide acceptable pressure under specific demands, while sizing of control valves is based on the desired flow conditions, velocity, and pressure differential. A related need is to ensure that pressure fluctuations associated with surge conditions are kept below an acceptable limit. Excessive pressure surges generate high fluid velocity fluctuations and may cause resuspension of settled particles as well as biofilm detachment.

A second element of hydraulic integrity is the *reliability of supply*, which refers to the ability of the system to maintain the desirable flow rate even when components are out of service (e.g., facility outage, pipe break) and is normally accomplished by providing redundancy in the system. Examples include looping of the pipe network and the development of backup sources to ensure multiple delivery points to all areas.

Many water quality parameters change with *length of time in the distribution system*, a factor directly related to the hydraulic design of the system. For example, chlorine residuals decrease with the increasing age of water and may be completely lost, and trihalomethanes concentrations may increase with time. In addition, higher concentrations of substances may leach from pipe materials and linings if the contact time with the water is increased. Low velocities in pipes create long travel times, resulting in pipe sections where sediments can collect and accumulate and microbes can grow and be protected from disinfectants. Furthermore, sediment deposition will result in rougher pipes with reduced hydraulic capacity. If peak velocity is increased or flow reverses in these pipe sections due to any operational change or shock loading, such as tank filling or draining, valve opening or closing, pump going on- or off-line, unexpected higher system pressure, or hydrant flushing, there is a risk that deposits will be suspended and carried to consumers. Long detention times can also greatly reduce corrosion control effectiveness by effecting phosphate inhibitors and pH management. Thus, reducing residence time is an important hydraulic issue both in pipes and in storage facilities.

A final component of hydraulic integrity is *maintaining sufficient mixing and turnover rates in storage facilities*. Insufficient turnover rates and incomplete (uneven) hydraulic mixing in reservoirs can allow short-circuiting between the tank inlet and outlet and generate pockets of stagnant water with depleted disinfectant residual. This can lead to bacterial regrowth and other biological changes in the water, including nitrification and taste and odor problems.

This chapter discusses the factors that can cause the loss of hydraulic integrity, the consequences of losing hydraulic integrity, how to detect loss of hydraulic integrity, techniques for maintaining hydraulic integrity, and how to recover system hydraulic integrity once it is lost.

FACTORS CAUSING LOSS OF HYDRAULIC INTEGRITY

There are many different ways that a water distribution system can lose its hydraulic integrity, such that water quality becomes impaired. A loss of hydraulic integrity implies a loss of positive line pressures, flow reversals, rapid changes in velocity, a reduction in hydraulic capacity, a detrimental increase in water residence time, or a combination of these events. Factors causing a loss of system hydraulic integrity include (1) pipe leaks and breaks, (2) rapid changes in pressure and flow conditions, (3) planned maintenance activities and emergencies, (4) tuberculation and scale formation in pipes, and (5) improper operational control.

Pipe Deterioration

Pipe deterioration resulting in leaks or breaks can lead to a loss of hydraulic integrity because adequate pressures can no longer be maintained. As discussed in detail in Chapter 4, all pipe materials are vulnerable to some kind of chemical or physical deterioration, and all water mains will require rehabilitation and eventual replacement. Aging pipe infrastructure and chronic water main breaks are a common problem for many water utilities. Analysis of water industry data showed that on average, main breaks occur 700 times per day in the United States (Cromwell et al., 2001). The condition of distribution system pipes is influenced by material type and age, line pressure, type of soil, installation procedures, and many other factors, making it difficult to predict where breaks and leaks will occur. Chapter 4 discusses the roles of leak detection and condition assessment in determining the current condition of distribution system infrastructure.

Pressure Transients and Changes in Flow Regime

Rapid changes in pressure and flow caused by events such as rapid valve closures or pump stoppages and hydrant flushing can create pressure surges of excessive magnitude. These transient pressures, which are superimposed on the

The after effects of a water main break that occurred beneath the side walk of an urban street. Photo courtesy of Bureau of Laboratory Services, Philadelphia Water Department.

normal static pressures present in the water line at the time the transient occurs, can strain the system leading to increased leakage and decreased system reliability, equipment failure, and even pipe rupture in extreme cases. High-flow velocities can remove protective scale and tubercles, which will increase the rate of corrosion. Uncontrolled pump shutdown can lead to the undesirable occurrence of water-column separation, which can result in catastrophic pipeline failures due to severe pressure rises following the collapse of the vapor cavities. Vacuum conditions can create high stresses and strains that are much greater than those occurring during normal operating regimes. They can cause the collapse of thin-walled pipes or reinforced concrete sections, particularly if these sections were not designed to withstand such strains. In less drastic cases, strong pressure surges may cause cracks in internal lining, damage connections between pipe sections, and destroy or cause deformation to equipment such as pipeline valves, air valves, or other surge protection devices. Sometimes the damage is not realized at the time, but may cause the pipeline to collapse in the future, especially if combined with repeated transients.

Transient pressure and flow regimes are inevitable. All systems will, at some time, be started up, switched off, or undergo rapid flow changes such as those caused by hydrant flushing, and they will likely experience the effects of human errors, equipment breakdowns, earthquakes, or other risky disturbances (Wood et al., 2005). Figure 5-1 illustrates typical hydraulic events following a pump trip.

Low pressure transients may promote the collapse of water mains, leakage into the pipes at joints and seals under sub-atmospheric pressures, and backsiphonage (see Chapter 4). There is also evidence that pressure transients can lead to the intrusion of contaminants into the distribution system. LeChevallier et al. (2003) reported the existence of low and negative pressure transients in a number of distribution systems. Gullick et al. (2004) studied intrusion occurrences in distribution systems and observed 15 surge events that resulted in a negative pressure. Most were caused by the sudden shutdown of pumps at a pump station because of either unintentional (e.g., power outages) or intentional (e.g., pump stoppage or startup tests) circumstances. Friedman et al. (2004) confirmed that negative pressure transients can occur in the distribution system and that the intruded water can travel downstream from the site of entry. Locations with the highest potential for intrusion were sites experiencing leaks and breaks, areas of high water table, and flooded air-vacuum valve vaults.

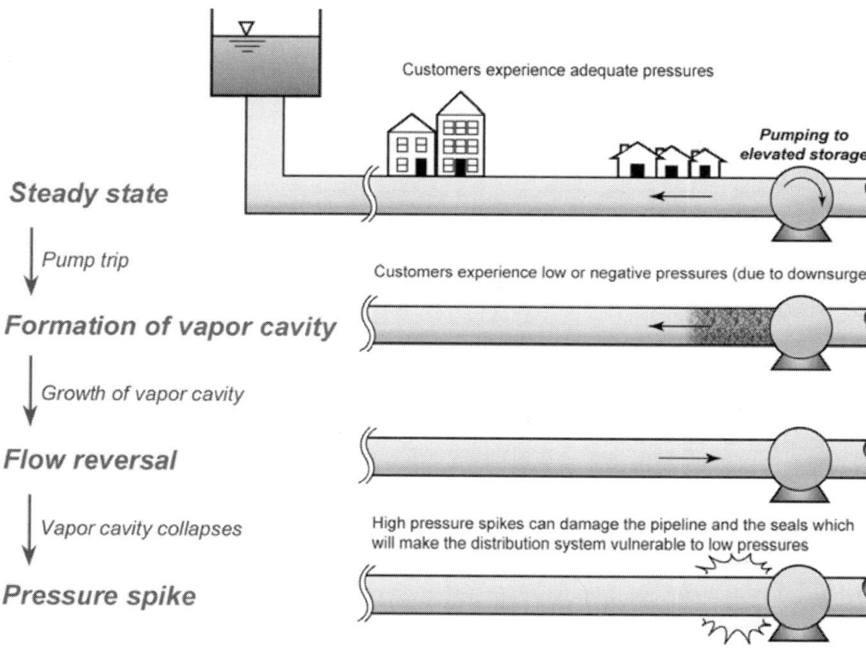

FIGURE 5-1 Hydraulic events following a pump trip. The system is pumping drinking water to an elevated storage tank while serving the intermediate customers with adequate pressures. Due to an unexpected power failure, the pump quickly runs down (loses speed). This will create a negative pressure wave (downsurge) that will propagate into the distribution system, putting the customers at a potential intrusion risk due to negative pressures. In addition, it is possible that the pressure drops to the point that a vapor pocket forms adjacent to the pump. Subsequently, this cavity will collapse and produce a large pressure spike that can damage the pipeline and the seals which will make the system even more vulnerable to low pressure events.

Hydraulic Changes during Maintenance and Emergencies

Water distribution systems are occasionally subject to emergencies or planned maintenance activities in which certain components become inoperable and the system can no longer provide the minimum level of service to customers (AWWA, 2005). Planned maintenance activities include supplies going off line (e.g., stopping the treatment plant or shutting down a well); reservoir shutdown for inspection, cleaning, or repairs; installation of new pipe connections; pipe rehabilitation or break repairs; and transmission main valve repairs. Examples of emergency situations include earthquakes, hurricanes, power failures, equipment failures, or transmission main failures. All these activities can result in a reduction in system capacity and supply pressure and changes to the flow paths of water within the distribution system.

Tuberculation and Scale

The hydraulic capacity of distribution systems can be compromised by deposits on the internal surface of the pipelines. The deposition of corrosion products in the form of tubercles and other types of scales on the interior of the pipes can seriously clog water lines and thus restrict the flow of water. Scales may also form because metal salts such as calcium carbonate, aluminum silicate, etc. (see Chapter 6) in treated water entering the network are supersaturated, leading to their precipitation on the pipe walls. Excessive pressure may be necessary to deliver the required flow of water in pipes with tuberculation and scales, further weakening aging pipes. The reduction in hydraulic capacity is caused by the increases in head loss due to the roughness of the deposits and to the decrease in pipe diameter that they cause.

Inadequate Operational Control

Historically, utilities have focused on the quality of water leaving the treatment plant, because of regulatory drivers, and on the quantity of water supplied by the distribution system, because of their mission to satisfy water demand and maintain system pressure. Thus, it is not surprising that distribution system operations at many utilities and their associated professionals (designers, builders, plumbers, inspectors, etc.) have been water quantity focused rather than water quality focused.

There is now greater recognition of the water quality effects of how long water is retained in the various elements of the distribution system. Retention time or water age is strongly related to the characteristics of the system and its operation. For example pipe roughness, which affects water flow and residence time, may be modified by repair or rehabilitation. Operational activities, such as

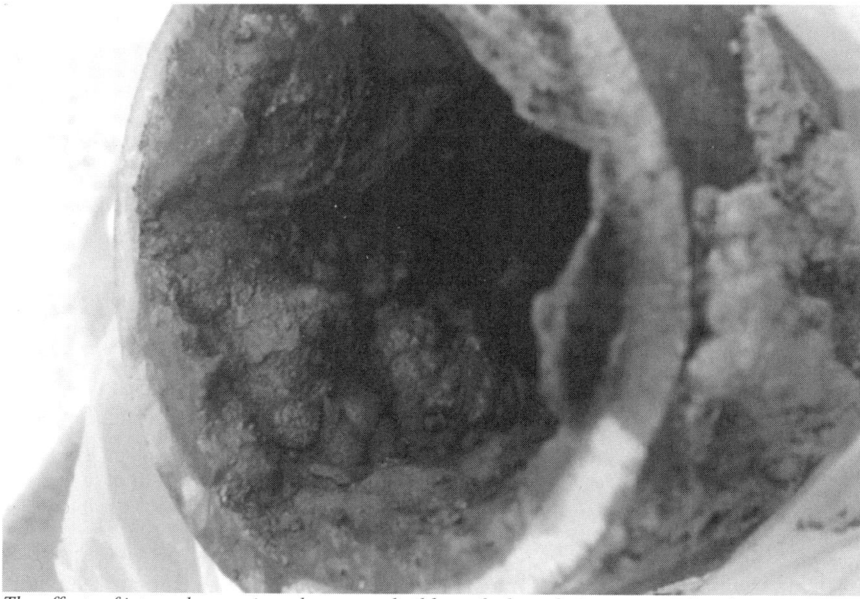

The effects of internal corrosion, shown as a build up of tuberculation, on an unlined cast iron water main. Photo courtesy of Bureau of Laboratory Services, Philadelphia Water Department.

pump scheduling and planned maintenance, or unplanned effects, such as unexpected changes in demand, will all have an effect on water age. A particularly important issue that demonstrates the interaction of system operation and water quality is the ability or inability of utilities to ensure adequate mixing intensity and time in storage tanks to minimize short circuiting and to limit residence times to be within acceptable limits. Interestingly, the design of tanks to ensure adequate turnover is required in only 15 of 34 states that responded to a survey of drinking water programs conducted by the Association of State Drinking Water Administrators in March 2003 (see Table 2-4). Dealing with these issues is discussed in the context of system operation later in this chapter.

CONSEQUENCES OF A LOSS IN HYDRAULIC INTEGRITY

There are several detrimental consequences of losing system hydraulic integrity, including contamination of the distribution system via intrusion, sedimentation, a reduction in hydraulic capacity, loosening of scale, and extended water age. Each of these has attendant water quality implications, as described below.

External Contamination

A distribution system can become contaminated by the external environment for several reasons. The most well documented contamination events are backflow and direct contamination at breaks and repair sites, discussed in Chapter 4. A specific type of backflow event related to a loss of hydraulic integrity is called intrusion, which refers to the entrance of contamination into the water distribution system through leaks (caused by corroded areas, cracks, and loose joints) because of sustained low or negative pressures or a pressure transient. When a section of the distribution system is depressurized due to a normal shutdown, failure of a main or a pump, routine flushing, or emergency fire-fighting water drawdown, contaminated water can be pulled into the main. For example, during a large fire, a pump is connected to a hydrant. High flows pumped out of the distribution system can result in a significantly reduced water pressure around the withdrawal point. A partial vacuum is created in the system, which can cause suction of contaminated water into the potable water system through nearby leaks. During such conditions, it is possible for water to be withdrawn from nonpotable sources into the distribution system and subsequently distributed to homes and buildings located near the fire. The same conditions can be caused by a water main break.

Sustained low pressure events and transient pressure events that lead to intrusion of contaminated water have the potential for substantial water quality and health implications. The potential for intrusion of contaminated groundwater into pipes with leaky joints or cracks seems greatest in systems with pipes below the water table and where pathogens or chemicals are in close proximity to the pipe. As discussed in Chapter 4, two recent studies (Kirmeyer et al., 2001; Karim et al., 2003) have established that soil and water samples collected immediately adjacent to pipelines can contain high fecal coliform concentrations and viruses. In the event of a large intrusion of pathogens, the disinfectant residual normally sustained in drinking water distribution systems may be insufficient to neutralize contaminated water (see Chapter 6 discussion on Adequate Disinfectant Residual). Transient events can also generate high intensities of fluid shear and may cause resuspension of settled particles as well as biofilm detachment.

Sedimentation

When water is moving slowly through a pipe, particles suspended in the water may settle out into the pipe. The accumulated sediment reduces the pipe's hydraulic capacity. They also serve as a food source for bacteria and create a hospitable environment for microbial growth. If not removed these materials may cause water quality deterioration, taste and odor problems, or discoloration of the water. This is particularly evident if the sediments are disturbed (stirred up) by changes in the flow of water, such as when a main break occurs, a service reservoir is filling or draining, a pump is going on or off line, or during normal hydrant flushing activities. The normal flow of water through the system will reduce some but not all sediment accumulation over time, and supplemental measures are periodically needed to clear out the system.

Reduction in Hydraulic Capacity and Associated Increase in Pumping Costs

As metal pipes age their roughness tends to increase and their cross sectional area tends to decrease due to encrustation and tuberculation of corrosion products on the pipe walls. This increase in hydraulic roughness and decrease in effective diameter will increase the resistance to flow and reduce the hydraulic capacity of the affected mains. Other deposits such as microbial slimes can also result in a significant decrease in the hydraulic capacity of water mains. The reduction in the hydraulic capacity can lead to a subsequent unwanted reduction in system pressure due to the higher head loss. The loss in system pressure can result in a water system that cannot deliver the necessary fire flows and, in the extreme, it provides the potential for backflow of contaminants.

A flooded transmission main and metering chamber. This is a prime location for intrusion to occur in the event of low or negative pressure transients. Photo courtesy of Philadelphia Water Department's Bureau of Laboratory Services.

In order to meet demand in such systems, higher pumping rates are needed to overcome the higher head losses and to avoid or postpone the replacement, duplication, or rehabilitation of tuberculated mains. This can overload motors and result in a significant increase in energy consumption and operational and maintenance costs of a water utility. Furthermore, the additional pumping can over-pressurize certain portions of the distribution system, thereby increasing leaks and breaks, and it can lead to ineffective utilization of storage tanks and reservoirs because high pressure in the mains prevents outflow from the reservoirs. If these reservoirs are subsequently put back into service during peak times when consumption is high, this may result in the provision of "old" (poor quality) water.

Poor Water Quality from Sediment Suspension and Removal of Scales

Changes in flow (magnitude and direction) within the water distribution system as a result of hydrant flushing and valve and pump operation can scour sediments, tubercles, and scales from the interior pipe walls and degrade water quality. For example, when the water velocity is increased or flow direction is reversed, sediment deposited on the pipe walls during periods of low flow may be re-suspended and scales may detach. These materials may cause the water to be colored, turbid, and sometimes odorous. Also, it is possible that these particles have adsorbed contaminants such as arsenic and other metals that originated in the source water, as discussed in Chapter 6.

Hydraulic Integrity and Water Age

As distribution system water ages, its quality degrades, such that delivering "younger" water is a desirable operational goal for water utilities. However, the concept of water age is complex. Water age at a given location and time in a water distribution system is actually a mixture of water parcels that have traveled along different paths through the distribution system with correspondingly different travel times. Therefore, the age of water at a given point in the distribution network is not a single value, but rather a distribution of values, termed a residence time distribution (Levenspeil, 2002). This concept is illustrated in Figure 5-2, which shows the results of a study conducted by EPA in collaboration with the Greater Cincinnati Water Works in which a calcium chloride tracer was introduced into an isolated portion of the distribution system (Clark et al., 2004; Panguluri et al., 2005). Figure 5-2 shows the field results from 34 continuously recording specific conductivity meters that were deployed at various nodes in the system, with an EPANET modeling prediction superimposed on the data. The three concentration peaks represent the different parcels of water that have taken different routes to the monitoring point, resulting in a residence time distribution at that monitoring station at the time the data were collected.

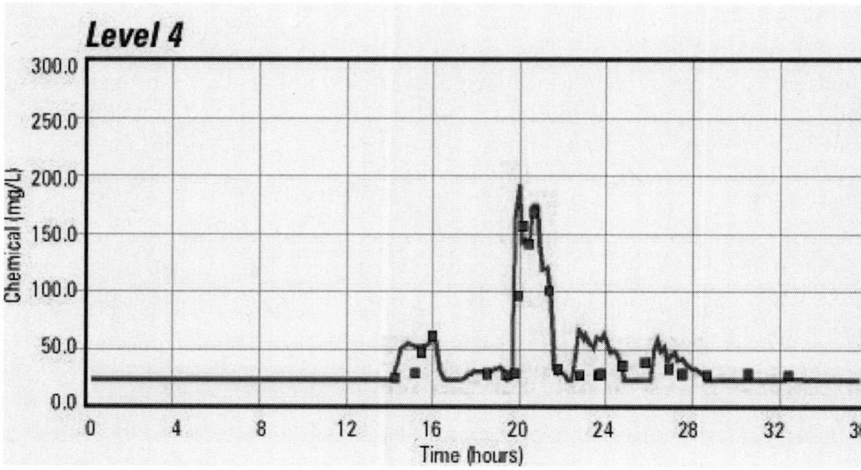

FIGURE 5-2 Field results of monitoring at location CM-18 from 34 continuously recording specific conductivity meters, with a detailed all-pipe (non-skeletonized) EPANET modeling prediction superimposed on the data. SOURCE: Panguluri et al. (2005).

For the purposes of this report, water age at a specific point in the distribution system is assumed to be the mean of the residence time distribution. The report uses the term "water age" as a surrogate for water quality. However, it should be noted that while water quality may depend on the age of the water, it may also depend on the specific residence time distribution at that point in the network or on one of its statistics (such as its variance). These complexities are infrequently considered in studies where water age is measured, making this an area ripe for additional research.

In addition to water age at any one point in the network being a distribution of values, the age of water delivered to all consumers is also a distribution of values, the shape of which depends on the location of the consumer, seasonality, whether the network is looped versus one way, the existence of storage facilities, etc. A typical system may deliver water to consumers that has resided in the network for a few days, but many systems have some portion of the network where residence time is much longer. For example, in Blacksburg, Virginia, 97 percent of the water in the main distribution system has a water age of less than 7 days, but 1 percent of the system has a residence time longer than 28 days. Premise plumbing adds another layer of complexity that is addressed in Chapter 8.

Hydraulically, increased water age is a consequence of many factors, including the inevitable loss of carrying capacity as pipes age. However, system design and operation have the most significant impact on water age, particularly where water storage facilities are concerned. For example, high residence time in these facilities can allow the disinfectant residual to be completely depleted, thereby preventing the protection of finished water from additional microbial contaminants that may be present in the distribution system downstream of the facilities. A survey of water utilities found that bacterial regrowth became a problem in free chlorine residual systems when water age reached three days whereas in monochloramine residual systems regrowth was not a problem until water age reached or exceeded seven days (Baribeau et al., 2005). Other negative consequences of increased water age are discussed in Chapter 6.

DETECTING LOSS OF HYDRAULIC INTEGRITY

Ideally, the verification of hydraulic integrity should involve real-time monitoring of pressure, flow direction, and velocity based on telemetry data. This type of data can be transmitted electronically from permanently installed measurement devices in the field. Typical measurement locations should include treatment plants and wells, pump and booster stations, reservoirs, valves, and other critical points in the system such as elevated sites.

An effective system-wide monitoring program can capture local variations in hydraulic behavior (e.g., pressure, flow) at a specific point in a water distribution system but cannot provide an overall understanding of the spatial and temporal changes, complex flow pathways, and interactions among the various water system characteristics. Thus, water distribution system network models are attractive supplements to monitoring for evaluating hydraulic and water quality changes throughout the distribution system. By combining telemetry data and modeling information, water utilities can gain a more complete and accurate picture of their systems hydraulic and water quality operation and performance capabilities. For example, the North Marin Water Authority in North Marin, California, draws its water from two sources, one of very poor quality with high levels of natural organic matter and another source of very high quality. Because of demand variations there is a great deal of mixing between the water sources at various nodes in the system leading to wide variations in trihalomethane (THM) values over a given day. On the surface these variations in THM concentration were unexplainable until hydraulic modeling techniques were applied which clearly showed that these variations were the result of the mixing effect from the two sources of water (Clark and Buchberger, 2004). Hydraulic integrity is best measured by monitoring and modeling of the system hydraulic parameters, as discussed below.

System-Wide Monitoring

Monitoring the operation of water distribution system components yields data used to detect the system hydraulic integrity. This can be accomplished in real-time by means of a Supervisory Control and Data Acquisition (SCADA) system, which provides local and remote (supervisory) real-time control and monitoring of selected process equipment and parameters at strategic locations throughout the water distribution system. Any parameter with a proper sensor and transmitter that can produce an analog signal (e.g., 4-20 mADC) proportional to the variation of the measured parameter can be monitored in real-time or historically via the SCADA system. The acquired data can be viewed on a real-time basis and also archived in a database for historical evaluation at a later date. The data generated from the sensors and transmitters is conveyed to the central control system using various communication media such as telephone lines, fiber optic cables, or radio and cellular systems. The amount of data collected is determined by the polling frequency of the SCADA system.

To detect changes in hydraulic integrity, certain hydraulic characteristics of water system components should be monitored continually in the distribution system via SCADA. These include reservoir inflow/outflow rates, water volumes and levels (used to calculate daily volume turnover), pump station operation such as status and speed settings, pump discharge flows and pressures, valve positions, regulating valve downstream (and /or upstream) pressures, pipe flow rates, and pressures at strategic sites. In addition, disinfectant residual, temperature, conductivity, turbidity, dissolved oxygen, and pH can be continuously monitored at the treatment plant. Temperature in storage tanks and reservoirs could also be monitored via SCADA to detect thermal stratification that results from poor mixing characteristics. Temperature differential between the inflow and the bulk water in the reservoir can result in density gradients inside the storage facility and cause stratification and poor hydraulic mixing and, thus, the greatest potential for water quality deterioration (Mahmood et al., 2005).

Continuous system-wide monitoring provides insight into the patterns of operational characteristics throughout the distribution system. An analysis of these patterns can directly determine if the system hydraulic integrity is not compromised and the system is operating as designed, or detect any unanticipated operational anomalies. For example, high night-time flows in specific areas could be an indicator of high leakage. Sonic leak detection equipment (discussed in Chapter 4) can be used to pinpoint the exact location of those leaks, which can then be isolated and repaired. Similarly, unexpected low pressure readings, excessive pumping, or a drop in reservoir levels in a specific area could indicate a large main break that may increase the potential for backflow.

Another function of SCADA is the ability to monitor and remotely control local conditions of water system components based on any desired range of operating conditions or set points. For example, a pump can be set to turn on or off automatically when the pressure at a critical location or the water level in a reservoir drops to a specified lower limit or goes above a specified upper limit. Alarms can be set to alert operators when a fault within the system equipment (e.g., equipment operating out of its normal range or overheating of a pump) and any breach in the system hydraulic integrity is detected. For example, extreme fluctuations in pressure and flow readings could result from pressure surges generated from a power failure at a pump station. SCADA could then divert water to the affected region from a different pump station, thus ensuring adequate supply and fire flow protection.

SCADA systems also contain pertinent system operational information required for water distribution network modeling (Cesario, 1995), such as the boundary conditions (e.g., tank water levels, valve and pump statuses and settings) for the network model as well as local flow and pressure conditions. These data can be used for calibrating network models (the process of adjusting model parameters so that modeled values reasonably match with measured data), confirming normal system operation, verifying daily variation in total system demands (based on a mass balance of the flows from the treatment plant and wells and in and out of the reservoirs), estimating water losses during main breaks, and investigating and solving operational problems. Operating data can be time specific or represent several consecutive points in time for comprehensive dynamic (extended period simulation) network modeling (e.g., 24-hour simulation) (see Chapter 7 for details on modeling). Clark et al. (2004) list many benefits of remote monitoring and network modeling for water security protection.

Beyond remote controlled, real-time monitoring provided by SCADA, actual field measurements can be made to detect any potential loss of system hydraulic integrity. Hydrant tests are performed to determine if fire flow requirements are met as an indicator of the hydraulic strength of the water system. Head loss tests are conducted to determine the hydraulic capacity of pipes as an indication of system hydraulic performance capability. Pump efficiency tests can be used to determine whether or not pump performance (e.g., overall system efficiency, electrical motor performance, and pump hydraulics) is degrading with time and if replacement or upgrading of equipment is warranted. Hydraulic grade line tests of a pipeline profile (stretches of pipes) help locate partially closed valves and deteriorated pipes with poor hydraulic capacity (high roughness). Field measurements of pressure, flow conditions, velocity, and other water system characteristics can also be carried out using a variety of measurement devices at any facility to verify questionable SCADA readings.

Network Modeling

Computer based mathematical models provide an effective and viable means of analyzing hydraulic and water quality conditions in distribution systems (see Clark and Grayman, 1998; Lansey and Boulos, 2005; Panguluri et al., 2005; Boulos et al., 2006). They can calculate the spatial and temporal variations of flow, pressure, velocity, reservoir level, water age, source contribution, disinfectant concentration, and other hydraulic and water quality parameters throughout the distribution system. These predictive capabilities are useful for detecting a loss of system hydraulic integrity. For example, model results can help identify areas of low pressures, excessive head losses, and high water age; compute water losses; locate partially closed valves; verify that the replacement or addition of a new supply source (e.g., emergency service connections or adding a new reservoir or well) will have little or no effect on the flow, velocity, and pressure patterns and residence times; estimate filling and draining cycles of reservoirs; detect oversized facilities; calculate interzone water transfers; and determine the adequacy of the system to supply fire flows under a variety of demand loading and operating conditions.

A few specific models are of particular importance to maintaining hydraulic integrity. First, surge models can be used to assess the hydraulic adequacy of the system under various transient conditions, identify weak spots, and evaluate the efficacy of surge control devices. These models could be instrumental in future research to better understand the potential for intrusion to contaminate distribution systems. Second, computational fluid dynamics (CFD) modeling has potential for investigating hydraulic mixing and transport characteristics in storage facilities and pipes for a wide range of designs and system operational conditions (Panguluri et al., 2005). CFD models predict flow patterns, heat transfer, and chemical reactions via the solution of partial differential equations that describe conservation of mass, momentum, and energy in a two- or three-dimensional grid that approximates the pipe or tank geometry. CFD models are used to simulate temperature profiles, unsteady hydraulic and water quality conditions, and decay of constituents in bulk flow and in storage facilities. However CFD modeling requires experienced and skilled programmers for effective application (Panguluri et al., 2005). Such network modeling applications greatly enhance the ability of water utilities to effectively manage, operate, and maintain their water distribution systems and deliver an adequate level of service to their customers.

MAINTAINING HYDRAULIC INTEGRITY

Water utilities often find themselves choosing between two approaches to preserve system hydraulic integrity: (1) reacting only to emergencies or (2) acting to prevent problems from occurring. The desirable approach is to develop an active program to prevent future problems and service interruptions.

To maintain the hydraulic integrity of water distribution systems and ensure the highest possible water quality, travel times in the system should be kept as short as possible and large fluctuations in the hydraulic regime and low flow and pressure conditions should be avoided. This can be accomplished by implementing best design, management, operational, and maintenance practices, as discussed below. Hydraulic modeling, discussed in the previous section, is also a critical component that can be used to identify problems areas within the distribution system and to develop design and operational alternatives that address the deficiencies. Those practices necessary to maintain both physical and hydraulic integrity, such as preventing the formation of leaks and cracks in pipe mains and using backflow prevention devices, are discussed in the previous chapter.

System Redundancy

Reliability of water distribution systems, which is necessary to minimize outages, is provided by building redundancy in the system in the form of looping and backup sources. A looped (as opposed to branched) multi-source system has the hydraulic advantage of carrying water to any location from more than one direction when a high rate is required (e.g., a fire flow demand) or when a pipe or source is out of service (see Chapter 1). Sufficient interconnections between the distribution mains are necessary to improve the ability of the system to maintain the normal supply by re-routing the water when a breakdown occurs. Dead-end distribution lines should be avoided. A fire-flow demand or large water use on a dead-end main can only draw water through a single pipe, with the maximum flow dictated by the size and length of the pipe. In addition, during scheduled maintenance or repairs on dead-end mains both the supplied customers and available fire flows will be affected. Availability of back-up power (e.g., generators in pump stations), extra pumps, additional reservoirs, standby wells, and emergency interconnections with other systems will provide the necessary redundant sources.

Redundancy can also be facilitated by ensuring an adequate number of operable valves and hydrants, as well as their strategic placement to allow for control of the system and for shutdown of sections for emergency repair and planned maintenance (Male and Walski, 1991).

Management of Pressure Zones

Water distribution systems work best with minimal fluctuations in pressure. The pressure differential range, which specifies the operating values for maximum and minimum pressure to be maintained, is based on local engineering standards and conditions. Many states have established requirements for the design, construction, operation, and maintenance of drinking water distribution systems that relate to hydraulic parameters. For example, 32 of 34 responding state require that distribution systems be designed for an operational pressure of at least 20 psi under all flow conditions (see Table 2-3). Further, nine of 34 require both a minimum and maximum velocity through pipes. These requirements determine the maximum and minimum ground elevations that can be supplied. The minimum pressure establishes the highest ground elevation that can be supplied, and the maximum pressure defines the lowest ground elevation. The former criterion ensures that the highest customers will be supplied with at least the minimum pressure, while the latter ensures that the lowest customers will not experience objectionably high pressures.

To supply water at acceptable pressure, the distribution system is thus divided into a number of distinct pressure zones. The maximum change in elevation across each zone is determined by the difference between the maximum and minimum design pressure values. Adding new pressure zones or adjusting existing pressure zone boundaries is needed when pressure differentials are outside their desirable values. Pressure zone boundaries are delineated through the use of closed valves. To improve reliability, pressure-regulating valves (or pumps) are normally installed between the zones (along the pressure zone boundaries), and stretches of new pipe are added to eliminate dead ends.

Pressure zoning is desirable but requires careful planning and design (for details, see Boulos et al., 2006). Proper design of pressure zones will reduce leaks (because leakage normally varies exponentially with pressure and will be reduced with a fall in system pressures), breaks, and pumping costs; improve reservoir turnover rates; and avoid over-pressurizing the system. Existing facilities (e.g., reservoirs, pumps, pressure regulating valves) and natural (e.g., rives, lakes) or political boundaries (e.g., city limits, county and state boundaries) will influence the design and modification of pressure zones (Cesario, 1995).

Surge Protection

Pressure events or surges that can allow intrusion to occur are caused by sudden changes in water velocity due to loss of power, sudden valve or hydrant closure or opening, a main break, fire flow, or an uncontrolled change in on/off pump status (Boyd et al., 2004). Intrusion can be minimized by knowing the causes of pressure surges, defining the system's response to surges, and estimating the system's susceptibility to contamination when surges occur (Friedman et al., 2004). Pressure transients in distribution systems are usually most severe at pump stations and control valves, in high-elevation areas, in locations with low static pressures, and in remote locations that are distanced from overhead storage (Friedman et al., 2005).

A number of devices can be used for controlling transients in pipeline systems (Boulos et al., 2005, 2006; Wood et al., 2005). The general principles of pressure surge control devices are to store water or otherwise delay the change of flow or to discharge water from the line so that rapid or extreme fluctuations in the flow regime are minimized. Devices such as pressure-relief valves, surge anticipation valves, surge vessels, surge tanks, pump bypass lines, or any combination thereof are commonly used to control maximum pressures. Storage tanks with a free water surface can be effective in controlling surges. Minimum pressures can be controlled by increasing pump inertia or by adding surge vessels, surge tanks, air-release/vacuum valves, pump bypass lines, or any combination of these components. The overriding objective is to reduce the rate at which flow changes occur. Figure 5-3 illustrates typical locations for the various surge protection devices in a water distribution system.

Because no two distribution systems are hydraulically the same, there are no general rules or universally applicable guidelines for eliminating objectionable pressures in distribution systems. Any surge protection device must be chosen accordingly. The final choice will be based on the initial cause and location of the transient disturbance(s), the system itself, the consequences if

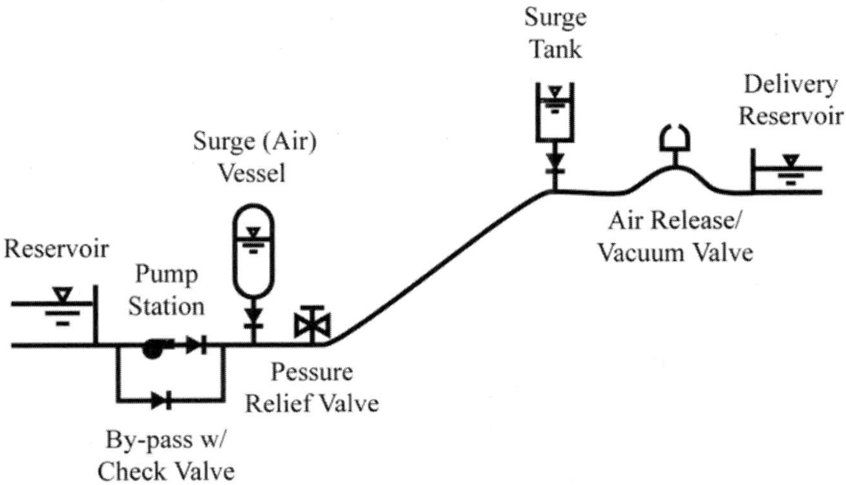

FIGURE 5-3 Typical locations for various surge protection devices. SOURCE: Reprinted, with permission, from Boulos et al. (2006). © 2006 by MWH Soft Pub.

remedial action is not taken, and the cost of the protection measures. A combination of devices may prove to be the most effective and economical. Final determination of the adequacy and efficacy of the proposed measure should be checked and validated using detailed surge modeling. Boulos et al. (2005, 2006) provide a detailed transient flow chart that offers a comprehensive guide to the selection of components for surge control and suppression in distribution systems. Good maintenance, pressure management, an adequate disinfectant residual, and routine monitoring programs are also essential components of transient protection.

Flushing Water Mains

Flushing is one of the most ubiquitous activities of water utilities for both maintaining and recovering the integrity of distribution systems because it is the primary means by which to remove contaminated water from the system. It was discussed briefly in Chapter 4 in association with the cleaning and disinfection of water mains following pipe installation, repair, and replacement. It is a topic of Chapter 6, which focuses on water quality integrity, because flushing is routine in areas with repeat customer complaints about color, taste, or odor; in dead ends mains; and in storage facilities. Its importance with respect to maintaining hydraulic integrity is that flushing removes accumulated sediment and corrosion products that reduce the hydraulic capacity of the pipe, improving the flow of water through the distribution system.

Flushing (discussed in greater detail in a subsequent section) is performed by isolating sections of the distribution system and opening fire hydrants (or flushing valves) to cause a large volume of flow to pass through the isolated pipelines so that a scouring action is created. Water is then discharged through a hydrant, which in turn removes any material buildup from the pipe. When flushing pipes, it is important to ensure that the flushing velocity is sufficient to suspend loose sediments. Flushing should continue until the water has cleared and disinfectant residual has reached normal expected levels. To minimize any negative environmental impacts (as flushed water may be high in suspended solids and other contaminants that can harm waterbodies), flushed water is normally discharged into sanitary or combined sewers or storm water management facilities. It is important to optimize flushing programs, as excessive flushing can waste significant volumes of water.

Operation and Design for Water Age Minimization

As discussed in Chapter 1, a primary reason for water quality problems within distribution systems is the advanced water age necessitated by the provision of adequate standby fire flow and redundant capacity. This requires that utilities use standpipes, elevated tanks, and large storage reservoirs, as well as

Hydrant flushing. Photo courtesy of Bureau of Laboratory Services, Philadelphia Water Department.

larger-sized pipes than would otherwise be necessary. The effect of designing and operating a system to maintain adequate fire flow and redundant capacity can result in long travel times and low velocities between the treatment plant and the consumer, which can be detrimental to water quality.

Brandt et al. (2004, 2006) have recently completed a two-volume study sponsored by the American Water Works Association Research Foundation to suggest ways to minimize water age (retention times) while at the same time controlling water quality degradation and providing the pressure and quantity constraints that are required to maintain water service. In particular, Brandt et al. (2006) have developed a diagnostic methodology by which a water utility can assess and then minimize water quality problems associated with excessive retention times. Best management practices for controlling retention time can generally be categorized into storage and network methods. Storage methods include adjusting pump schedules, reducing the operational top water level, removing storage tanks from service, and reconfiguring reservoir and storage tanks to avoid dead zones. Network methods include altering network valving patterns, installing time actuated valves, flushing (manual and automated), and abandoning and downsizing mains (Brandt, 2006).

An important aspect of hydraulic integrity maintenance is to ensure sufficient mixing and to minimize water age in storage facilities—issues which if not addressed can generate pockets of stagnant water with depleted disinfectant residual and associated water quality problems. Mixing will eliminate internal dead zones within a storage facility and prevent short-circuiting between the tank inlet and outlet. Completely mixed flow can be achieved by using a turbulent (high velocity) inlet jet, mechanical mixers, or hydraulic circulation systems. Controlling pumping rates and fill and discharge rates can also provide adequate intensities to achieve complete mixing. For example, Grayman et al. (2000) recommend that to avoid stratification in distribution storage facilities, the fill time should exceed the mixing time. A utility's SCADA system can be used to monitor the real-time mixing intensity within a storage facility, and as such is useful for process control. It should be noted that utilities may be constrained in their ability to provide complete mixing due to the increased energy requirements.

Both poor mixing and improper tank discharge management can increase the residence time of water in a service reservoir. To combat this potential problem, frequent exercising of reservoirs (i.e., continuously mixing the water and making sure that fresh water replaces stagnant water) is required. Grayman et al. (2000) used various modeling techniques to develop a set of general guidelines for reducing water quality deterioration associated with inadequate mixing and excessive water age in distribution storage facilities. They reviewed the application of CFD, compartment, and physical scale models. A stand-alone model called CompTank is presented which provides a wide range of alternatives and allows the user to model water age and the concentrations of reactive or conservative substances over long time periods.

***

There is limited information about how to operationally reduce water age in an existing system while taking into account larger issues such as minimizing operational costs and maintaining the other aspects of hydraulic integrity, such as reliability of supply and adequate pressure for all water uses. At the present time, there is so much variability in the system parameters affecting distribution system operation that it is not possible, for example, to quantify the tradeoff between the risk of running out of water and the risk of delivering water of poor quality. This quandary is manifested in our inability to optimally maintain and operate storage facilities. The benefits of large storage tanks are not clear, nor is it easy to determine whether to remove a tank from service or reduce its volume. Answering such questions will require research that quantifies how various actions (such as removing a storage tank from service) will affect other aspects of hydraulic integrity (such providing fire flow and minimizing water age) within a given distribution system.

RECOVERING HYDRAULIC INTEGRITY

When a distribution system experiences high head losses, inadequate pressures or flows, high turbidity from scale loosening or resuspension of sediment, or low disinfectant residual and high bacterial counts from advanced water, there are several steps the utility should take. One of the first steps is to consider one or more of the standard techniques available to remove any loose sediment, biofilm, and tubercles that may be the cause of the problem. These procedures can restore most of the pipes' original hydraulic capacity, and include conventional and unidirectional flushing, air scouring, swabbing, abrasive pigging, chemical cleaning, mechanical cleaning and lining (nonstructural, cement or epoxy applied linings), and structural lining. If the problems persist even after the application of these techniques, replacement of the pipes should be considered (see Chapter 4). A brief discussion of each technique follows.

It should be noted that to overcome increasing head losses and local deficiencies in system pressure and to increase the carrying capacity of water mains, increased levels of pumping are usually needed. This will result in increases in energy consumption and increased operational costs for a water utility.

Conventional Flushing

Conventional flushing generally involves opening hydrants in a specific area of the distribution system until the water visually runs clear. While effective in quickly removing loose particles, this type of flushing is usually not effective in dislodging well-attached deposits and cannot remove scales and tuberculation. Because in a looped system the water will flow to the hydrant from multiple mains and directions, it becomes very difficult to achieve the high-velocity flushing required to scour and remove deposits (as shown in Figure 5-4). As a result, some sediment and biofilm may not be removed, and the cleanup method requires a substantial quantity of water. In addition, because the dynamics of the entire distribution system are not considered, it is possible that the water used to the flush the system may come from a component that has not been previously cleaned. Therefore, sediments, detached biofilm, etc., may simply be transported from one part of the distribution system to another.

Unidirectional Flushing

Unidirectional flushing involves the closure of valves and opening of hydrants to create a one-way flow in the water mains (see Figure 5-5). This increases the speed of the water flow so that the shear velocity near the pipe wall is maximized, producing a scouring action in the mains, effectively removing sediment deposits and biofilm. Flushing should start at a clean water source (e.g., pump station) and proceed outward in the system so that flushing water is drawn from previously flushed reaches. This ensures that clean water is always used to flush the mains. No special equipment is needed; however, substantial planning time is required to define the flushing zones, the valves and hydrants to be operated, the duration of the flush for each zone, the required velocities, and the sequence of operation. A hydraulic model of the distribution system can

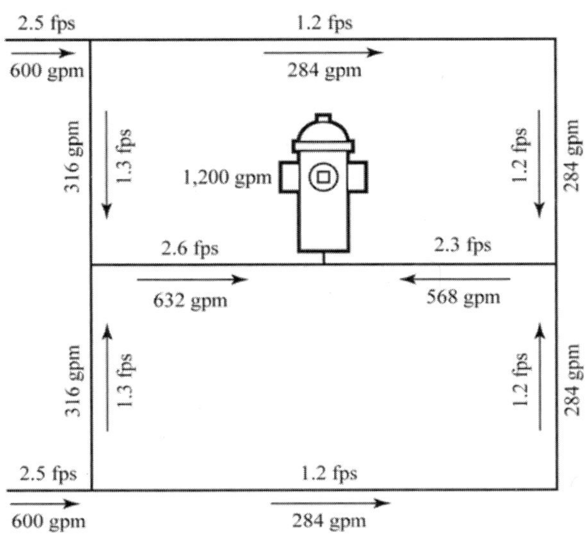

FIGURE 5-4 Conventional flushing in a looped system results in water flowing toward the hydrant from all directions generating lower velocity and less scouring of the mains. SOURCE: Reprinted, with permission, from Boulos et al. (2006). © 2006 by MWH Soft Pub.

greatly simplify and expedite the planning process, especially for estimating pipe flow rates, velocities, and flushing times. While more costly and time consuming than conventional flushing, unidirectional flushing is more effective and uses less water (Hasit et al., 2004). There are often long-term water quality benefits because deposits and water of questionable quality are actually removed rather than being re-routed to other parts of the distribution system.

Work done by Slaats (2001) demonstrated the velocities needed to entrain sediments, and these were within the range of velocities used for flushing. Carriere et al. (2002) showed that loose deposits could be removed by unidirectional flushing as a function of time, pipe material, and water characteristics. Gauthier et al. (1997) showed that loose deposits in a French system removed by flushing contained organisms including invertebrates, protozoa, and bacteria (although it should be noted that French distribution systems maintain no disinfectant residual such that their ecology is not representative of U.S. distribution systems). The abiotic constituents were primarily iron, volatile solids, calcium, aluminum, and other insoluble materials. Deposits flushed from four systems in the United Kingdom were all high in iron and manganese (Marshall, 2000).

Not all systems can or will routinely flush. There may be water restrictions that preclude flushing, and customers may be upset if they see water being

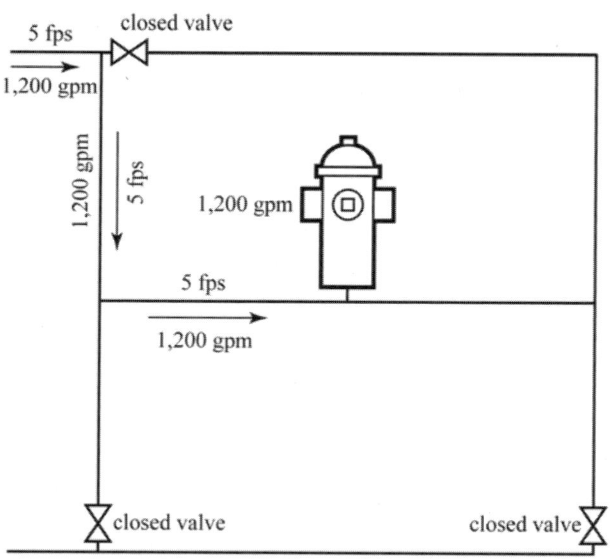

FIGURE 5-5 Unidirectional flushing results in water flowing toward the hydrant in only one direction resulting in higher velocity, more scouring, and better cleaning of the mains with less water use. SOURCE: Reprinted, with permission, from Boulos et al. (2006). © 2006 by MWH Soft Pub.

"wasted" while they are being told to conserve. Additionally, there is often a requirement that disinfectant residuals in the flushed water be neutralized, and this may be more complicated if chloramines are present compared to chlorine.

Air Scouring, Swabbing, and Abrasive Pigging

There is a long history of cleaning pipelines in order to remove accumulated material resulting from corrosion, improper pH adjustment, post precipitation of water treatment chemicals, and biofilm growth. Cleaning usually is a precursor to another process like lining or insertion rather than a process onto itself. This is due to the fact that cleaning potentially exposes unprotected metal pipe which would result in additional water quality problems.

Scouring, swabbing, and abrasive pigging are progressively more aggressive cleanup techniques that involve more specialized equipment and specialized skills. Although a few water utilities have implemented these methods using their own staff, typically these methods are contracted to specialty firms. Air scouring involves the continuous injection of filtered, compressed air into the pipe, along with a continuous but smaller flow of water. Given a continuous supply of water and air in the right proportions, discrete "slugs" of water are formed in the pipe and driven along by the compressed air at high velocity. The high velocity slugs tend to remove silt, sediment, loose matter, and debris from the base of the pipe. No disassembly of the pipe is necessary. Water scouring involves the insertion of a high-pressure water jet into the pipe to remove deposited materials. The water jet pressure can be adjusted to remove the deposits without damaging the piping material. The jet will back flush the deposited material to the insertion point in the pipeline. While jetting is very effective, it is limited to the length of the jetting equipment, which will result in frequent insertion points, and to small diameter pipes.

More aggressive techniques, such as swabbing and abrasive pigging, work to varying degrees in removing heavy sediment, biofilm, adherent material, tuberculation, and even very hard scale (Ellison et al., 2003). Swabbing involves driving cylindrical foam sponges (known as swabs) through pipes using water pressure. The swabs travel along the water main and scrub the scale encrustations and slime build-up from the inner pipe walls. Loosened debris and swabs are eventually flushed out at an exit point. Currently, pigging is used primarily if there are hydraulic problems in the water mains, i.e., to improve the "C" factor (roughness coefficient) of the pipes. It involves isolating a segment of the distribution system and passing a fluid-propelled object through the pipe. A styrofoam plug is often used as the "pig," which is normally the same or slightly bigger in diameter than the water main and is shaped like a torpedo (Deb, 1990). Increasing sizes of pigs are passed through the pipeline to gradually remove deposits within the pipeline. The abrasiveness of the pig results in varying quality of water being discharged during the cleaning process. For instance, higher concentrations of suspended solids normally follow the more abrasive pigs, as these scour the inner lining of the pipe.

Both pigging and swabbing can be difficult to implement because they require the removal of hydrants or the installation of new pipeline appurtenances (e.g., pig launching and receiving stations). Few water utilities have implemented these methods using their own staff, such that these methods are usually contracted to specialty firms.

Chemical Cleaning, Mechanical Cleaning, and Lining

Chemical cleaning to restore old pipes involves the recirculation in an isolated pipe section of proprietary acids and surfactants to remove scale and deposit, while mechanical cleaning is accomplished by dragged scrapers. Scrapers are devices that use springs to force blades against the wall of the pipe. As the device moves through the pipe, the blades scrape the material off the walls which can then be flushed from the pipe. These techniques are typically applied in the rehabilitation of older unlined cast iron pipes that have become scaled and tuberculated. In another example, a process using a cleansing solution of an organic oxide scavenger and muriatic acid circulated through an isolated section of distribution main worked effectively for small diameter pipelines (Estrand, 1995). Compared to air scouring and pigging, chemical cleaning is infrequently used due to the cost of chemicals and their proper disposal after cleaning.

It is common practice to reline a cleaned pipeline to protect the newly exposed metallic pipeline material. The most common technique is to use concrete mortar applied to the internal surface, a technology that has been used for over 50 years. Spray-on epoxy lining is a newer method that is especially useful when the water is low in hardness, which can cause a cement lining to deteriorate. Most recently, polyurethane lining is becoming a competitive alternative to concrete mortar lining especially in long pipelines with few service connections. This type of chemical lining on the inner surface of the pipe is referred to as nonstructural lining and does not increase the pipe's structural integrity.

CONCLUSIONS AND RECOMMENDATIONS

Maintaining the hydraulic integrity of distribution systems is vital to ensuring that water of acceptable quality is delivered in acceptable amounts. The most critical element of hydraulic integrity is adequate water pressure inside the pipes. The loss of water pressure resulting from pipe breaks, significant leakage, excessive head loss at the pipe walls, pump or valve failures, or pressure surges can impair water delivery and increase the risk of contamination of the water supply via intrusion. In addition, slow moving water or changes in the flow regime (including flow reversals) and advanced water age can negatively impact finished water quality. Proper system design, operation, and maintenance, along with monitoring and modeling, can help water utilities achieve a high degree of hydraulic integrity and reliability and extend the life of their distribution systems. The following conclusions and recommendations focus on the highest priority issues.

Water residence times in pipes, storage facilities, and premise plumbing should be minimized. Excessive residence times can lead to low disinfectant residuals and leave certain service areas with a less protected drinking water supply. In addition, long residence times can promote microbial regrowth and the formation of disinfection byproducts. From an operational viewpoint it may be challenging to reduce residence time where the existing physical infrastructure and energy considerations constrain a utility's options. Furthermore, limited understanding of the stochastic nature of water demand and water age makes it difficult to assess the water quality benefits of reduced residence time. Research is needed to investigate such questions, as well as how to achieve minimization of water residence time while maintaining other facets of hydraulic integrity (such as adequate pressure and reliability of supply).

Positive water pressure should be maintained. Low pressures in the distribution system can result not only in insufficient fire fighting capacity but can also constitute a major health concern resulting from potential intrusion of contaminants from the surrounding external environment. A minimum residual pressure of 20 psi under all operating conditions and at all locations (including at the system extremities) should be maintained. The minimum value could be adjusted based on site specific conditions.

Where feasible, surge protection devices should be installed. Because these devices provide the only practical opportunity to prevent intrusion of contaminants due to low or negative pressure events, surge tanks should be considered at all pump stations (to dampen negative pressure waves) and other surge control devices at vulnerable locations in the system such as high points. This can be aided by a comprehensive surge analysis on a representative network model of the distribution system to select, locate, and size the most effective combination of surge protection devices. Although looped networks are generally less susceptible to objectionable pressure transients than single long transmission main systems, they must still be protected against low or negative pressure transients.

Distribution system monitoring and modeling are critical to maintaining hydraulic integrity. Hydraulic parameters to be monitored should include inflows/outflows and water levels for all storage tanks, discharge flows and pressures for all pumps, flows and/or pressure for all regulating valves, and pressures at critical points. An analysis of these patterns can directly determine if the system hydraulic integrity is compromised or if the system is operating as designed, or detect any unexpected operational anomalies. Calibrated distribution system models can calculate the spatial and temporal variations of flow, pressure, velocity, reservoir level, water age, and other hydraulic and water quality parameters throughout the distribution system. Such results can, for example, help identify areas of low or negative pressure and high water age, estimate filling and draining cycles of storage facilities, and determine the adequacy of the system to supply fire flows under a variety of demand loading and operating conditions.

REFERENCES

American Water Works Association (AWWA). 1986. Introduction to Water Distribution Principles and Practices of Water Supply Operations. Denver, CO: AWWA.

AWWA. 2005. AWWA Manual M32: Computer modeling of water distribution systems. Denver, CO: AWWA.

Baribeau, H., N. L. Pozos, L. Boulos, G. F. Crozes, G. A. Gagnon, S. Rutledge, D. Skinner, Z. Hu, R. Hofmann, R. C. Andrews, L. Wojcicka, Z. Alam, C. Chauret, S. A. Andrews, R. Dumancis, and E. Warn. 2005. Impact of Distribution System Water Quality on Disinfection Efficacy. Denver, CO: AwwaRF.

Boulos, P. F., B. W. Karney, D. J. Wood, and S. Lingireddy. 2005. Hydraulic transient guidelines for protecting water distribution systems. J. Amer. Water Works Assoc. 97(5):111–124.

Boulos, P. F., K. E. Lansey, and B. W. Karney. 2006. Comprehensive Water Distribution Systems Analysis Handbook for Engineers and Planners. Second edition. Pasadena, CA: MWH Soft Pub.

Boyd, G. R., H. Wang, M. D. Britton, D. C. Howie, D. J. Wood, J. E. Funk, and M. J. Friedman. 2004. Intrusion within a simulated water distribution system due to hydraulic transients. 1: Description of test rig and chemical tracer method. J. Environ. Eng. 130(7):774–783.

Brandt, M., J. Clement, J. Powell, R. Casey, D. Holt, N. Harris, and C. T. Ta. 2004. Managing Distribution Retention Time to Improve Water Quality—Phase I. Denver CO: AwwaRF.

Brandt, M., J. Clement, J. Powell, R. Casey, D. Holt, N. Harris, and C. T. Ta. 2004. Managing Distribution Retention Time to Improve Water Quality—Phase II. Denver CO: AwwaRF.

Buchberger, S. G., J. T. Carter, Y. H. Lee, and T. G. Schade. 2003. Random Demands, Travel Times and Water Quality in Deadends. Denver, CO:AWWA.

Buchberger, S. G., and G. J. Wells. 1996. Intensity, duration and frequency of residential water demand. J. Water Resources Planning and Management 122(1):11–19.

Buchberger, S. G., and L. Wu. 1995. A model for instantaneous water demand. J. Hydraulic Eng. 121(3):232–246.

Carriere, A., B. Barbeau, V. Gauthier, C. Morissette, R. Millette and A.Lalumiere. 2002. Unidirectional flushing: loose deposits characterization in the test zones of four Canadian distribution systems. *In:* Proceedings of the AWWA Water Quality Technology Conference. Denver, CO: AWWA.

Cesario, L. 1995. Modeling, analysis and design of water distribution systems. Denver, CO: AWWA.

Clark, R. M., and S. G. Buchberger. 2004 Responding to a contamination threat in a drinking water network: the potential for modeling and monitoring. Pp 9.1-9.26 *In:* Water Supply Systems Security. L. W. Mays (ed.). New York: McGraw-Hill.

Clark, R. M., W. M. Grayman, S. G. Bucberger, Y. Lee, and D. J. Hartman. 2004. Drinking water distribution systems: an overview. Pp 4.1-4.2 *In:* Water Supply Systems Security. L. W. Mays (ed.). New York: McGraw-Hill.

Clark, R. M., and W. M. Grayman. 1998. Modeling water quality in drinking water distribution systems. Denver, CO: AWWA.

Clark, R. M., S. Panguluri, and R. C. Haught. 2004. Remote monitoring and network models: their potential for protecting U.S. water supplies. Pp. 14.1–14.22 *In:* Water Supply Systems Security. Mays, L. W. (ed). New York: Mc Graw-Hill.

Cromwell, J., G. Nestel, and R. Albani. 2001. Financial and economic optimization of water main replacement programs. Denver, CO: AwwaRF.

Deb, A. K, J. K. Snyder, J. J. Chelius, and D. K. O'Day. 1990. Assessment of Existing and Developing Water Main Rehabilitation Practices. Denver, CO: AwwaRF.

Ellison, D., S. J. Duranceau, S. Ancel, G. Deagle, and R. McCoy. 2003. Investigation of Pipe Cleaning Methods. Denver, CO: AwwaRF.

Estrand, C., A. Hicatt, and J. Ludwidg. 1995. Chemical cleaning process for water pipe systems. *In*: Proceedings of the Hydraulics of Pipelines Conference, ASCE, Phoenix, AZ.

Friedman, M., L. Radder, S. Harrison, D. Howie, M. Britton, G. Boyd, H. Wang, R. Gullick, D. Wood and J. Funk. 2004. Verification and Control of Pressure Transients and Intrusion in Distribution Systems. Denver, CO: AwwaRF.

Gauthier, V., C. Rosin, L. Mathieu, J. M. Portal, J. C. Block, P. Chaix, and D. Gatel. 1997. Characterization of the loose deposits in drinking water distribution systems. *In:* Proceedings of the AWWA Water Quality Technology Conference. Denver, CO: AWWA.

Grayman, W. M., L. A. Rossman, C. Arnold, R. A. Deininger, C. Smith, J. F. Smith, and R. Schnipke. 2000. Water quality modeling of distribution system storage facilities. Denver, CO: AwwaRF.

Gullick, R. W., M. W. LeChevallier, R. C. Svindland, and M. J. Friedman. 2004. Occurrence of transient low and negative pressures in distribution systems. J. Amer. Water Works Assoc. 96(11):52–66.

Hasit, Y. J., A. J. DeNadai, H. M. Gorill, S. B. McCammon, R. S. Raucher, and J. Whitcomb. 2004. Cost and Benefit Analysis of Flushing. Denver, CO: AwwaRF.

Karim, M., Abbaszadegan, M. and M. W. LeChevallier. 2003. Potential for pathogen intrusion during pressure transients. J. Amer. Water Works Assoc. 95(5):134–146.

Kirmeyer, G. J., M. Friedman, K. Martel, D. Howie, M. LeChevallier, M. Abbaszadegan, M. Karim, J. Funk, and J. Harbour. 2001. Pathogen intrusion into the distribution system. Report No. 90835. Denver, CO: AwwaRF and AWWA.

Lansey, K. E., and P. F. Boulos. 2005. Comprehensive Handbook on Water Quality Analysis for Distribution Systems. Pasadena, CA: MWH Soft Pub.

LeChevallier, M. W., R. W. Gullick, M. R. Karim, M. Friedman, and J. E. Funk. 2003. The potential for health risks from intrusion of contaminants into distribution systems from pressure transients. Jour. Water Health 1(1):3–14.

Levenspeil, O. 2002. Modeling in chemical engineering. Chemical Engineering Science 57: 4691–4696.

Mahmood, F., J. G. Pimblett, N. O. Grace, and W. M. Grayman. 2005. Evaluation of water mixing characteristics in distribution system storage tanks. J. Amer. Water Works Assoc. 97(3):74–88.

Marshall, G. P. 2000. Understanding and Preventing Discolored Water. UKWIR Report #01/DW/03/17. London: UKWIR Ltd.

Panguluri, S., W. M. Grayman, and R. M. Clark. 2005. Distribution system water quality report: a guide to the assessment and management of drinking water quality in distribution systems. Cincinnati, OH: EPA Office of Research and Development.

Slaats, N. (ed.). 2001. Processes Involved in the Generation of Discolored Water. Nieuwegein, the Netherlands: KIWA.

Walski, T. M., D. V. Chase, D. A. Savic, W. M. Grayman, S. Beckwith and E. Koelle. 2003. Advanced Water Distribution Modeling and Management. Waterbury, CT: Heastad Press.

Wood, D. J., S. Lingireddy, and P. F. Boulos. 2005. Pressure Wave Analysis of Transient Flow in Pipe Distribution Systems. Pasadena, CA: MWH Soft Pub.

6
Water Quality Integrity

As discussed in Chapters 4 and 5, breaches in physical and hydraulic integrity can lead to the influx of contaminants across pipe walls, through breaks, and via cross connections. These external contamination events can act as a source of inoculum, introduce nutrients and sediments, or decrease disinfectant concentrations within the distribution system, resulting in a degradation of water quality. Even in the absence of external contamination, however, there are situations where water quality is degraded due to transformations that take place within piping, tanks, and premise plumbing. Most measurements of water quality taken within the distribution system cannot differentiate between the deterioration caused by externally vs. internally derived sources. For example, decreases in disinfectant concentrations with travel time through the distribution system could be the result of demand from an external contamination event or it could be due to disinfectant reactions with pipe walls and natural organic matter remaining after treatment.

This chapter deals with the various internal processes or events occurring within a distribution system that lead to degradation of water quality, the consequences of those processes, methods for detecting the loss of water quality, operational procedures for preventing these events, and finally, how to restore water quality integrity if it is lost. In many cases, the detection methods and recovery remedies are similar to those discussed in previous chapters.

FACTORS CAUSING LOSS OF WATER QUALITY INTEGRITY AND THEIR CONSEQUENCES

For water quality integrity to be compromised, specific reactions must occur that introduce undesirable compounds or microbes into the bulk fluid of the distribution system. These reactions can occur either at the solid–liquid interface of the pipe wall or in solution. Obvious microbial examples include the growth of biofilms and detachment of these bacteria within distribution system pipes and the proliferation of nitrifying organisms. Important chemical reactions include the leaching of toxic compounds from pipe materials, internal corrosion, scale formation and dissolution, and the decay of disinfectant residual that occurs over time as water moves through the distribution system. All these interactions are governed by a suite of chemical and physical parameters including temperature, pH, flow regime, concentration and type of disinfectant, the nature and abundance of natural organic matter, pipe materials, etc. Many of these variables may be linked in distribution systems; for example, seasonal increases in temperature may be accompanied by changes in organic matter, flow regimes, and disinfectant concentrations. As a consequence, attempting to correlate the occurrence of a given event (such as corrosion, microbial growth, disinfectant decay, or DBP formation) within distribution systems to a single variable (such as temperature) is difficult.

Biofilm Growth

One way in which water quality can be degraded in the distribution system is due to the growth of bacteria on surfaces as biofilms. Virtually every water distribution system is prone to the formation of biofilms regardless of the purity of the water, type of pipe material, or disinfectant used. The extent of biofilm formation and growth, the microbial ecology that develops, and the subsequent water quality changes depend on surface-mediated reactions (e.g., corrosion, disinfectant demand, immobilization of substrates for bacterial growth), mass transfer and mass transport processes, and bulk fluid properties (concentration and type of disinfectants, general water chemistry, organic concentration, etc.). These interactions can be exceedingly complex, which typically means that the mechanisms leading to biofilm growth may not be obvious and are often system specific.

Bacteria growing in biofilms can subsequently detach from the pipe walls. Because these organisms must survive in the presence of the disinfectant residual present in the distribution system, the interaction between the suspended organisms and residual is critical. If the residual has decayed due to reactions with compounds in the water or with the pipe wall, intrusion, or other sufficient external contamination, it is possible for attached bacteria to be released into water that contains insufficient disinfectant to cause their inactivation. The potential for this to occur is higher in premise plumbing, which generally has longer water residence times that may lead to very low disinfectant concentrations.

Pathogenic Microorganisms

An obvious risk to public health from distribution system biofilms is the release of pathogenic bacteria. As discussed in Chapter 3, there are instances where opportunistic pathogens have been detected in biofilms, including *Legionella*, *Aeromonas* spp., and *Mycobacterium* spp. Assessing risk from these organisms in biofilms is complicated by the potential for two modes of transmission. *Aeromonas* spp. causes disease by ingestion, while the other two organisms cause the most severe forms of disease after inhalation. In the case of *Aeromonas* spp., which is included as one of the unregulated "contaminants" to be tested for in the Contaminant Candidate List, it has been shown that drinking water isolates carry virulence factors directly involved in pathogenesis (Sen and Rogers, 2004).

Coliforms and Heterotrophs

Another consequence of biofilms is their potential to support the growth and release of organisms of regulatory concern, especially coliforms. Coliforms released from biofilms may result in elevated coliform detection even though physical integrity (i.e., breaches in the distribution system) and disinfectant residual have been maintained (Characklis, 1988; Haudidier et al., 1988; Smith et al., 1990). It should be noted that coliforms arising from biofilms are generally considered to be low risk (see Chapter 2), which is also inferred by EPA's variance to the Total Coliform Rule for coliforms emanating from biofilms (see page 208). However, coliform regrowth may indirectly present a risk by masking the presence of bacteria introduced in a simultaneous contamination event. If repeated occurrences of coliforms in the distribution system force a utility to notify the public, there can be a loss of consumer confidence and trust in the utility.

The regrowth of heterotrophs in biofilms can also be of concern, especially for European communities that are required to monitor their presence. Some U.S. utilities routinely monitor heterotrophs using heterotrophic plate counts (HPC) as a general indicator of microbial quality, and may be required to assess their numbers if chlorine residuals are too low. In general, heterotrophic bacteria are usually not of public health concern, but with the growing immunocompromised population many utilities are interested in minimizing the presence of these organisms in their water.

Corrosion and Other Effects

In addition to the regrowth issue, biofilms in distribution systems can cause other negative effects on finished water quality. The processes listed here do not require that the organisms detach from the surfaces, since the changes in water quality are due to their metabolic activities as they grow on the surfaces.

Bacterial biofilms may contribute to the corrosion of pipe surfaces and their eventual deterioration. Although a considerable amount of corrosion internal to the pipe can be mediated by abiotic factors, it is known that bacteria can both directly and indirectly influence corrosion of metal surfaces. Of particular concern is the pitting of copper that can lead to pinhole leaks in premise plumbing. Geesey et al. (1993) reported that pitting of copper plumbing in four hospitals around the world was likely attributable to bacterial activity. Wagner et al. (1997) have said that biologically produced polymers typical of biofilms create high and low chloride concentration cells, and consequently localized corrosion cells, leading to increased copper corrosion. Laboratory studies have shown that the presence of bacteria on copper surfaces could accelerate corrosion when compared to an abiotic system (Webster et al., 2000). In other studies, specific organisms were correlated with copper corrosion and could be isolated from pits (Bremer and Geesey, 1991; Bremer et al., 1992). However, other research has shown that organisms alone did not cause copper pitting, and that particulate matter was also required (Walker et al., 1998).

Microbes may also influence iron surfaces in distribution systems. Iron bacteria can grow on ferrous metal surfaces (Ridgway et al., 1981), and by virtue of their metabolism may modify the local chemistry at the metal surface which in turn promotes localized corrosion (Victoreen, 1974). As stated by McNeill and Edwards (2001), there are many possible effects of bacterial action and biofilm formation on iron corrosion. These include the production of differential aeration cells (Lee et al., 1980), soluble metal uptake by biofilm polymers (Tuovinen et al., 1980), changes in iron speciation by oxidation or reduction (Shair, 1975; Denisov et al., 1981; Kovalenko et al., 1982; Okereke and Stevens, 1991; Chapelle and Lovely, 1992; Nemati and Webb, 1997), and the production of pH gradients (Tuovinen et al., 1980) or corrosive hydrogen sulfide (Tuovinen et al., 1980; DeAraujo-Jorge et al., 1992). All of these factors can contribute to increased localized corrosion and the deterioration of the pipe material, as well as influencing water quality by causing the release of metal ions or corrosion products and associated problems with water color.

Other effects of biofilms are worth noting. As demonstrated in the wastewater industry, it is possible to have nitrifying bacteria present in biofilms, and these organisms could result in nitrification episodes in distribution systems where chloramine is used (Wolfe et al., 1990, and see the section below). *Actinomycetes* or fungi present in biofilms may result in taste and odor problems (Burman, 1965, 1973; Olson, 1982), which then lead to consumer complaints. Excess biofilm growth can result in the loss of hydraulic capacity by increasing fluid frictional resistance at the pipe wall (see examples in Characklis et al., 1990). Finally, growth of biofilms and the associated organics can create a chlorine demand at the pipe wall.

Biologically Stable Water

Because this report focuses on distribution system events, it does not delve into failures or breaches at the treatment plant that might allow a breakthrough of contaminated water. Nonetheless, a brief discussion of biologically stable water is warranted, given its potential to reduce the growth of bacteria in the distribution system. Drinking water is generally considered to be biologically stable if it does not support the growth of bacteria in the distribution system. In its broadest sense, biologically stable water restricts growth because it lacks an essential nutrient (nitrogen or phosphorus), is sufficiently low in utilizable organic carbon, or contains adequate disinfectant. Although all of these parameters may influence biofilm growth, the U.S. drinking water industry has typically viewed biologically stable water as sufficiently low in organic carbon as to limit the proliferation of heterotrophic bacteria. In this context, the general concepts of microbial stable water and maximum regrowth potential are relatively well understood (Rittman and Snoeyink, 1984; Sathasivan et al., 1997).

Another mechanism for ensuring biological stability is the maintenance of an adequate disinfectant residual. However, since disinfectants decay in the distribution system, reliance on a residual to ensure biological stability may not be entirely feasible. Within distal portions of the distribution system or within stagnant portions of premise plumbing, disinfectants disappear via reactions with pipe or bulk water or via nitrification. At these locations, any available organics can then be freely utilized by the bacteria present.

The reduction of organic carbon to control microbial growth may allow utilities to decrease their reliance on disinfectants. This approach also has the advantage of decreasing the potential for the production of disinfectant by-products (DBPs). Organic carbon removal is most often accomplished through enhanced coagulation, granular activated carbon filtration, or biological filtration. Although there is controversy surrounding target concentrations of organics that will limit regrowth, some recommendations have been made. van der Kooij et al. (1989) and van der Kooij and Hijnen (1990) showed a correlation between assimilable organic carbon (AOC) and regrowth in a non-disinfected distribution system, and provided evidence for biological stability in the Netherlands when the AOC concentration (*Pseudomonas fluorescens* P17 + *Spirillum* NOX) is reduced to 10 μg acetate C eq/L (van der Kooij 1992). LeChevallier et al. (1991) have suggested that coliform regrowth may be controlled by influent AOC levels (P17 + NOX) below 50 μg acetate C eq/L. Based on a field study, LeChevallier et al. (1996) subsequently recommended a level below 100 μg C/L to control regrowth. Servais et al. (1991) have associated biological stability with a biodegradable dissolved organic carbon (BDOC) level of 0.2 mg/L, but Joret et al. (1994) have stated that the value is 0.15 mg/L at 20° C and 0.30 mg/L at 15° C.

It should also be noted that organic carbon may not be the limiting nutrient. In Japan and Finland, evidence supports the concept that phosphorus is limiting (Miettinen et al., 1997; Sathasivan et al., 1997; Sathasivan and Ohgaki, 1999; Lehtola and Miettinen, 2001; Keinanen et al., 2002; Lehtola et al., 2002a,b, 2004). In these cases, the addition of phosphate-based corrosion inhibitors may decrease the biological stability of the water and allow for regrowth (Miettinen et al., 1997).

This discussion illustrates that the best strategy for creating and maintaining biologically stable water is most likely to be system specific. Each water utility should identify the limiting nutrient and best practices to attain and then maintain biological stability. Changing water quality goals should then keep these factors in mind. For example, the dosing of ammonia during a switch to chloramination would relieve nitrogen limitations to regrowth, whereas dosing of phosphate corrosion inhibitors can relieve phosphate limitations.

Nitrification

Biological nitrification is a process in which bacteria oxidize reduced nitrogen compounds (e.g., ammonia) to nitrite and then nitrate. It is associated with nitrifying bacteria in distribution systems and long retention times in water supply systems practicing chloramination. One of the most important problems exacerbated by nitrification is loss of the chloramine disinfectant residual. This occurs because a reduction in ammonia results in an increased ratio of chlorine to ammonia nitrogen. This ratio controls the stability of monochloramine, which is governed by a complex set of reactions (Jafvert and Valentine, 1992; also see following section on loss of disinfectant residual). As the ratio approaches 1.5 on a molar basis, a rapid loss of monochloramine occurs attributable to the eventual oxidation of N(III) to primarily nitrogen gas and the release of more ammonia. The released ammonia can then be further oxidized by the nitrifying organisms, establishing what amounts to a positive feedback loop. Furthermore, the loss of disinfectant residual removes one of the controls on the activity of nitrifiers, and it may also lead to the increased occurrence of microorganisms such as coliforms (Wolfe et al., 1988, 1990) and heterotrophic bacteria.

As discussed in NRC (2005), the loss of chloramine residual is the most significant health threat that can result from nitrification. It should be noted, however, that there are other lesser health effects of nitrification that may be important for certain populations. Nitrite and nitrate have been shown to cause methemoglobinemia (blue baby syndrome), an acute response to nitrite that results in a blockage of oxygen transport (Bouchard et al., 1992). Methemoglobinemia affects primarily infants below six months of age, but it may occur in adults of certain ethnic groups (Navajos, Eskimos) and those suffering from a genetic deficiency of certain enzymes (Bitton, 1994). Pregnant women may also be at a higher risk of methemoglobinemia than the general population (Bouchard et al., 1992). A second concern is that nitrate may be reduced to nitrite in the low pH environment of the stomach, reacting with amines and amides to form N-nitroso compounds (Bouchard et al., 1992; De Roos et al., 2003). Nitrosamines and nitrosamides have been linked to different types of cancer, but the intake of nitrate from drinking water and its causal relation to the risk of cancer is still a matter of debate (Bouchard et al., 1992). A study by Gulis et al. (2002) in Slovakia related increased colorectal cancer and non-Hodgkin's lymphoma to medium (10.1–20 mg/l) and high (20.1–50 mg/l) concentrations of nitrate nitrogen in drinking waters. Similarly, Sandor et al. (2001) showed a correlation between the consumption of waters containing greater than 88 mg/l nitrate nitrogen and gastric cancer. Despite numerous papers (Sandor et al., 2001; Gulis et al., 2002; Kumar et al., 2002; De Roos et al., 2003; Coss et al., 2004; Fewtrell, 2004), the concentration at which nitrate nitrogen in drinking waters presents a health risk is unclear (Fewtrell, 2004). Finally, a lesser but still significant water quality effect of nitrification is a reduction in alkalinity and pH in low alkalinity waters. This may cause the pH to decrease to the point that corrosion of lead or copper becomes a problem.

It is important to recognize that nitrate and nitrite may come from sources other than nitrification. van der Leeden et al. (1990) found that 93 percent of all U.S. water supplies contain less than 5 mg/l nitrate, but noted that these values may be changing as a result of the increased use of nitrate-containing fertilizers. Increased use of chloramination (up to 50 percent of the surface water systems in the United States may use chloramination in the near future as a result of the Stage 1 Disinfectants/Disinfection Byproducts Rule; EPA, 2003) may result in higher levels of nitrate in drinking waters (Bryant et al., 1992), but the increment in nitrate plus nitrite nitrogen from this source would typically be less than 1 mg/L, which is well below the current maximum contaminant level (MCL). Thus, as stated earlier the concern may be predominantly for more susceptible populations (pregnant women, infants, some ethnic groups).

Interestingly, although nitrification is a recognized potential problem in water systems practicing chloramination, nitrification control is required or encouraged in only 11 of 34 states that responded to a survey of drinking water programs conducted by the Association of State Drinking Water Administrators in March 2003 (see Table 2-5). This illustrates the need for state agencies to recognize the potential issues associated with chloramination and nitrification, and thereby prepare their utilities to deal with this potentially problematic issue.

Leaching

All materials in the water distribution system, including pipes, fittings, linings, other materials used in joining or sealing pipes, and internal coatings leach substances into the water. The processes that account for this include corrosion, dissolution, diffusion, and detachment. Taste and odor problems (Burlingame et al., 1994; Khiari et al., 2002) are the most likely outcome of leaching because most substances leaching into water from materials in the distribution system are non-toxic, present only at trace levels, or are in a form unlikely to cause health problems.

There are however, a few situations in which leaching may present a substantial health risk. By far the most significant is the leaching of lead from lead pipe, lead-containing solder, and lead service connections. Monitoring of lead in tap water and replacement of these lines are important components of the Lead and Copper Rule. Other materials used in distribution systems that have the potential for leaching include PVC pipes manufactured before about 1977. These are known to leach carcinogenic vinyl chloride into water at levels above the MCL (AWWA and EES, Inc., 2002). Cement materials have, under unusual circumstances, leached aluminum into drinking water at concentrations that caused death in hemodialysis and other susceptible patients (Berend et al., 2001). Because levels of aluminum normally present in drinking water can also threaten this population, the FDA has issued guidance for water purification pre-treatments in the U.S. for dialysis and other patients (*http://www.gewater.com/library/ tp/1111_Water_The.jsp*). Asbestos fibers may also be released from asbestos cement; the content of asbestos in water is regulated with an MCL, although utilities are not required to monitor for asbestos in the distribution system. Finally, excessive leaching of organic substances from linings, joints, and sealing materials have occasionally been noted. Some of these substances may support the growth of biofilms (Shoenen, 1986), such that their use should be limited.

For new materials, NSF International establishes levels of allowable contaminant leaching through ANSI/NSF Standard 61 (see Chapter 2). However, this standard, which establishes minimum health effect requirements for chemical contaminants and impurities, does not establish performance, taste and odor, or microbial growth support requirements for distribution system components. This is unfortunate because research has shown that distribution system components can significantly impact the microbial quality of drinking water via leaching. Procedures are available to evaluate growth stimulation potential of different materials (Bellen et al., 1993), but these tests are not applied in the United States by ANSI/NSF.

Internal Corrosion

Internal corrosion manifests as (1) the destruction of metal pipe interiors by both uniform and pitting corrosion (see Chapter 4) and (2) the buildup of scales of corrosion products on the internal pipe wall that hamper the flow of water (see Chapter 5). A large number of water quality parameters such as disinfectant residual, temperature, redox potential, alkalinity, calcium concentration, total dissolved solids concentration, and pH play an important role both in the internal corrosion of pipe materials and the subsequent release of iron. The products of corrosion may appear in water as dissolved and particulate metals, and the particles may cause aesthetic problems because of their color and turbidity if they are present in sufficient concentration. Metals such as lead and copper in tap water are governed by the Lead and Copper Rule; asbestos particles and iron particles with adsorbed chemicals such as arsenic (Lytle et al., 2004) are of concern because of possible health effects. The quality of distributed water must be controlled so that both corrosion and metal release do not cause water quality problems.

Scale Formation and Dissolution

Scale on pipe surfaces may form in distribution systems for a variety of reasons including precipitation of residual aluminum coagulant after filtration, precipitation of corrosion products, precipitation of corrosion inhibitors, and precipitation of calcium carbonate and silicate minerals. Scale that forms in a thin, smooth coat that protects the metal pipe by reducing the rate of corrosion is generally desirable, whereas uncontrolled precipitation can reduce the effective diameters of distribution pipes and can create rough surfaces, both of which reduce the hydraulic capacity of the system (as discussed in Chapter 5) and increase the cost of distributing water.

In terms of internal contamination events, rough surfaces and scales with reduced metals such as ferrous iron can increase problems with biofilms (Camper et al., 2003). That is, ferrous iron reacts with chlorine and monochloramine, reducing the effective concentration of disinfectant in the vicinity of biofilms. Furthermore, rough surfaces contain niches where microbes can grow without exposure to hydraulic shear. If the scale material is loosely attached to the pipe wall, such as some aluminum precipitates, hydraulic surges can result in substantial increases in the turbidity of tap water. Scales are also important because they can dissolve under some water quality conditions and release metals to the water in the distribution system. For example, Sarin et al. (2003, 2004) showed that iron scales release iron during flow stagnation, which then causes turbid and colored water. Dodge et al. (2002), Valentine and Stearns (1994), and Lytle et al. (2002) showed that uranium, radium-226, and arsenic, respectively, could be adsorbed to iron corrosion scales found in distribution systems. (In order for these metals to accumulate they must be present in the source water.) Lytle et al. (2002) showed that arsenic would accumulate on iron solids in distribution systems even when present in water at concentrations less than 10 μg/L. Aluminum and manganese solids can also adsorb metal contaminants and may subsequently release them because of changes in water quality. Research is needed to fully characterize this potential source of contamination related to internal corrosion and scale dissolution and to find ways to control it.

Other Chemical Reactions that Occur as Water Ages

Many water distribution systems in the United States experience long retention times or increased water age, in part due to the need to satisfy fire fighting requirements. Although not a specific degradative process, water age is a characteristic that affects water quality because many deleterious effects are time dependent. The most important for consideration here are (1) the loss of disinfectant residuals and (2) the formation of DBPs. The importance of water age is recognized in part by the survey of state drinking water programs where nearly all states that responded to the survey either required or encouraged utilities to minimize dead ends and to have proper flushing devices at remaining dead ends (Table 2-3).

Loss of Disinfectant Residual

Maintenance of a disinfectant residual throughout a distribution system is considered an important element in a multiple barrier strategy aimed at maintaining the integrity of a distribution system. It is generally assumed that the presence of a disinfectant is desirable because it may kill pathogenic organisms, and therefore the lack of a disinfectant is an undesirable situation. The absence of a disinfectant residual when one is expected may also indicate that the integrity of the system has been compromised, possibly by intrusion or nitrification. If the disinfectant is chloramine, its decay will produce free ammonia that could promote the onset of nitrification. Understanding the nature of the processes leading to disinfectant losses, especially when those processes lead to excessive decay rates, is important in managing water quality.

Loss of disinfectants in distribution systems is typically due to reduction reactions in the bulk water phase and at the pipe–water interface that reduce disinfectant concentration over time, although nitrification (in the case of chloramine) can also play a role. Dissolved constituents that can act as reductants in the aqueous phase include natural organic matter (NOM) and ferrous Fe(II) and manganous Mn(II) ions. These substances may occur in the water either as a result of incomplete removal during treatment, from the corrosion of pipe material (e.g., cast iron), or from the reduction of existing insoluble iron and manganese deposits. Disinfectants may also readily react with reduced forms of iron and manganese oxides typically found on the surface of cast iron pipes as well as with adsorbed NOM (Tuovinen et al., 1980, 1984; Sarin et al., 2001, 2004). Benjamin et al (1996) found that the accumulation of iron corrosion products at the pipe wall and the release of these products into the bulk water led to a deterioration of water quality. There have been several reports that the loss of chlorine residuals in corroded unlined metallic pipes (particularly cast iron) increases with increasing velocity (Powell, 1998; Powell et al., 2000; Grayman et al., 2002; Doshi et al., 2003). Correlative evidence for the role of corrosion in reducing disinfectant residuals was produced by Camper et al. (2003), who studied the interactions between pipe materials, organic carbon levels, and disinfectants using annular reactors with ductile–iron, polyvinyl chloride (PVC), epoxy, and cement-lined coupons at four field sites. They found that iron surfaces supported much higher bacterial populations than other materials.

Modeling efforts to understand disinfectant decay have been primarily empirical in nature or semi-mechanistic, and they have mostly addressed non-biological reactions. The primary purpose of these types of models is to serve as a predictive tool in managing water quality. Most modeling research has targeted the relatively fast reactions of free chlorine in the aqueous phase, predicting free chlorine decay versus hydraulic residence time using single system-specific decay coefficients. For example, Vasconcelos et al. (1996) developed several simple empirical mathematical models to describe free chlorine decay. Clark (1998) proposed a chlorine decay and TTHM formation model based on a competitive reaction between free chlorine and NOM. The model was validated against the Vasconcelos et al. (1996) data sets and found to be as good or better (based on r^2 values) than the models examined by Vasconcelos et al. (1996).

More sophisticated models have improved predictive management capabilities and are also useful as research tools in the elucidation of fundamental processes. Rossman et al. (1994) developed a chlorine decay model that includes first-order bulk phase and reaction-limited wall demand coefficients; this model is incorporated into EPANET[1]. The model developed by Clark (1998) was extended to include a rapid and slow reaction component and to study the effect of variables such as temperature and pH (Clark and Sivaganesan, 2001). Further extensions included the formation of brominated byproducts (Clark et al., 2001). McClellan et al. (2000) modeled the aqueous-phase loss of free chlorine due to reactions with NOM by partitioning the NOM into reactive and non-reactive fractions. Other models have incorporated reactions with reactive pipe surfaces that may dominate the loss pathways (Lu et al., 1995; Vasconcelos et al., 1997) as well as bulk phase reactions. Clark and Haught (2005) were able to predict free chlorine loss in corroded, unlined metallic pipes subject to changes in velocity by modeling the phenomena as being governed by mass transfer to the pipe wall where the chlorine was rapidly reduced.

Less studied has been the loss of monochloramine in distribution systems. Monochloramine, while generally less reactive than free chlorine, is inherently unstable because it undergoes autodecomposition. While autodecomposition occurs via a complex set of reactions, the net loss of monochloramine occurs according to the stoichiometry:

$$3 \cdot NH_2Cl \rightarrow N_2 + NH_3 + 3\,Cl^- + 3\,H^+ \qquad (1)$$

This reaction has been reasonably well studied (Valentine et al., 1998; Vikesland et al., 2000) and can be approximated (in the absence of other reactions) by a simple second-order relationship:

$$1/[NH_2Cl] - 1/[NH_2Cl]_o = k_{vcsc}\, t \qquad (2)$$

where k_{vcsc} is a rate constant describing the second order loss of monochloramine (Valentine et al., 1998). Its derivation involves the simplifying assumption that monochloramine decays by a mechanism involving the rate limiting formation of dichloramine that then rapidly decays. As such, k_{vcsc} is a combination of several fundamental rate constants and the Cl/N ratio. It can be simply calculated and used to predict monochloramine decay in the aqueous phase in the absence of other demand reactions. It should be pointed out that

---

[1] EPANET is a model developed by EPA that performs an extended period simulation of hydraulic and water quality behavior within pressurized pipe networks (see Chapter 7).

chloramine will decay more rapidly than predicted by this approach if significant amounts of demand substances other than ammonia are present in solution or if reactions with pipe walls are considered. Other demand substances can include NOM and reduced metals in the aqueous phase, as well as Fe(II) in pipe deposits.

Wilczak (2001) found that a sequential first-order empirical model best fit the East Bay Municipal Utility District's chloramine decay data. Palacios and Smith (2002) found that chloramine decay in San Francisco's water was consistent with a sequential first-order model, but that a simple first-order decay rate could be applied to the data due to the low organic matter concentrations. Duirk et al. (2002) developed a comprehensive aqueous-phase chloramine reaction model that accounts for both monochloramine autodecomposition as well as reduction by NOM that is similar in structure to that proposed by McClellan et al. (2000). Reaction of trace levels of free chlorine that equilibrate with monochloramine was a key mechanism accounting for slow monochloramine loss due to reaction with NOM. As a consequence, loss of chloramine should decrease with increasing pH because both autodecomposition and its reaction with NOM become slower. Table 6-1 summarizes the mechanisms for loss of a chloramine residual.

Disinfection Byproduct Formation

Formation of DBPs in distribution systems is attributable to reactions of chemical disinfectants with NOM either in bulk solution or associated with pipe deposits (Rossman et al., 2001). The importance of NOM associated with pipe deposits is based largely on evidence from controlled lab studies and is open to speculation, and must certainly be very system specific.

TABLE 6-1 Reactions that Reduce Chloramine Residual

Reaction	Stoichiometry
Chloramine auto-decomposition	$3\ NH_2Cl \rightarrow N_2 + NH_4^+ + 3\ Cl^- + 2\ H^+$
Oxidation of organic matter by chloramine	$0.1\ C_5H_7O_2N + NH_2Cl + 0.9\ H_2O \rightarrow$ $0.4\ CO_2 + 0.1\ HCO_3^- + 1.1\ NH_4^+ + Cl^-$
Reaction of chloramine with corrosion products at pipe wall	$0.5\ NH_2Cl + H^+ + Fe^{2+} \rightarrow$ $Fe^{3+} + 0.5\ NH^{4+} + 0.5\ Cl^-$
Oxidation of nitrite by chloramine	$NH_2Cl + NO_2^- + H_2O \rightarrow$ $NH_3 + NO_3^- + HCl$

SOURCE: Reprinted, with permission, from Woolschlager et al. (2001). © 2001 by IWA Publishing.

Most studies have focused on formation of halogenated DBPs produced from reactions of NOM with free chlorine, especially those DBPs that are currently regulated by the EPA—trihalomethanes (THMs) and haloacetic acids (HAAs). However, over 600 potentially harmful DBPs have been identified (Richardson, 1998) including both chlorinated and brominated compounds. Brominated compounds arise from the oxidation of bromide which can be an important factor in determining DBP speciation even when found at the sub-milligram per liter level. Many of the DBPs formed in chloraminated systems are the same as those observed in systems practicing chlorination (Figure 6-1). This may be a consequence of similar formation mechanisms involving free chlorine or attributable to the practice of prechlorination prior to ammonia addition and subsequent chloramine formation. However, the rates of formation of most DBPs are much slower in chloraminated systems, resulting in the reduced formation of many DBPs, especially THMs.

Table 6-2 lists some of the DBPs rated as high priority and observed in a recent comprehensive survey of 12 full-scale treatment plants in the United States in 2000. The halogenated DBPs detected in this study have included mono-, di-, tri-, and/or tetra- species of halomethanes (HMs) (including iodinated species); haloacetonitriles (HANs); haloketones (HKs); haloacetaldehydes (HAs); and halonitromethanes (HNMs). The presence of bromide resulted in a shift in speciation for the trihalomethanes (THMs) and haloacetic acids (HAAs). Brominated DBPs for other classes of DBPs (HANs, HKs, HAs, HNMs) were also detected. Chloramination formed certain dihalogen-substituted DBPs (HAAs, HAs) preferentially over related trihalogenated species. In addition, chlorine dioxide produced dihalogenated HAAs (Richardson et al., 2004). Recently several DBPs have been identified as unique to chloraminated systems. These include N-nitrosodimethylamine (NDMA) (Choi and Valentine, 2002 a,b, Mitch and Sedlak, 2002), cyanogen chloride, and several iodohaloacetic acids, none of which are currently regulated at the federal level. The state of California has, however, established a notification level of 10 ppb in drinking water for NDMA, a potent carcinogen (*http://www.dhs.ca.gov/ps/ddwem/chemicals/NDMA/NDMAindex.htm*). NDMA seems to be a relatively widespread DBP and may become more prevalent as the use of chloramination increases.

Given the relatively high reactivity of free chlorine with NOM, it is not surprising that a significant amount of DBPs is formed in the water treatment plant as the result of primary disinfection (i.e., disinfection at the treatment plant to meet CT requirements). DBP formation, however, continues in the distribution system, as shown in Figure 6-2. Based on an evaluation of data from utilities that participated in the Information Collection Rule (ICR) and that use surface water as their source, TTHMs increased through distribution systems on average about 50 percent when chlorine was used to maintain the distribution system residual (McGuire and Graziano, 2002). Similar results were obtained for chloraminated distribution systems, mainly because these systems had water with higher TTHM precursors than those utilities that were using free chlorine. Chloramine-specific DBPs (like N-nitrosodimethylamine cyanogen chloride,

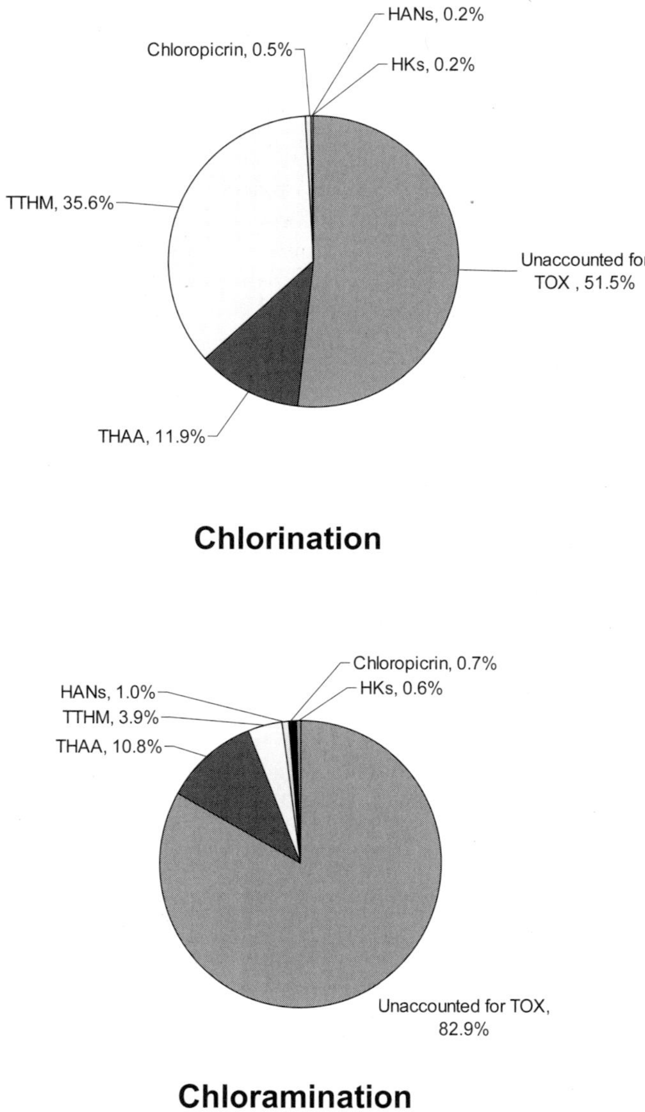

FIGURE 6-1 Comparison of Halogenated DBPs. SOURCE: Richardson et al. (2004).

TABLE 6-2 Chlorination and Chloramination Disinfection Byproducts

Chloroform	Dichloroacetonitrile	Acetaldehyde	
Bromodichloromethane	Bromochloroacetonitrile	Propionaldehyde	**THMs** -$CHCl_3$
Chlorodibromomethane	Tribromoacetic acid	Butyraldehyde	**HAAs** -$CHCl_2COOH$
Bromoform	Trichloroacetonitrile	Valeraldehyde	**HANs** -$CHCl_2CN$
Chloroacetic acid	Dichloroacetonitrile	Hexanal	**Chloropicrin (halonitro-methanes)** -CCl_3NO_2
Bromoacetic acid	Bromochloroacetonitrile	Heptanal	
Dichloroacetic acid	Dibromoacetonitrile	Octyl aldehyde	**Cyanogen Halides** -CNCl
Trichloroacetic acid	Dichloropropanone	Benzaldehyde	
Bromochloroacetic acid	Trichloropropanone	Nonyl aldehyde	
Bromodichloroacetic acid	Chloropicrin (chloronitromethane)	Decryl aldehyde Formaldehyde	
Dibromoacetic acid	TOX, TOCl, TOBr (as Cl)	Glyoxal	
Chlorodibromoacetic acid	Cyanogen Chloride	Methylglyoxal	
Tribromoacetic acid	Dichloroaldehyde	Acetaldehyde	
Trichloroacetonitrile		Propionaldehyde	

Note: this table is not comprehensive, as new DBPs are discovered on a regular basis.
SOURCE: Reprinted, with permission, from Richardson and Krasner (2003).

iodohaloacetic acids) are expected to form primarily in distribution systems since chloramine is not usually used during primary disinfection (although ammonia is sometimes added in the treatment plant to stop THM and HAA formation). Finally, haloacetic acid levels are also expected to increase in the distribution system, but not to the same degree as THMs.

It should be noted that processes may occur in distribution systems that cause a loss of DBPs. For example Baribeau et. al. (2006) showed that the formation of several haloacetic acids did not increase with water age in a chlorinated distribution system (Figure 6-3). Speight and Singer (2005) correlated HAA reduction to a reduction in chlorine and suggested that the observed HAA loss was due to biodegradation that was otherwise inhibited in the presence of

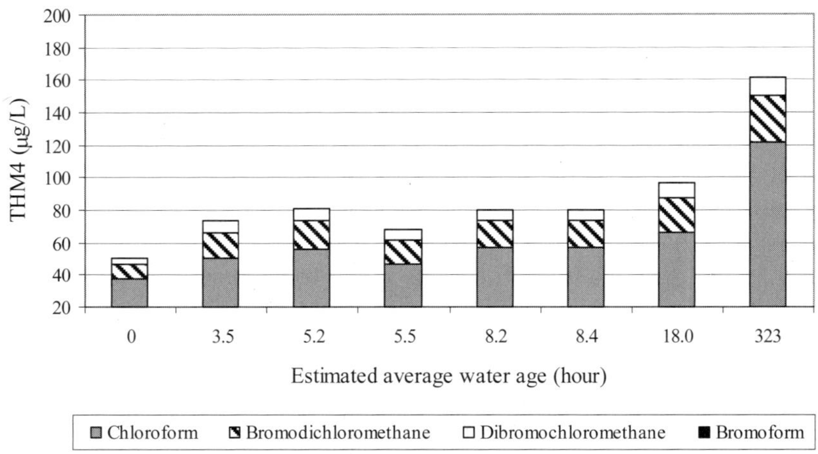

FIGURE 6-2 Changes in total trihalomethanes in a system with free chlorine with water age. SOURCE: Reprinted, with permission, from Baribeau et al. (2006). © 2006 by American Water Works Association.

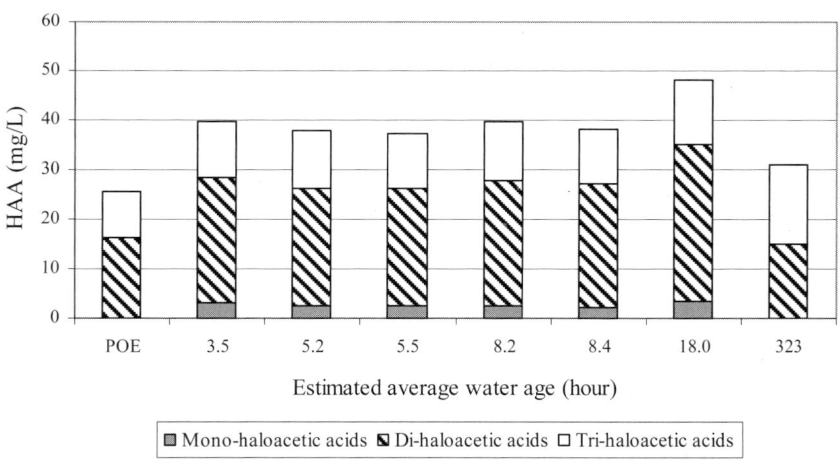

FIGURE 6-3 HAA concentrations in a chlorinated system as a function of water age. SOURCE: Reprinted, with permission, from Baribeau et al. (2006). © 2006 by American Water Works Association.

chlorine. However, interpreting field data can be difficult, and changes in DBP concentrations may alternatively be attributable to changes in treatment plant operation (Pereira et al., 2004). The nature of DBP decay processes are not well established in actual distribution systems, but laboratory studies suggest that these may include biodegradation (Baribeau et al., 2005a), hydrolysis (Zhang and Minear, 2002), and reduction by reduced forms of iron (Chun et al., 2005; Zhang et al., 2004).

DBP modeling efforts can be categorized as empirical and semi-mechanistic. Motivation for modeling includes estimating the extent of the problem from easily measured parameters, predicting the influence of treatment practices aimed at reducing DBPs, and as a tool in establishing fundamental mechanisms. Amy et al. (1987) correlated DBP formation to a number of important variables that include chlorine dosages, DOC and bromide concentrations, temperature, and contact time. Harrington et al. (1992) used a similar approach to develop an empirical model to simulate THM and HAA formation during water treatment. More recently, semi-mechanistic kinetic models have been developed that couple disinfectant loss to the formation of selected DBPs. McClellan et al. (2000) proposed a model for the formation of THMs in chlorinated water assuming a fixed number of chlorine-consuming and THM-forming sites per mg C in the aqueous phase. Duirk and Valentine (2002) used a similar approach to model dichloroacetic acid formation from the reaction of monochloramine with NOM. As already stated, no efforts have as yet included the specific role of NOM on deposit/pipe surfaces that may be required to adequately model DBP formation in distribution systems. In spite of the limitations, considerable progress has been made in using models to explain observations and make simple predictions about the influence of treatment practices and distribution system residence time. Continued effort is needed to refine these models by including unifying principles that are not system specific and an improved description of all pertinent phenomena.

It should be pointed out that while chlorine demand may be a useful measurement to correlate with DBP formation in the bulk water (Gang et al., 2002), this would not be the case if the demand were governed by inorganic constituents or by reactions with deposit materials (corrosion products, etc.).

DETECTING LOSS OF WATER QUALITY INTEGRITY

Distinguishing the loss of water quality integrity due to internal changes as opposed to changes brought about by external events is very difficult because there are few parameters that can be conclusively linked to internal contamination. Routinely monitored parameters such as temperature, pH, disinfectant residual, and even microbial constituents cannot differentiate between external and internal contamination. Other less routinely monitored constituents, including dissolved metals, turbidity, total organic carbon, synthetic organic compounds, or nuisance organisms such as invertebrates may also not be definitive for external vs. internal sources. However, if utilities have a good understanding of their distribution system and its points of vulnerability, value judgments as to the potential source of the detected contaminants can be made. With these limitations in mind, the following sections on detecting physical, chemical, and biological changes are presented. When there is a clear distinction between the ability of the monitoring method to distinguish between internal and external contamination, these methods are thereby identified.

Detection of Physical and Chemical Changes

Taste and Odor

Tastes and odors detectable by the consumer are a common indication of a loss of water quality integrity (McGuire, 1995). In fact consumers may only complain about the loss of water quality if they detect taste and odors (Watson, 2004). Fortunately, methods exist to directly evaluate the flavor and odor of tap water (Krasner et al., 1985; Dietrich et al., 2004; APHA, 2005) and a guide exists to determine possible sources within the distribution system and customers' premises (McGuire et al., 2004).

Because most drinking waters in the United States contain a total chlorine residual, the taste and odor of tap water might be described as "chlorinous." Whether this is noticeable to the water-consuming public depends on the chlorine species present, the concentration of the residual, and the temperature of the tap water. Other causative agents of tastes and odors in drinking water are usually metals, volatile organic chemicals, and microbial activity, with the latter being the most prevalent (APHA, 2005). A very common cause is open storage reservoirs where algae have been allowed to grow within the water as well as along the sides of the basins. These algae can produce earthy, musty, grassy, fishy, decaying vegetation and similar odors. Watson and Ridal (2004) credited taste and odors to periphyton and more specifically to certain cyanobacteria present in biofilms, as well as to the presence of dreissenid mussels in the Great Lake region. Skjevrak et al. (2004) detected the presence of ectocarpene, dictyipterenes, beta-ionone, menthol, menthone, and other VOCs in biofilms within distribution systems. Furthermore, bacteria such as actinomycetes can give rise to geosmin, and other microorganisms such as certain fungi have been associated with consumer complaints about taste and odor.

Much of the biological activity that causes taste and odor problems is indirect. Taste and odors problems may arise as a result of bacterial processes in certain types of pipes, such as iron, copper, and lead (Geldreich, 1996). In water systems with chlorophenols or bromophenols, biological activity (particularly fungal) can convert these compounds to very odorous chloro/bromo-anisoles that have much lower thresholds of odor detection than the original compounds (Bruchet, 1999; Montiel et al., 1999). It is also possible that other chlorinated and oxidant-derived byproducts can be produced or allowed to increase in the distribution system to the point where they begin to be detectable by customers. Finally, if a source water contributes sulfur or iron to the distribution system (such as from a groundwater supply), biological activity in the distribution system can produce compounds that change the taste of the water.

Tastes and odors may be associated with external contamination events, such as permeation and intrusion. Among the compounds most likely to present a taste and odor problem stemming from an external contamination event are gasoline additives or constituents, soluble components of soil, and compounds found in sewage.

Changes in taste and odor can occur anywhere in the distribution system that the chlorine residual deteriorates and the water becomes stagnant, such as in storage tanks, at dead-end water mains, and behind closed valves. Also in stagnant areas of the distribution system where corrosion conditions release iron into the water, the iron may be detected by customers both visually and by taste. Interestingly, most nuisance tastes and odors that cause customer complaints originate within customers' premises (except for those that come from source water such as geosmin, 2-methylisoborneol, and certain chemical spills) (Suffet et al., 1995; Khiari et al., 2002). Common causes are stagnant plumbing (musty odors from biological growth), backflow events (various types of chemical odors), hot water heater odors (hydrogen sulfide from biological activity in hot water tanks), and corrosion of plumbing materials (release of copper, zinc, and iron). New plastic pipe can leach odors for a period of time.

Within the main distribution system, new pipe and facilities need to be checked for their contributions to potential off-odors before they are released for use. Ductile iron pipe that is lined with cement-mortar might have an asphaltic coating that can leach volatile organic chemicals into the water if it has not cured sufficiently. New pipe joint lubricant can also impart aldehyde-type odors to the water. New linings of storage tanks also need to be cured adequately before being placed into service.

Finally, the stability of the chlorine or chloramine residual is important to controlling undesirable tastes and odors. Blending of source waters or boosting of disinfectants that is not well controlled can produce dichloramine (which is more odorous than monochloramine). It has also been shown that when a system uses chlorine dioxide as a primary oxidant, chlorite in the distribution system can react with free chlorine to reform chlorine dioxide that (1) can give a strong chlorinous odor at the tap and (2) can be released into the air of a home and react with volatile organic chemicals (such as from new carpet or paneling) to create cat urine or kerosene type odors (Dietrich and Hoehn, 1991).

Color and Turbid Water

Colored and turbid water at the tap is a strong indication that corrosion of iron, iron release from scales, and post precipitation of aluminum salts are not being controlled. The presence of colored or turbid water is therefore indicative of changes in quality due to internal contamination. Iron released from the pipe wall as Fe^{2+} will diffuse into the bulk water where it is oxidized by oxygen or disinfectant, and then precipitates as ferric oxyhydroxide particles that cause color and turbidity (Lytle and Snoeyink, 2002; Sarin et al., 2003). The effect of this process may be made worse if the particles settle during periods of low flow and are then resuspended by hydraulic surges. Post precipitation of aluminum may result in particles that are loosely attached to the pipe wall and are suspended during hydraulic surges. This type of aluminum precipitate can be the cause of turbid or dirty water.

Dissolved and Particulate Metal Concentrations

If water leaving the treatment plant has metal concentrations that meet regulatory requirements, and if these levels increase in transit through the distribution system (typically iron) or in premise plumbing (lead, copper), then it may be assumed that leaching and internal corrosion are occurring. Although it is beyond the scope of this report to discuss the details of the Lead and Copper Rule, this is the only example of a regulation that specifically addresses the internal degradation of water quality in a distribution system. In the case of iron, elevated levels are more likely to be associated with secondary standards and aesthetic concerns rather than with a specific public health threat.

Disinfectant Residual and Disinfection Byproduct Measurements

Measurements of disinfectant residuals and DBP concentrations often accompany one another and are routinely practiced using a number of standard analytical methods. Sudden temporal increases in disinfectant loss indicate a sudden change in water quality or system characteristics. For example, this might be due to significant input of a reactive contaminant due to a cross-connection or rapid onset of nitrification in the system. Unexpected spatial losses point to problems associated with specific elements of the distribution system such as a zone where internal corrosion is excessive or where nitrification is occurring.

Identification of the causes of excessive disinfectant loss and DBP formation involves a combination of bench studies and field observations. The significance of what is considered excessive loss must be gauged against what is considered "normal" or not excessive. Free chlorine is stable for many days in water containing no reactive constituents such as NOM. The applied dose is then a bench mark for comparison. Simulated Distribution System (SDS) jar testing using water obtained from the point of entry into a distribution system or at other points can be used to determine the rate of disinfectant loss attributable to bulk phase reactions as well as DBP formation. This requires only measurements of the free chlorine and DBP concentrations as a function of contact time. The rate of chlorine loss and DBP formation in the SDS can then be compared with losses observed in the system between points of known hydraulic residence time (acknowledging that residence time at a given point is actually a distribution of values and not easy to determine—see Chapter 5). Differences between the lab and field must be attributable to processes occurring inside the distribution system, most likely at the pipe–water interface.

Determining if chloramine loss is "excessive" is more complicated. The SDS jar test will be indicative of how fast the chloramine disappears in the *bulk* phase but will not by itself reveal the mechanism if the loss is unexpectedly high. This must also be compared to the rates of loss from autodecomposition, which can be predicted using the second order relationship previously discussed or perhaps measured in "distilled water" at the same pH as the system of interest. If the bulk water reactions are much higher than expected in clean water or than predicted, then one can presume the presence of significant amounts of reactive substances such as ferrous iron or NOM. The loss rate in the bulk phase can then be compared to values determined by measuring chloramine concentrations in the system at points of known hydraulic residence times. If the rate of loss determined by the field measurements are much higher than the bulk loss rates, then presumably this is due to reactions at the pipe–water interface. These include biological nitrification and reaction with reduced iron.

Indicators of Nitrification

Smith (2006) recently summarized important parameters (Table 6-3) that can be used as indicators of biological nitrification—one phenomenon that is directly associated with internal changes in water quality. The most important indicators (after loss of residual) are formation of nitrite and nitrate, loss of ammonia, and a decrease in pH. An increase in heterotrophic plate count may also

TABLE 6-3 Usefulness of Water Quality Parameters for Distribution System Nitrification Monitoring

Parameter/Usefulness		
Very Useful	**Useful**	**Limited Usefulness**
Total Chlorine	Nitrate	Dissolved Oxygen
Nitrite-N	Total ammonia-N	TOC
Free ammonia-N	HPC-R2A	Hardness
Temperature	pH	Alkalinity
Free Chlorine*		AOB**

* Very useful during breakpoint chlorination (not for routine monitoring)
** AOB=Ammonia oxidizing bacteria. Of limited usefulness until rapid, inexpensive enumeration methods become available.
SOURCE: Reprinted, with permission, from Smith (2006). © 2006 by American Water Works Association.

indicate the growth of nitrifying organisms in the system. However, since biological nitrification can be associated with biofilms (Regan et al., 2003), an absence of nitrifying organisms in the bulk phase does not necessarily indicate the absence of nitrification.

Detection of Biological Changes

Several biological constituents, some of which are part of compliance monitoring, can be used to detect the loss of water quality integrity due to both internal and external contamination events. The applicability of each group of organisms for assessing internal changes in water quality is described in this section.

Heterotrophic Plate Counts

Since the end of the 19th century, heterotrophic plate counts (HPC) have been used as an indicator of the proper functioning of treatment processes (Bartram et al., 2003). By extension, HPC have also been used as an index of water quality and safety in the distribution system, and the method continues to be used in many countries as an index of regrowth (an internal event) of microorganisms within the distribution system. Although it is difficult to establish the exact contribution of suspended bacteria vs. proliferation and release of biofilm cells if increases in HPC are observed, there is evidence that biofilm growth and detachment can be the source of elevated bacterial numbers. Published accounts by van der Wende et al. (1989) and LeChevallier et al. (1990) demonstrated that elevated bacterial counts in water could not be attributed to replication of suspended cells, but rather was due to biofilm growth on pipe surfaces. Accordingly, the Surface Water Treatment Rule allows HPC levels to be used as a surrogate for a "detectable" residual for regulatory compliance purposes, provided that HPC is less than or equal to 500 colony forming units (CFU)/ml. Although other countries do not set specific numerical limits for HPC (Robertson and Brooks, 2003), the European Union has a recommendation of 100 CFU/ml. In addition, the World Health Organization is presently debating whether or not HPC counts should be included in their regulations on water quality.

Linking changes in HPC with a meaningful water quality variable can be very difficult. Many conditions, such as an increase or decrease in the organic carbon concentration, stagnation, loss of disinfectant residual, and/or nitrification, will result in an increase in HPC. Another cause of observed rises in HPC might be that chorine-injured organisms regain culturability, even though their actual numbers have not changed. It is also possible that increased HPC is not due to bacterial growth in the distribution system but may originate from an external contamination event. In a study where HPC was analyzed in a distribution system with groundwater as its source, the concentrations of HPC varied with distance in the distribution system (Pepper et al., 2004), suggesting that most of the HPC bacteria originated from the distribution system rather than from the source water, although it could not be determined with certainty. Intrusion could account for increased HPC, either by fomenting bacterial regrowth as a result of nutrient intrusion, or simply by increasing HPC as a result of bacterial intrusion. Therefore, differentiating between possible causes for changes in HPC cannot be done without thorough monitoring of the system over a long period of time, and without having a thorough knowledge of the microbial diversity, the physiology, and the ecology of the microbiota being detected (Szewsyk et al., 2000).

A further complication is that the concentrations of HPC bacteria in a water sample are dependent on the media being used for their enumeration. R2A agar has been shown to give the highest numbers; however, the importance of this with regards to the use of different media vis-à-vis the detection of anomalies in the system is yet to be determined. Because of the variety of incubation temperatures and media used, it is difficult to compare within or between systems.

Regardless of the detection method employed, it is possible to use HPC bacteria as a general indicator of distribution system hygiene and performance. This approach requires that samples be taken at regular spatial intervals along the distribution system at time points that reflect the hydraulic residence time of the water in that section of the pipe. If increased bacterial numbers are seen in the same "packet" of water in a plug flow system, it is evidence that deterioration in water quality has occurred. With reasonable forensic investigation, the utility can then determine if the increased counts are due to internal vs. external events. This type of system monitoring is already performed by industries (other than water supply) that rely on high-quality water for manufacturing purposes.

Coliforms

Under ideal circumstances, the presence of coliforms in a drinking water sample should indicate external fecal contamination of the water supply, which is the main premise behind the current Total Coliform Rule. Although this concept has served the industry reasonably well, it is not without flaws. Methods may not be sufficiently sensitive for detection, and sample collection may give false positive (e.g., contaminated faucet screens) or false negative (disinfectant residual not neutralized) results. Studies of coliform presence in distribution systems indicate that coliforms may be introduced via treatment breakthrough as well as by intrusion events, main breaks, and other external contamination events (Besner et al., 2002). Furthermore, on occasion, coliforms have been shown to multiply in biofilms, contributing to their detection in drinking waters (LeChevallier et al., 1996). These same authors indicated that there was a correlation between coliform occurrence and variables such as temperature, AOC levels, and disinfectant type being used. Therefore, it is often not possible to determine if a coliform-positive sample is the result of an external contamination event vs. regrowth of the organism within a biofilm. This problem is acknowledged in the Total Coliform Rule, in which the EPA allows for a variance from the regulation (40CFR, Code of Federal Regulations, 1993) "for systems that demonstrate to the State that the violation of the total coliform MCL is due to a persistent growth of total coliforms in the distribution system rather than fecal or pathogenic contamination, a treatment lapse or deficiency, or a problem in the operation or maintenance of the distribution system."

Although the term coliform is used, it should not be forgotten that the group includes several different genera (see Table 3-2) which may survive/regrow differently under different conditions. When *E. coli* is found in drinking water, it is generally believed to be associated with an external contamination event (see references in Tallon et al., 2005) and linked to fecal contamination rather than an internal/biofilm source. Consequently, *E. coli* is one monitoring tool that is used to distinguish between internal and external contamination. However, it should be noted that *E. coli* is less resistant to disinfectants than some other pathogenic bacteria, viruses, and the protozoan cysts/oocysts. Thus, its absence does not indicate an absence of pathogens.

Other Indicators of Fecal Contamination

There is a great deal of interest in identifying alternative indicators for waterborne pathogens, as evidenced by the recent publication of the NRC report *Indicators for Waterborne Pathogens* that summarizes the most recent insights on the topic (NRC, 2004). An overview of some of these organisms (*Clostridium perfringens*, *Enterococci* and fecal streptococci, *Bacteroides spp.*, *Bacillus subtilis* spores, *Pseudomonas spp.*, *Aeromonas spp.*, *Staphylococci*, HPC bacteria, hydrogen sulfide producers, and bacteriophages) is given in Table III of Tallon et al. (2005). As noted in the table, most of these organisms are not entirely specific to fecal contamination and/or suffer from difficulties in detection. As a case in point, the spores of *Clostridium perfringens* have been proposed as an indicator of fecal contamination of water and have been used as surrogates for assessing the efficacy of water treatment processes designed to remove viruses and the cysts/oocysts of *Giardia* and *Cryptosporidium spp.* (Payment and Franco, 1993; Venczel et al., 1997). Because the spores are more resistant to disinfection and the environment, their responses to these stresses are less pronounced than vegetative bacteria. This indicator is not always specific for fecal contamination, however, because it can be found in soils and sediments as part of the natural flora. Nonetheless, to date it has not been identified as a part of the natural flora of drinking water distribution systems, and as such, may be a reasonable indicator of external contamination regardless of whether it arises from fecal contamination or the soil.

Direct Detection of Pathogens

It would be optimal to directly measure pathogenic microorganisms or molecules specific for them instead of relying on indicator organisms. At the present time, however, this approach is not realistic for several reasons. There is a wide diversity of potential pathogens, ranging from viruses to bacteria to fungi to protozoa, and each group of organisms represents a unique challenge for detection. For many viruses and protozoa, there are no appropriate lab-based culturing methods. If culturing methods are possible, selective media can reduce the chances for recovering stressed organisms. Organisms can be present in such low numbers that direct sampling is not sufficiently sensitive, and concentration methods are also prone to error. With the advent of improved molecular methods in the future, some of these limitations may be overcome. A review of the issues associated with implementing these novel methods for pathogen detection as well as problems associated with conventional approaches for assessing microbial water quality have been published previously (NRC, 2004). This report also points out that even if these new methods show promise, standardization and validation are critical if the methods are to be used in a regulatory context.

Summary

Water quality integrity needs to be evaluated rapidly and, if at all possible, using in-line, real time methods (as discussed in Chapter 7). Unfortunately, current microbial detection methods do not lend themselves to this approach, given the many types of microbes possible, the limitations of individual indicators, and the rapidity and sensitivity of certain methods. HPC can be a useful parameter, but only if frequent monitoring is carried out and only if anomalous levels are detected (Robertson and Brooks, 2003). The levels that have been proposed as action levels or guidelines by various industries are too site-, season-, and method-specific to be generally applied. In any case, HPC counts do not lend themselves to on-line monitoring because it may take up to seven days for results, depending on the media being used. The concentration of HPC within the distribution system may be an indirect measure of AOC and BDOC in the water as organic carbon may be a reason for HPC regrowth (Robertson and Brooks, 2003). It may be tempting to suggest that AOC and/or BDOC be measured in place of HPC, but these assays are not as easily applied by most utilities and often take longer than the incubation period required for the HPC measurements.

Because many of the organisms present within distribution systems cannot be detected using the plate count methods typical for HPC (Block, 1992; Leclerc, 2003), efforts have been made to use rapid and relatively inexpensive microscopic techniques to visualize all microbes present employing fluorescent DNA stains, such as acridine orange direct counts (AODC), 4'-6-diamidino-2-phenylindole (DAPI), SYBER® Green, propidium iodide, etc.). Problems associated with these methods are that both dead and live cells are detected, there is interference from inorganic constituents, background autofluorescence can occur, and the level of detection may not be sufficiently low. Other dyes have been used by some investigators (McFeters et al., 1999), but their usefulness for routine monitoring remains to be seen. It is also possible to use specialized equipment such as the ChemScan RDI/ Scan RDI™ or flow cytometers to quantify fluorescently stained organisms. In these cases, the equipment is expensive, and for flow cytometry, extensive optimization may be required.

Indirect methods could be used to determine water quality integrity; for example adenosine triphosphate (ATP) methods have been proposed, but they are rather expensive and do not lend themselves to routine monitoring. Although hand-held ATP photometers have been developed and are currently being used in the food industry, the usefulness of these methods in the water industry remains in question, given their low level of sensitivity and the relatively dilute nature of finished water.

Molecular methods for the detection of microbes are still far from routine. Although promising, PCR-based methods employing specific primers for a suite of targeted organisms or for 16S ribosomal DNA may suffer from the same problems as HPC and coliform counts. New analytical methods have been used for early detection of chemical agents in water, and many approaches are being developed in response to the need to detect chemical bioterrorism and warfare agents. Calles et al. (2005) describe photoionization and quadrupole ion trap, time-of-flight mass spectrometry as a means of detection of certain hazardous compounds in a fast and sensitive manner. Similar approaches for biological contaminants and indicator organisms are possible, but considerably more research and development will be needed to ensure that the methods are reliable. Even with the possibility for increased sensitivity and results that can be obtained more quickly, the overall limitations on the use of coliform bacteria to signify fecal contamination still exist.

Indeed, it is improbable that one indicator (such as coliforms) will serve all needs because of the varied sources of contamination, the different types of organisms involved, the impact these events have on public health, and the changing regulatory climate. NRC (2004) proposes a tiered approach for microbial monitoring. The first level is routine monitoring of common indicators to provide an early warning of a health risk or a change from background conditions that could pose a health risk. These methods should be rapid, reasonably inexpensive, and low in cost. If a potential problem is identified at this level, it should be followed by more detailed studies to assess the extent of public health risk. This might involve expanded sampling and using a more tailored detection method for indicators or even direct measurement of pathogens with molecular methods. The third level is a detailed investigation of the source of contamination so that it can be ameliorated. The need for standardized methods decreases as the investigation moves through the three phases, such that at the third level, specialized research tools may be required.

MAINTAINING WATER QUALITY INTEGRITY

Maintenance of water quality in the distribution system requires diligent attention by treatment plant personnel and those in charge of the distribution system. A delicate balance must be achieved in order to comply with relevant regulations while taking into account the detrimental effects that may occur as water travels through miles of pipes to the consumer. As regulations become increasingly more complicated and stringent, it is more difficult for utilities to balance the requirements. For example, there may be a need to bolster disinfectant residuals at various points throughout the distribution system, but this may lead to unacceptable levels of DBPs. The issue is even more complicated in premise plumbing where long periods of stagnation ultimately influence the water quality that the consumer receives. The following section describes methods and processes for maintaining the water quality integrity of potable water. The committee believes that nitrification control is best accomplished by maintaining an adequate disinfection residual and by booster disinfection, both of which are discussed below. The reader is referred to AWWA (2006) for further details on nitrification control.

Adequate Disinfection Residual

Maintenance of a disinfectant residual in the distribution system is required under the Surface Water Treatment Rule and has been designated as the best available technology for compliance with the Total Coliform Rule. The practice of carrying a disinfectant residual through the distribution system is also integral to the control of biofilms. This residual is intended to act as a prophylactic in the event of intrusion or backflow of a contaminant within the distribution system. With regard to the latter, the difficulty arises in determining what constitutes an adequate residual. There are examples of disease outbreaks caused by external contamination of a distribution system with a virus (Levy et al., 1998) and *Giardia* (Craun and Calderon, 2001); in both cases, a disinfectant residual was present or required. A few studies have examined the persistence of pathogens introduced into water carrying a disinfectant residual. Using sewage at various concentrations, Snead et al. (1980) demonstrated that there was no pathogen inactivation by chorine at 0.2 mg/L when 0.01 percent sewage was added. Payment (1999) reported inactivation of indigenous coliforms in sewage only if the chlorine concentration was greater than 0.6 mg/L. These low levels of chlorine are typical in sections of distribution systems with more advanced water age and in premise plumbing. More recent work has involved modeling of potential intrusion events to obtain insight on how chlorine and monochloramine inactivate organisms under relevant scenarios. Propato and Uber (2004), whose approach incorporates the variable factors of intrusion location along with mixing and contact time prior to consumption to simulate inactivation of pathogens in the distribution system, showed that monochloramine did not provide any protection against contamination by *Giardia* under realistic conditions. In another modeling study, chloramine and chlorine were evaluated for their ability to inactivate *Giardia* and *E. coli* O157:H7 under a range of water quality conditions in the distribution system (Baribeau et al., 2005b). This group demonstrated that chlorine at a level of 0.5 mg/L would not inactivate *Giardia*, but was sufficient to disinfect *E. coli* when a simulated sewage intrusion event was 0.2 percent of the total flow. In contrast, monochloramine under the same conditions performed poorly in reducing the *E. coli* counts in a reasonable amount of time. In both of these modeling studies, there were inherent assumptions that remain to be verified under field conditions. However, the insights obtained, along with the laboratory and disease outbreak data, demonstrate that criteria for disinfecting organisms introduced during external contamination events is not well understood.

Federal regulations regarding the maintenance of a distribution system residual require, for large systems, a detectable free or combined residual in 95 percent of the sample results analyzed during a one-month period, or demonstration of a heterotrophic plate count less than 500 CFU/mL. Smaller systems have reduced monitoring requirements. Some states have chosen to define "detectable residual" including specific requirements for chlorinated and chloraminated distribution systems. For example, Texas requires 0.2 mg/L for chlorinated water and 0.5 mg/L for chloraminated water (TECQ, 2005). The North Carolina regulations are even more stringent (NCDENR, 2004), in that when chlorine is the single applied disinfectant, the residual disinfectant in the distribution system must be at least 0.2 mg/l as free chlorine in at least 95 percent of the samples each month. When ammonia and chlorine are applied together as disinfectants, the residual disinfectant must be at least 2.0 mg/l as combined chlorine in at least 95 percent of the samples each month.

In addition to the state regulations mentioned above, the literature implicates and in some cases makes suggestions for appropriate disinfectant levels. For example, systems that maintained dead-end free chlorine levels of < 0.2 mg/liter or monochloramine levels of < 0.5 mg/liter had substantially more coliform occurrences than systems that maintained higher disinfectant residuals (LeChevallier et al., 1996). This committee did not reach consensus on recommending specific numbers, given the need for additional research on the level of protection provided by maintenance of a disinfectant residual and the large variability in contact time between points of contaminant entry and consumers, both within an individual distribution system and between systems. To date, most studies have examined rather large amounts of contamination (1 percent or more), and studies have not been done in flowing pipes, where the hydrodynamics would be important. It is not clear what level of microbial inactivation would be required during an event of a given magnitude, nor how that might vary depending on the type of organisms in the vicinity of the distribution system. Given that current federal regulations for surface water systems require a "detectable" disinfectant level within the distribution system, each utility should set targets depending on the expected loss of residual in the system. This loss will depend on (1) the extent of treatment to minimize disinfectant demand in the bulk water and (2) distribution system operational practices that minimize the disinfectant demand of the pipe walls and minimize water age (such as turnover of stored water).

Booster Disinfection

Reactions can reduce disinfectant residuals within distribution systems, such that some utilities have chosen to use booster chlorination or booster chloramination to increase residuals at susceptible locations. Using additional points of disinfectant application in the distribution system can reduce the amount of chlorine added at a treatment plant for the purpose of maintaining the distribution system residual. This, in turn, has the potential to limit DBP formation and subsequent exposure of those consumers' drinking water from taps close to the initial source. The booster disinfection simultaneously increases protection (in terms of the presence of a residual) for those drinking water from taps with longer hydraulic residence times.

An important consideration for implementing booster chlorination or chloramination is the proper location of facilities. Kirmeyer et al. (2000) list the following criteria in selecting the location of booster stations:

- The location should be such that a relatively large volume of water can be disinfected.
- The water to be treated travels in one direction.
- The chlorine residual in the water has begun to decrease, but has not totally dissipated.
- The chlorine can be applied uniformly into the water.
- The location is acceptable by neighbors and is easily accessible for chemical delivery vehicles with room for chemical storage and feed equipment.
- Power is readily available.
- Communications systems are readily available for the SCADA system.
- Flow and/or residual pacing can be used.
- Safety concerns can be addressed.
- For a common inlet/outlet line, chlorine should be injected as the storage facility is filling, although mixing the chlorine throughout the contents may be difficult.

Booster Chlorination

Booster chlorination is an alternative for maintaining a residual in drinking water systems where substantial disinfectant degradation occurs with travel through the system. When Uber (2003) surveyed 4,000 drinking water utilities, 15 percent of the respondents reported currently using booster chlorination for

(1) disinfectant residual maintenance, (2) prevention of biological regrowth, and (3) disinfection after open reservoirs. Utilities practicing booster disinfection ranged in size from 0.14 to 830 MGD, with an average of 55 MGD. Most (55 percent) of the booster stations operated with a constant delivery dose, although 35 percent used flow-pacing or residual pacing to adjust dose. A few stations used a time-dependent set-point regime. Fifty-seven (57) percent of the stations were controlled manually, 33 percent were automated, and ten percent were controlled remotely with the aid of SCADA. Half of the stations with automatic control also had remote alarms. Examples provided in the report show that incorporation of decay rate and THM formation data is fundamental to predicting whether there will be any net gain in maintenance of residual and formation of THMs when disinfectant application is changed from a single location to multiple locations. Important products of the study were the Booster Disinfection Design and Analysis software and network models, which aid in the placement and operation of booster disinfection systems. The software is capable of providing information such as (1) setting the dosing schedules given the locations are provided; (2) selecting of booster dose schedules and location; and (3) heuristic screening of potential booster locations.

Booster Chloramination

Approximately 12 percent of the respondents to the survey published in Uber (2003) practice chloramination, by one of two methods. Most reportedly use chlorine to bind excess ammonia—a useful approach if chloramine decay results in the excess ammonia or if sufficient ammonia remained during the initial formation of chloramine. Three booster stations were identified in the survey using a second method in which both chlorine and ammonia were applied at the same location. This approach is used when there is a need to increase the overall concentration of chloramine present. Wilczak et al. (2003) reviewed operating practices at utilities employing booster chloramination with the addition of free chlorine (Martin and Cummings, 1993; Cohen, 1998; Ireland and Knudson, 1998) or both chlorine and ammonia (Potts et al., 2001). Monitoring of chlorine and ammonia was practiced by all utilities, in the majority of cases with on-line combined chlorine analyzers. Nitrite, pH, and on-line free ammonia analyzers were also employed by some of the utilities. Process control options included manual dose control, dosage determined by the flow, dosage determined by the flow along with measurements of the chlorine residual, and dosage set by the flow and controlled by a desired feed level set point. In all cases, operators could manually alter the chemical doses depending on water quality results. The goal of all utilities was to maintain a total chlorine residual of at least 2.0 mg/L.

***

Several factors should be investigated before implementing booster disinfection. Hatcher et al. (2004) described recommendations made to the Sweetwater Authority, which could be made without capital expenditure in the distribution system. Among these were (1) optimizing corrosion control, (2) improving the biostability of the water by reducing the free ammonia concentrations at the treatment plant effluent to prevent nitrification (thereby avoiding disinfectant loss), and (3) conversion of the last of three plants to chloramine to avoid chlorine-chloramine blending in the system. Grayman et al. (2004) described the methods to characterize and improve mixing within a storage reservoir so that disinfectant decay in the reservoir could be minimized. If operations such as these (improved tank mixing, optimized chloramine formation at the plant, improved corrosion control to reduce disinfectant demand by pipe surfaces) can improve the maintenance of the total chlorine residual, boosting may not be necessary.

Corrosion Control

As discussed in Chapter 4, there are measures that can be taken to control both internal corrosion and metal release, including materials selection for the distribution system, addition of a corrosion inhibitor such as phosphates, control of the chemistry of the water being distributed, or some combination of these approaches. Materials selection is important because materials that are not subject to corrosion can be used when it is very difficult to control water composition. The use of phosphate inhibitors to control problems with iron, lead, and copper is widespread, but care must be taken to ensure that the added phosphate does not decrease the biological stability of the water or cause a problem in municipal wastewater treatment plant discharges. Water stagnation is an important cause of many iron release problems, so distribution systems must be designed to maintain flowing water conditions to the extent possible (Sarin et al., 2004). Also, the use of proper pH control and maintenance of an acceptable alkalinity concentration are also effective ways to control both corrosion and metal release (Sarin et al., 2003).

It should be noted that internal corrosion control can positively influence the effectiveness of chlorine-based disinfectants for inactivation of bacteria in biofilms. Corrosion products react with residual chlorine, preventing the biocide from penetrating the biofilm and controlling coliform growth. Studies have shown that free chlorine is impacted to a greater extent than monochloramine, although the effectiveness of both disinfectants is impaired if corrosion rates are not controlled (LeChevallier et al., 1990, 1993). Increasing the phosphate-based corrosion inhibitor dose, especially during the summer months, can help reduce corrosion rates.

Materials Specification to Control Leaching

Chapter 4 discusses how the quality of materials is critical to minimizing the potential for external contamination to enter into the potable water system via leaks, breaks, and permeation. Internal contamination processes like leaching and corrosion are also related to the quality of the materials used in the distribution system. Most substances leaching into water from materials in the distribution system are non-toxic and unlikely to cause health problems. However, PVC pipes manufactured before about 1977, cement materials, and the excessive leaching of organic substances from linings and joining and sealing materials have occasionally been noted. In addition to the direct effect of leaching, new pipe materials can have a significant indirect effect on internal water quality. For example, they may exert a chlorine demand that can reduce the residual disinfectant in the distribution system and hence degrade the microbial quality of the drinking water (Haas et al., 2002). Standards for manufacture of materials and guidance for specifying materials need to be updated to address water quality issues (e.g., leaching of emerging chemicals, biogrowth promoting potential, leaching of non-health related but taste/odor-related chemicals, susceptibility to permeation).

RECOVERING WATER QUALITY INTEGRITY

Recovering water quality integrity following an internal contamination event in the distribution system revolves around a few specific activities. In many cases, the same method is used to address several issues including colored/turbid water, loss of residual, nitrification, and elevated microbial counts. Options are limited to (1) flushing to remove the taste/odor/color/ turbidity or to restore disinfectant concentrations, (2) permanently switching disinfectants to maintain a residual, typically from free chlorine to chloramine, (3) periodically changing from chloramine to free chlorine to mitigate nitrification, (4) implementing corrosion control to reduce corrosion and leaching, or (5) changing water sources. These approaches are discussed in more detail below. It should be noted that other distribution system maintenance and repair options such as cleaning, relining, replacement, and localized disinfection can also alleviate internally derived water quality problems; these methods are described in Chapters 4 and 5.

Flushing

As shown in previous chapters, water main flushing is an operational activity that involves moving water through the distribution system, often at a rate that facilitates scouring of the surfaces, and discharging it through hydrants or blow-off ports. Many researchers and utility managers have suggested that optimized flushing is important in maintaining or recovering water quality (Pattison, 1980; Emde et al., 1995, 1997; Smith et al., 1996; Antoun et al., 1997; Barbeau et al., 1999). Disinfectant residuals may be regained or maintained by moving out "old" water and replacing it with water containing a measurable residual.

Flushing can also be used to remove deposits as well as discolored water resulting from suspended material, which can be an effective method for combating biofilm growth. Gauthier et al. (1997) showed that loose deposits in a French system removed by flushing contained organisms including invertebrates, protozoa, and bacteria. Ackers et al. (2001) characterized the sediments removed by flushing as corrosion products, components of the pipe lining, treatment breakthrough, animal and biomatter, and calcium deposits. Antoun et al. (1997) recommended that flushing be used to reduce the potential for total coliform positive samples in a distribution system, and the approach seemed to alleviate the problem. Similarly, Emde et al. (1995 and 1997) suggested that flushing at sufficient velocities could remove biomass from pipe surfaces, therefore controlling biofilms. This approach was utilized by the Zurich Water Supply in Switzerland to control regrowth in a distribution system supplying water without a secondary disinfectant (Klein and Forster, 1998). In another situation, temporary control of invertebrates through a program including flushing was advocated (van Lieverloo et al., 1998).

A recent survey (Friedman et al., 2003) showed that of 23 U.S. utilities that responded, 20 have regularly scheduled flushing programs, while the remaining three flushed on an as-needed basis. In order of frequency cited, the objectives used for flushing were to eliminate colored water, restore disinfectant residual, reduce turbidity, eliminate tastes and odors, reduce the number of bacteria, reduce DBP precursors, remove sediment, comply with regulations, maintain water quality, decrease chlorine demand, respond to customer complaints, reduce corrosion inhibitor build-up, eliminate stale water, respond to animal activity, and eliminate lime deposits. Another survey (summarized in Table 2-5) suggests that flushing is variously supported by state agencies. Of 34 responding states, in only 11 states are flushing/cleaning/pigging required, with 20 others encouraging the practices by utilities. One of the difficulties in assessing the efficacy of flushing for restoring or maintaining water quality is the lack of data collection by utilities. A nation-wide survey (Chadderton et al., 1992) showed that utilities typically implemented flushing in response to consumer complaints. In this study they also found that less than half of the utilities collected water samples during the flushing process for analysis of chlorine, turbidity, bacteria, or other parameters. In most cases, flushing proceeded until the water was visually clear. In light of the benefits that can be attained by properly conducted flushing (to minimize the amount of water wasted and appropriately discharge the waste), more attention should be given to this approach for maintaining quality or resolving problems.

Change Disinfectant

For nearly 100 years, drinking water utilities in the United States and Canada have converted to chloramine for improved ability to maintain a residual at dead ends or other areas with high hydraulic residence time due to the lower decay rates of chloramine versus free chlorine. Systems also have converted to chloramine for taste, odor, or DBP control (Kirmeyer et al., 1993). Lowering of the THM standard from 100 to 80 μg/L in 2001 and the forthcoming requirement to meet this value at each monitoring location have encouraged additional utilities to switch to chloramine.

In recent years the ability of chloramine to penetrate biofilms and control *Legionella* has received more attention and helped those who have already converted for other reasons (e.g., maintenance of residual, taste and odor, or DBP control) justify their decision. Preliminary results show that although *Legionella* is better controlled with chloramination (see Chapter 3 and Pryor et al., 2004), it is possible that other organisms of public health concern such as *Mycobacteria* could have a selective advantage under these conditions (Pryor et al., 2004). The ability of chloramine to control biofilms has been documented by several utilities that find it easier to meet the requirement of the Total Coliform Rule using chloramine (Norton et al., 1997; Richard Mann, Metropolitan Water District of Southern California, Los Angeles, CA, personal communication, 1998) although this increased control is not universal (Muylwyk et al., 2001). In a multi-year study in pipe loops with a three-day residence time, both chlorine and chloramine were found to be effective disinfectants but chloramine persisted longer (Clark et al., 1994). Other laboratory studies have not demonstrated an advantage of monochloramine over free chlorine (Ollos et al., 1998; Clark and Sivaganesan, 1999; Camper et al., 2003).

Meanwhile, issues regarding the disturbance of bacteria or metallic oxides on the pipe walls during the disinfectant switch and their influence on water quality have been and continue to be a concern. A recent example of how the disturbance of pipe walls can affect water quality is described by Edwards and Dudi (2004) regarding the District of Columbia Water and Sewer Authority's drinking water lead levels after the conversion from free chlorine to chloramine. During the time when chlorine was used as the disinfectant in the distribution system, lead dioxide formed on the lead pipes. When the redox potential decreased because of the conversion from chlorine to monochloramine, the lead dioxide was converted to a more soluble lead (+II) compound, and this caused an increase in the lead concentration in the bulk water. While it is likely that iron release and elevated coliform levels concomitant with the conversion to chloramine will resolve as the distribution systems reequilibrate, in some cases utilities have intervened with flushing to accelerate the transition, but noted increases in water quality problems during the flushing event.

In addition to permanent changes in disinfectant, there are instances where short-term switches are practiced. Drinking water utilities using chloramine as a disinfectant residual sometimes temporarily switch to free chlorine both as a preventative and control measure for nitrification. This can be accomplished system-wide by simply turning off the ammonia feed pumps. To switch to free chlorine in an isolated pressure zone or storage facility, enough chlorine must be added to pass the breakpoint and achieve a free chlorine residual. There are differing views about the practice of periodic chlorination in chloraminated systems. Indeed, the practice has been abandoned by some utilities in Southern California as a response to legitimate concerns over short-term exposure to elevated DBP levels (Hatcher et al., 2004).

When nitrification occurs in storage tanks, and other strategies such as adjustments in chlorine to ammonia ratio, increased turnover, or flushing have not solved the problem, breakpoint chlorination is a common response (Skadsen, and Cohen, 2006). In this case chlorine is added to oxidize all of the nitrogen species (ammonia and nitrite) and achieve a free chlorine residual. This strategy is effective because ammonia oxidizing bacteria are sensitive to free chlorine (Baribeau, 2006), but temporary since their populations will recover after the weaker disinfectant (chloramine) is reestablished in the facility or system. This is demonstrated by a need for repeated breakpoint chlorination of reservoirs especially in summer months.

Adequate mixing is important for efficient breakpoint events and for routine maintenance of the disinfectant residual within storage facilities. Figure 6-4 shows the variability of a chloramine (total chlorine) residual at the sample tap on a common inlet/outlet pipe to a storage tank. As the tank fills, water from the distribution system with a total chlorine residual of about 2 mg/L enters the tank;

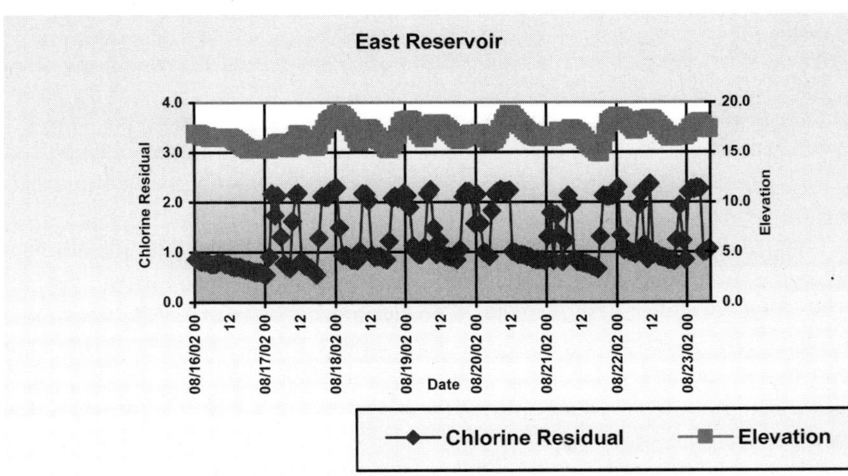

FIGURE 6-4 Variability in storage tank chlorine residual as a function of fill and drain cycle. SOURCE: Reprinted, with permission, from Guistino (2003). © 2003 by Joe Guistino.

as the tank drains the water is observed to have a total chlorine residual of about 1 mg/L. Some intermediate points are seen on the figure that are characteristic of distribution water mixing with tank water (Guistino, 2003). Grayman et al. (2004) have described design and operational problems that lead to poor mixing within storage facilities. Figure 6-4 also illustrates the utility of using continuous disinfectant analyzers on storage facilities. The data reveal that the disinfectant residual is a function of the fill and drain cycle (or elevation) of the tank more than the water quality of the storage facility and distribution system.

Regardless of whether disinfectant changes are long- or short-term, utilities should be aware that these changes may have implications for protecting public health, especially during an intrusion event. Chloramine may be inadequate for protection against microorganisms that enter the distribution system during intrusion, as discussed previously for *Giardia* cysts but also for enteric viruses with less susceptibility to chloramine than chlorine. Karim et al. (2003) showed that over half of soil samples collected during pipe replacements tested positive for enteric viruses.

Change Treatment/Corrosion Control

If corrosion or metal release is identified as a problem, one of the measures listed in the previous section on corrosion control should be undertaken. Identification of the cause of the problem is most important relative to selection of the best approach. For example, if the cause of the problem is variable pH, treatment to control pH and possibly to add alkalinity for the purpose of increasing buffer intensity is probably necessary. Adding alkalinity also benefits corrosion control via the addition of carbonate (from the standpoint of developing a stable scale).

CONCLUSIONS AND RECOMMENDATIONS

Beyond contamination that enters potable water from external sources (intrusion, cross connections, etc.), there are processes within the distribution system that contribute to degradation of water quality. The large surface area to volume ratio of pipe surfaces, reactive pipe materials, advanced water ages, and bulk water reactions all contribute to deleterious changes in water quality from the treatment plant to the consumer. Maintaining water quality integrity in the distribution system is challenging because of the complexity of the system. There are interactions between the type and concentration of disinfectants, corrosion control schemes, operational practices (e.g., flow characteristics, water age, flushing practices), the materials used for pipes and plumbing, the biological stability of the water, and the efficacy of treatment. In some cases, changes to improve water quality may be reasonably easy, while others may be extremely difficult. The following conclusions and recommendations are made.

Prior to distribution, the quality of treated water should be adjusted to minimize deterioration of water quality. For example, appropriate use of phosphate inhibitors and control of pH and alkalinity can be used to minimize both internal corrosion of lead, copper, and iron pipes and the formation of colored water owing to the release of iron from corrosion scales. Coagulation with aluminum salts should be done in a way that minimizes the residual aluminum concentration in filtered water, thereby reducing the amount of aluminum that precipitates in the distribution system. Ensuring that water is biologically stable via removal of organic carbon will have a positive impact on the preservation of water quality as it travels from the treatment plant to the consumer. It should be kept in mind that other chemical adjustments may have an impact on the biological stability of the water.

Microbial growth and biofilm development in distribution systems should be minimized. Even though general heterotrophs are not likely to be of public health concern, their activity can promote the production of tastes and odors, increase disinfectant demand, and may contribute to corrosion. Biofilms may also harbor opportunistic pathogens (those causing disease in the immunocompromised); this is of greatest importance in premise plumbing where long residence times contribute to disinfectant decay and subsequent bacterial growth and release. Coliforms may also proliferate in biofilms. With perhaps the exception of *E. coli*, coliforms from biofilms are indistinguishable from those arising from external contamination. When these coliforms are detected, it is difficult to determine if a contamination event of public health significance has occurred (fecal contamination) as opposed to the growth of indigenous organisms.

Residual disinfectant choices should be balanced to meet the overall goal of protecting public health. For free chlorine, the potential residual loss and DBP formation should be weighed against the problems that may introduced by chloramination, which include nitrification, lower disinfectant efficacy against suspended organisms, and the potential for deleterious corrosion problems. Although some systems have demonstrated increased biofilm control with chloramination, this response has not been universal. This ambiguity also exists for the control of opportunistic pathogens.

Current microbial monitoring is limited in its ability to indicate distribution system contamination events, such that new methods and strategies are needed. Current methods are not specific for the source of contamination and do not allow for good process control because of the substantial amount of time needed to generate results. There are limits to the effectiveness of indicators with respect to predicting the presence of pathogens. A tiered approach to microbial monitoring should be considered in which common indicators are used initially to provide an early warning of a potential health risk, followed by more detailed studies to assess the extent of the risk using more specific methods. In concert with this, techniques for monitoring for specific pathogens with known public health significance should be developed, ideally with results available on-line and in real time. The implementation of best practices to maintain water quality (see Chapter 2) is needed until better monitoring approaches can be developed.

Standards for materials used in distribution systems need to be updated to address their impact on water quality, and research is needed to develop new materials that will have minimal impacts. Materials standards have historically been designed to address physical/strength properties including the ability to handle pressure and stress. Testing of currently available materials should be expanded to include (1) the potential for permeation of contaminants, and (2) the potential for leaching of compounds of public health concern as well as those that contribute to tastes and odors and support biofilm growth. The results of these tests should be incorporated into the standards in a way that water quality deterioration attributable to distribution system materials is minimized. Also, research is needed to develop new materials that minimize adverse water quality effects such as the high concentrations of undesirable metals and deposits that result from corrosion and the destruction of disinfectant owing to interactions with pipe materials.

REFERENCES

Ackers, J., M. Brandt, and J. Powell. 2001. Hydraulic characterization of deposits and review of sediment modeling. Report REF. No. 01/DW/03/18. London: UKWIR Ltd.

American Public Health Association (APHA). 2005. Standard Methods for the Examination of Water and Wastewater, 21st edition. A. D. Eaton, L. S. Clesceri, E. W. Rice, and A. E. Greenberg (eds.). Washington, DC: APHA.

Amy, G. L., P. A. Chadik, and Z. K. Chowdhury. 1987. Developing models for predicting trihalomethane formation potential and kinetics. J. Amer. Water Works Assoc. 79(7):89–97.

Antoun, E. N., T. Tyson, and D. Hiltebrand. 1997. Unidirectional flushing: a remedy to water quality problems such as biologically mediated corrosion. *In:* Proceedings of the AWWA Annual Conference. Denver, CO: AWWA.

American Water Works Association (AWWA). 1990. Position statement on chlorine residual. Pp. 196 *In*: 1995–1996 AWWA Officers and Committee Directory. Denver, CO: AWWA.

AWWA. 2006. Fundamentals and Control of Nitrification in Chloraminated Drinking Water Distribution Systems. AWWA Manual M56. Denver, CO: AWWA.

AWWA and EES, Inc. 2002. Permeation and leaching. *http://www.epa.gov/safewater/tcr/pdf/permleach.pdf.* Washington, DC: EPA.

Barbeau, B., K. Julienne, V. Gauthier, R. Millette, and M. Provost. 1999. Dead-end flushing of a distribution system: short and long-term impacts on water quality. *In:* Proceedings of the AWWA Water Quality Technology Conference. Denver, CO: AWWA.

Baribeau, H., S. W. Krasner, R. Chinn, and P. C. Singer. 2005a. Impact of biomass on the stability of HAAs and THMs in a simulated distribution system. J. Amer. Water Works Assoc. 97:69–81.

Baribeau, H., N. L. Pozos, L. Boulos, G. F. Crozes, G. A. Gagnon, S. Rutledge, D. Skinner, Z. Hu, R. Hofmann, R. C. Andrews, L. Wojcicka, Z. Alam, C. Chauret, S. A. Andrews, R. Dumancis, and E. Warn. 2005b. Impact of Distribution System Water Quality on Disinfection Efficacy. Denver, CO: AwwaRF.

Baribeau, H., P. C. Singer, R. W. Gullick, S. L. Williams, R. L. Williams, S. A. Andrews, L. Boulos, H. Haileselassie, C. Nichols, S. A. Schlesinger, L. Fountleroy, E. Moffat, and G. F. Crozes. 2006. Formation and Decay of Disinfection By-Products in the Distribution System. Denver, CO: AWWA and AwwaRF.

Baribeau, H. 2006. Chapter 6: Growth and inactivation of nitrifying bacteria. *In:* Fundamentals and Control of Nitrification in Chloraminated Drinking Water Distribution Systems. AWWA Manual M56. Denver, CO: AWWA.

Bartram, J., J. Cotruvo, M. Exner, C. Fricker and A. Glasmacher. 2003. Heterotrophic Plate Counts and Drinking-Water Safety. London: IWA Publishing.

Bellen, G. E., S. H. Abrishami, P. M. Colucci, and C. Tremel. 1993. Methods for assessing the biological growth support potential of water contact materials. Denver, CO: AwwaRF.

Benjamin, M. M., H. Sontheimer, and P. Leroy. 1996. Corrosion of iron and steel. Pp. 29–77 *In:* Internal Corrosion of Water Distribution Systems. Denver, CO: AwwaRF/DVGW-Technologiezentrum Wasser.

Berend, K., G. Van Der Voet, and W. H. Boer. 2001. Acute aluminum encephalopathy in a dialysis center caused by a cement mortar water distribution pipe. Kidney International 59(2):746–753.

Besner, M.-C., V. Gauthier, P. Servais, and A. Camper. 2002. Explaining the occurrence of coliforms in distribution systems. J. Amer. Water Works Assoc. 94(8):95–109.

Bitton, G. 1994. Role of microorganisms in biogeochemical cycles. Pp. 51–73 *In*: Wastewater Microbiology. New York: John Wiley.

Block, J. C. 1992. Biofilms in drinking water distribution systems. *In:* Biofilms—Science and Technology. L. F. Melo, T. R. Bott, M. Fletcher, and B. Capdeville (eds.). Dordrecht, Netherlands: Kluwer Publishers.

Bouchard, D. C., M. K. Williams, and R. Y. Surampalli. 1992. Nitrate contamination of groundwater: sources and potential health effects. J. Amer. Water Works Assoc. 84(9):84–90.

Bremer, P. J., and G. G. Geesey. 1991. Laboratory-based model of microbially induced corrosion of copper. Appl. Environ. Microbiol. 57:1956–1962.

Bremer, P. J., G. G. Geesey, and B. Drake. 1992. Atomic force microscopy examination of the topography of a hydrated bacterial biofilm on a copper surface. Current Microbiol. 24:223–230.

Bruchet, A. 1999. Solved and unsolved cases of taste and odor episodes in the files of Inspector Cluzeau. Water Science and Technology 40(6):15–21.

Bryant, E. A., G. P. Fulton, and G. C. Budd. 1992. Chloramination. Pp. 128–170 *In:* Disinfection Alternatives for Safe Drinking Water. New York: Van Nostrand Reinhold.

Burlingame, G., J.Choi, M. Fadel, L. Gammie, J. Rahman, and J. Paran. 1994. Sniff new mains...before customers complain. Opflow 20(10):3.

Burman, N. P. 1965. Taste and odor due to stagnation and local warming in long lengths of piping. *In*: Proceedings of the Society for Water Treatment and Examination 14:125–131.

Burman, N. P. 1973. The occurrence and significance of actinomycetes in water supply. Pp. 219–230 *In*: *Actinomycetaeles*; Characteristics and Practical Importance. G. Sykes and F. A. Skinner (eds.). London and New York: Academic Press.

Calles, J., R. Gottler, M. Evans, and J. Syage. 2005. Early warning surveillance of drinking water by photoionization/mass spectrometry. J. Amer. Water Works Assoc. 97:62–73.

Camper, A. K., K. Brastrup, A. Sandvig, J. Clement, C. Spencer and A. J. Capuzzi. 2003. Impact of distribution system materials on bacterial regrowth. J. Amer. Water Works Assoc. 95(7):107–121.

Chadderton, R. A., G. L. Christensen, and P. Henry-Unrath. 1992. Implementation and Optimization of Distribution System Flushing Programs. Denver, CO: AWWA and AwwaRF.

Chapelle, F. H., and D. R. Lovely. 1992. Competitive exclusion of sulfate reduction by Fe(III)-reducing bacteria: a mechanism for producing discrete zones of high-iron ground water. Groundwater 30:29–34.

Characklis, W. G. 1988. Bacterial Regrowth in Distribution Systems. Denver, CO: AWWA and AwwaRF.

Characklis, W. G., M. H. Turakhia, and N. Zelver. 1990. Transport and Interfacial Transport Phenomena. Chapter 9 *In:* Biofilms. W. G. Characklis and K. C. Marshall (eds.). New York: John Wiley & Sons.

Choi, J., and R. L. Valentine. 2002a. A kinetic model of N-nitrosodimethylamine (NDMA) formation during water chlorination/chloramination. Wat. Sci. Tech. 6(3):65–71.

Choi, J. H., and R. L. Valentine. 2002b. Formation of N-nitrosodimethylamine (NDMA) from reaction of monochloramine: a new disinfection by-product. Water Res. 36(4):817–824.

Chun, C. L, R. M. Hozalski, and T. A. Arnold. 2005. Degradation of drinking water disinfection byproducts by synthetic goethite and magnetite. Environ. Sci. Technol. 39:8525–8532.

Clark, R. M., B. W. Lykins, Jr., J. C. Block, L. J. Wymer, and D. J. Reasoner. 1994. Water Quality Changes in a Simulated Distribution System. Journal of Water Supply Research and Technology—Aqua 43(6):263–277.

Clark, R. M. 1998. Chlorine demand and TTHM formation kinetics: a second- order model. J. Environ. Eng. 124:16–24.

Clark, R. M., and M. Sivaganesan. 2002. Predicting chlorine residuals in drinking water: a second order model. J. Water Resources Planning and Management 128:1–10.

Clark, R. M., R. Thurnau, M. Sivaganesan, and P. Ringhand. 2001. Predicting the formation of chlorinated and brominated by-products. J. Environ. Eng. 127:493–501.

Clark, R. M., and R. A. Haught. 2005. Characterizing pipe wall demand: implications for water quality. J. Water Resources Planning and Management 131:208–217.

Clark, R. M., and M. Sivaganesan. 1999. Characterizing the effect of chorine and chloramines on the formation of biofilm in a simulated drinking water distribution system. EPA/600/R-01/024. Washington, DC: EPA.

Cohen, Y. K. 1998. Forming chloramine and maintaining residual. Opflow 24(9):1.

Coss, A., K. P. Cantor, J. S. Reif, C. F. Lynch, and M. H. Ward. 2004. Pancreatic cancer and drinking water and dietary sources of nitrate and nitrite. Amer. J. Epidemiol. 159(7):693–701.

Craun, G. F., and R. L. Calderon. 2001. Waterborne disease outbreaks caused by distribution system deficiencies. J. Amer. Water Works Assoc. 93(11):65–75

DeAraujo-Jorge, T. C., C. M. L. Melo Coutinho, and E. V. De Aguiar. 1992. Sulfate - reducing bacteria associated with biocorrosion - a review. Memorias Instituto Oswaldo Cruz. 87:329.

Denisov, G. V., B. G. Kovrov, and T. F. Kovaleva. 1981. Effect of the pH and temperature of the medium on the rate of oxidation of Fe^{+2} to Fe^{+3} by a culture of *Thiobacillus ferrooxidans* and the coefficient of efficiency of biosynthesis (translated from Russian). Mikrobiologiya 50:696.

De Roos, A. J., M. H. Ward, C. F. Lynch, and K. P. Cantor. 2003. Nitrate in public water supplies and the risk of colon and rectum cancers. Epidemiology 14(6):640–649.

Dietrich, A. M., and R. C. Hoehn. 1991. Taste-and-odor problems associated with chlorine dioxide. Denver, CO: AwwaRF.

Dietrich, A. M., R. C. Hoehn, G. A. Burlingame, and T.Gittelman. 2004. Practical taste-and-odor methods for routine operations: decision tree. Denver, CO: AwwaRF.

Dodge, D. J., A. J. Francis, J. B. Gillow, G. P. Halada, and C. R. Clayton. 2002. Association of Uranium with Iron Oxides Typically Formed on Corroding Steel Surfaces. Environ. Sci. Technol. 36:3504.

Doshi, P. E., W. M. Grayman, and D. Guastella. 2003. Field testing the chlorine wall demand in distribution mains. Pp. 1–10 *In:* Proceedings of the Annual Conference of the American Water Works Association. Denver, CO: AWWA.

Duirk, S. and R. L. Valentine. 2002. Reaction of monochloramine with humic acid. J. Environ. Monitoring 4:85–89.

Edwards, M., and A. Dudi. 2004. Role of chlorine and chloramine in corrosion of lead-bearing plumbing materials. J. Amer. Water Works Assoc. 96(10):69–81.

Emde, K. M., K. Oberoi, and D. Smith. 1995. Evaluation of various methods for distribution system biofilm control. *In:* Proceedings of the AWWA Annual Conference. Denver, CO: AWWA.

Emde, K. M., K. Oberoi, and D. Smith. 1997. Maximizing distribution system maintenance activities. *In:* Proceedings of the AWWA Annual Conference. Denver, CO: AWWA.

EPA (Environmental Protection Agency). 1993. Total Coliform Rule. 40 CFR, Code of Federal Regulations. Washington, DC: National Archives and Records Administration.

EPA. 2003. Economic Analysis for Proposed Stage 2 DBPR. EPA 815-D-03-001. Washington, DC: EPA.

Friedman, M. J., K. Martel, A. Hill, D. Holt, S. Smith, T. Ta, C. Sherwin, D. Hiltebrand, P. Pommerenk, Z. Hinedi, and A. Camper. 2003. Establishing Site Specific Flushing Velocities. Denver, CO: AWWA and AwwaRF.

Gang, D. D., R. L. Segar, T. E. Clevenger, and S. K. Banerji. 2002. Using chlorine demand to predict TTHM and HAA9 formation. J. Amer. Water Works Assoc. 94(10):76–86.

Gauthier, V., C. Rosin, L. Mathieu, J. M. Portal, J. C. Block, P. Chaix, and D. Gatel. 1997. Characterization of the loose deposits in drinking water distribution systems. *In:* Proceedings of the AWWA Water Quality Technology Conference. Denver, CO: AWWA.

Geesey, G. G., P. J. Bremer, W. R. Fischer, D. Wagner, C. W. Keevil, J. Walker, A. H. L. Chamberlain, and P. Angell. 1993. Unusual types of pitting corrosion of copper tubes in potable water systems. *In*: Biocorrosion. London: CRC Press.

Geldreich, E. E. 1996. Creating microbial quality in drinking water. Pp. 39–101 *In:* Microbial Quality of Water Supply in Distribution Systems. Boca Raton, FL: Lewis Publishers.

Grayman, W. M., L. A. Rossman, M. A. Gill, Y. Li, and D. E. Guastella. 2002. Measuring and modeling disinfectant wall demand in metallic pipes. *In:* Proceedings of the EWRI Conference on Water Resources Planning and Management. Reston VA: ASCE.

Grayman,W. M., L. A. Rossman, R. A. Deininger, C. D. Smith, C. N. Arnold, and J. F. Smith. 2004. Mixing and aging of water in distribution system storage facilities. J. Amer. Water Works Assoc. 96(9):70–80.

Guistino, J. 2003. Are you getting a real distribution system sample? Source 16(4):10–11.

Gulis, G., M. Czompolyova, and J. R. Cerhan. 2002. An ecologic study of nitrate in municipal drinking water and cancer incidence in Trnava District, Slovakia. Environmental Research 88(3):182–187.

Haas, C. N., M. Gupta, R. Chitlluru, and G. Burlingame. 2002. Chlorine demand in disinfecting water mains. J. Amer. Water Works Assoc. 94(1):97–102.

Harrington, G. W., Z. K. Chowdhury, and D. M. Owen. 1992. Developing a computer model to simulate DBP formation during water treatment. J. Amer. Water Works Assoc. 84(11):78–87.

Hatcher, M., W. Grayman, C. Smith, and M. Mann. 2004. Monitoring and modeling of the Sweetwater Authority distribution system to assess water quality. *In:* Proceedings of the AWWA Annual Conference. Denver, CO: AWWA.

Haudidier, K., J. L. Paquin, T. Francais, P. Hartemann, G. Grapin, F. Colin, M. J. Jourdain, J. C. Block, J. Cheron, O. Pascal, Y. Levi, and J. Miazga. 1988. Biofilm growth in drinking water networks: a preliminary industrial pilot plant experiment. Water Sci. Technol. 20:109–115.

Ireland, C., and M. Knudson. 1998. Portland's experience with chloramine residual management. *In:* Proceedings of the Conference on Protecting Water Quality in the Distribution System: What is the Role of Disinfection Residuals? April 26–28, Philadelphia, PA.

Jafvert, C. T., and R. L. Valentine. 1992. Reaction scheme for the chlorination of ammoniacal water. Environ. Sci. Technol. 26:577–586.

Joret, J. C., C. Volk, R. Javadpoour, G. Randon, and P. Cote. 1994. Control of biodegradable organic matter during drinking water treatment. *In:* Proceedings of the International Seminar on Biodegradable Organic Matter. Montreal, Quebec.

Karim, M. R., M. Abbaszadegan, and M. W. LeChevallier. 2003. Potential for pathogen intrusion during pressure transients. J. Amer. Water Works Assoc. 95(5):134–146.

Keinanen, M. M., L. K. Korhohen, M. J. Lehtola, and I .T. Miettinen. 2002. The microbial community structure of drinking water biofilms can be affected by phosphorus availability. Appl. Environ. Microbiol. 68:434–439.

Khiari, D., S. Barrett, R. Chinn, A. Bruchet, P Piriou, L. Matia, F. Ventura, I. Suffet, T. Gittelman, and P. Leutweiler. 2002. Distribution generated taste-and-odor phenomena. Denver, CO: AwwaRF.

Kirmeyer, G. J., G. W. Foust, G. L. Pierson, J. J. Simmler, and M. W. Lechevallier. 1993. Optimizing Chloramines Treatment. Denver, CO: AWWA and AwwaRF.

Kirmeyer, G., M. Friedman, J. Clement, A. Sandvig, P. F. Noran, K. D. Martel, D. Smith, M. LeChevallier, C. Volk, J. Dyksen, and R. Cushing. 2000. Guidance Manual for Maintaining Distribution System Water Quality. Denver, CO: AwwaRF.

Klein, H., and R. Forster. 1998. Network operation without safety chlorination. Zurich Water Supply 16(3/4):165–174.

Kovalenko, T. V., G. I. Karavaiko, and V. P. Piskunov. 1982. Effect of Fe^{3+} ions on the oxidation of ferrous iron by *Thiobacillus ferroxidans* at various temperatures (translated from Russian). Mikrobiologiya 51:42.

Krasner, S. W., M. J. McGuire, and V. B. Ferguson. 1985. Tastes and odors: the flavor profile method. J. Amer. Water Works Assoc. 77:34–39.

LeChevallier M. W., B. H. Olson, and G. A. McFeters. 1990. Assessing and Controlling Bacterial Regrowth in Distribution Systems. Denver, CO: AWWA.

LeChevallier, M. W., W. Schulz, and R. G. Lee. 1991. Bacterial nutrients in drinking water. Appl. Environ. Microbiol. 57:857–862.

LeChevallier, M. W., C. D. Lowry, R. G. Lee, and D. L. Gibbon. 1993. Examining the relationship between iron corrosion and the disinfection of biofilm bacteria. J. Amer. Water Works Assoc. 85(7):111–123.

LeChevallier, M. W., N. J. Welch, and D. B. Smith. 1996. Full-scale studies of factors related to coliform regrowth in drinking water. Appl. Environ. Microbiol. 62:2201–2211.

Leclerc, H. 2003. Relationships between common water bacteria and pathogens in drinking-water. Pp. 80–118 *In*: Heterotrophic Plate Counts and Drinking-Water Safety. J. Bartram, J. Cotruvo, M. Exner, C. Fricker, and A. Glasmacher (eds.). London: IWA Publishing.

Lee, S. H., S. O'Connor, and B. K. Banerji. 1980. Biologically mediated corrosion and its effects on water quality in distribution systems. J. Amer. Water Works Assoc. 72:636–645.

Lehtola, M. J., and I. T. Miettinen. 2001. Microbially available organic carbon, phosphorus, and microbial growth in ozonated drinking water. Water Res. 35:1635–1640.

Lehtola, M. J., I. T. Miettinen, and P. J. Martikaninen. 2002a. Biofilm formation in drinking water affected by low concentrations of phosphorus. Can. J. Microbiol. 48:494–499.

Lehtola, M. J., I. T. Miettinen, T. Vartianen, and P. J. Martikaninen. 2002b. Changes in content of microbially available phosphorus, assimilable organic carbon and microbial growth potential during drinking water treatment processes. Water Res. 36:3681–3690.

Lehtola, M. J., I. T. Miettinen, M. M. Keinanen, T. K. Kekki, O. Laine, A. Hirvonen, T. Vartianen, and P. J. Martikaninen. 2004. Microbiology, chemistry and biofilm development in a drinking water distribution system with copper and plastic pipes. Water Res. 38:3769–3779.

Levy, D. A., M. S. Bens, G. F. Craun, R. L. Calderon, and B. L. Herwaldt. 1998. Surveillance for waterborne-disease outbreaks—United States, 1995–1996. MMWR 47(SS-5):1–34.

Lu, C., P. Biswas, and R. M. Clark. 1995. Simultaneous transport of substrates, disinfectants, and microorganisms in water pipes. Water Res. 29(3):881–894.

Lytle, D. A., and V. L. Snoeyink. 2002. Effect of ortho- and polyphosphates on the properties of iron particles and suspensions. J. Amer. Water Works Assoc. 94(10):87–99.

Lytle, D. A., T. J. Sorg, and C. Frietch. 2002. The significance of arsenic-bound solids in drinking water distribution systems. *In:* Proceedings of the AWWA Water Quality Conference. Seattle, WA.

Lytle, D. A, T. Sorg, and C. Frietch. 2004. The accumulation of arsenic in water distribution systems. Environ. Sci. Technol. 38:5365–5372.

Martin, P., and E. Cummings. 1993. Rechloramination in the distribution system. *In:* Proceedings of the AWWA Annual Conference. Denver, CO: AWWA.

McClellan, J. N., D. A. Reckhow, J. E. Tobiason, J. K. Edzwald, and D. B. Smith. 2000. A comprehensive kinetic model for chlorine decay and chlorination byproduct formation. Pp. 223–246 *In*: Natural Organic Matter and Disinfection By-Products: Characterization and Control in Drinking Water. S. W. Krasner, S. E. Barrett, and G. L. Amy (eds.). Washington, DC: American Chemical Society.

McFeters, G. A., B. H. Pyle, J. T. Lisle, and S. C. Broadaway. 1999. Rapid direct methods for enumeration of specific, active bacteria in water and biofilms. J. Appl. Microbiol. Symposium Suppl. 85:193S–200S.

McGuire, M. J. 1995. Off-flavor as the consumer's measure of drinking water safety. Water Sci. Technol. 31(11):1–8.

McGuire, M. J., N. and Graziano, N. 2002. Trihalomethanes in U.S. Drinking Water: NORS to ICR. *In:* Information Collection Rule Data Analysis. M. J. McGuire, J. L. McLain, and A. Obolensky (eds.). Denver, CO: AwwaRF.

McGuire, M., N. Graziano, L. Sullivan, R. Hund, and G. Burlingame. 2004. Water Utility Self-Assessment for the Management of Aesthetic Issues. Denver, CO: AWWA.

McNeill, L. S., and M. Edwards. 2001. Iron pipe corrosion in distribution systems. J. Amer. Water Works Assoc. 93(7):88–100.

Miettinen, L. T., T. Vartianinen, and P .J. Martikainen. 1997. Phosphorus and bacterial growth in drinking water. Appl. Environ. Microbiol. 63:3242–3245.

Mitch, W. A., and D. L. Sedlak. 2002. Formation of N-nitrosodimethylamine (NDMA) from dimethylamine during chlorination. Environ. Sci. Technol. 36(4):588–595.

Montiel, A., S. Rigal, and B. Welte. 1999. Study of the origin of musty taste in the drinking water supply. Water Science and Technology 40(6):171–177.

Muylwyk, Q., J. MacDonald, and M. Klawunn. 2001. Success! Switching from chloramines to chlorine in the distribution system: results from a one year full-scale trial. *In:* Proceedings of the Water Quality Technology Conference. Denver, CO: AWWA.

National Research Council (NRC). 2004. Indicators for Waterborne Pathogens. Washington, DC: National Academies Press.

NRC. 2005. Public Water Supply Distribution Systems: Assessing and Reducing Risks. Washington, DC: National Academies Press.

Nemati, M., and C. Webb. 1997. A kinetic model for biological oxidation of ferrous iron by *Thiobacillus ferroxidans*. Biotechnol. Bioeng. 53:478.

North Carolina Department of Environment and Natural Resources (NCDENR). 2004. Title 15A, Subchapter 18C of the North Carolina Administrative Codes. Section 2000—Filtration and Disinfection. Raleigh, NC: NCDENR.

Norton, C. D., and M. W. LeChevallier. 1997. Chloramination: its effect on distribution water quality. J. Amer. Water Works Assoc. 89(7):66–77.

Okereke, A., and S. E. Stevens, Jr. 1991. Kinetics of iron oxidation by *Thiobacillus ferroxidans*. Appl. Environ. Microbiol. 57:1052–1056.

Ollos, P. J., R. M. Slawson, and P. M. Huck. 1998. Bench scale investigations of bacterial regrowth in drinking water distribution systems. Water Sci. Technol. 38:275–282.

Olson, B. H. 1982. Assessment and implications of bacterial regrowth in water distribution systems. EPA-600/S2-82-072. Washington, DC: EPA.

Palacios, B., and C. D. Smith. 2002. The use of decay studies to plan San Francisco's conversion to chloramines. *In:* Proceedings of the Distribution and Plant Operators Conference. Denver, CO: AWWA.

Pattison, P. L. 1980. Conducting a regular main flushing program. J. Amer. Water Works Assoc. 72(2):88–90.

Payment, P., and E. Franco. 1993. *Clostridium perfringens* and somatic coliphages as indicators of the efficiency of drinking water treatment for viruses and protozoan cysts. Appl. Environ. Microbiol. 59:2418–2424.

Pepper, I. L., P. Rusin, D. R. Quintanar, C. Haney, K. L. Josephson, and C. P. Gerba. 2004. Tracking the concentration of heterotrophic plate count bacteria from the source to the consumer's tap. Int. J. Food Microbiol. 92:289–295.

Pereira, V. J., H. S. Weinberg, and P. C. Singer. 2004. Temporal and spatial variability of DBPs in a chloraminated distribution system. J. Amer. Water Works Assoc. 96:91–102.

Potts, D. E., W. G. Richards, and C. G. Hitz. 2001. A satellite chloramine booster station: design and water chemistry. *In:* Proceedings of the AWWA Distribution System Symposium. Denver, CO: AWWA.

Powell, J. C. 1998. Modeling Chlorine in Water Distribution Networks. Ph.D. Thesis. School of Civil Engineering. Faculty of Engineering. The University of Birmingham.

Powell, J. C., J. R. West, N. B. Hallam, C. F. Forster, and J. Simms. 2000. Performance of various kinetic models for chlorine decay. J. Water Resources Planning and Management ASCE 126:3–20.

Propato, M., and J. G. Uber. 2004. Vulnerability of water distribution systems to pathogen intrusion: how effective is a disinfectant residual? Environ. Sci. Technol. 38:3713–3722.

Pryor, M., S. Springthorpe, S. Riffard, T. Brooks, Y. Huo, G. Davis, and S. A. Satter. 2004. Investigation of opportunistic pathogens in municipal drinking water under different supply and treatment regimes. Water Sci. Technol. 50:83–90.

Regan, J. M., G. W. Harrington, H. Baribeau, R. DeLeon, and D. R. Noguera. 2003. Diversity of nitrifying bacteria in full-scale chloraminated distribution systems. Water Res. 37:197–205.

Richardson, S. D. 1998. Drinking Water Disinfection By-Products, Vol. 3. Pp. 1398–1421 *In:* Encyclopedia of Environmental Analysis and Remediation. R. A. Meyers (ed.). New York: John Wiley & Sons.

Richardson, S. D., and S. W. Krasner. 2003. Disinfection by-products and other emerging contaminants in drinking water. Trac-Trends in Analytical Chemistry 22(10):666–684.

Richardson, S. D., H. S. Weinberg, S. W. Krasner, and A. D. Thruston. 2004. Nationwide DBP Occurrence Study. EPA-600-R02-068. Athens, GA: EPA NERL.

Ridgway, H. F., E. G. Means, and B. H. Olson. 1981. Iron bacteria in drinking-water distribution systems: elemental analysis of *Gallionella* stalks using X-ray energy-dispersive microanalysis. Appl. Environ. Microbiol. 41:288–292.

Rittmann, B. E., and V. L. Snoeyink. 1984. Achieving microbially stable drinking water. J. Amer. Water Works Assoc. 76(10):106–114.

Robertson, W., and R. Brooks. 2003. The role of HPC in managing treatment and distribution of drinking-water. Pp. 233–244 *In*: Heterotrophic Plate Counts and Drinking-Water Safety. J. Bartram, J. Cotruvo, M. Exner, C. Fricker, and A. Glasmacher (eds.). London: IWA Publishing.

Rossman, L. A., R. M. Clark, and W. M. Grayman. 1994. Modeling chlorine residuals in drinking-water distribution systems. J. Environmental Engineering 120(4):803–820.

Rossman, L. A., R. A. Brown, P. C. Singer, and J. R. Nuckols. 2001. DBP formation kinetics in a simulated distribution system. Water Res. 35(14):3483–3489.

Sandor, J., I. Kiss, O. Farkas, and I. Ember. 2001. Association between gastric cancer mortality and nitrate content of drinking water: ecological study on small area inequalities. European Journal of Epidemiology 17(5):443–447.

Sarin, P., V. L. Snoeyink, D. A. Lytle, and W. M. Kriven. 2004. Iron corrosion scales: model for scale growth, iron release and colored water formation. J. Environmental Engineering 130(4):364–373.

Sarin, P., J. Clement, V. L. Snoeyink and W. M. Kriven. 2003. Iron release from corroded, unlined cast-iron pipe. J. Amer. Water Works Assoc. 95(11):85–96.

Sarin, P., V. L. Snoeyink, J. Bebee, W. M. Kriven, and J. A. Clement. 2001. Physicochemical characteristics of corrosion scales in old iron pipes. Water Res. 35(12):2961–2969.

Sathasivan, A., and S. Ohgaki. 1999. Application of new bacterial regrowth potential method for water distribution system – a clear evidence of phosphorus limitation. Water Res. 33:137–144.

Sathasivan, A., S. Ohgaki, K. Yamamoto, and N. Kamiko. 1997. Role of inorganic phosphorus in controlling regrowth in distribution systems. Water Res. Technol. 5:37–44.

Schoenen, D. 1986. Microbial growth due to materials used in drinking water systems. *In*: Biotechnology, Vol. 8. H. J. Rehm and G. Reed (eds.). Weinheim: VCH Verlagsgesellschaft.

Sen, K., and M. Rogers. 2004. Distribution of six virulence factors in *Aeromonas* species isolated from U.S. drinking water utilities: a PCR identification. J. Appl. Microbiol. 97:1077–1086.

Servais, P., G. Billen, C. Ventresque and G. P. Bablon. 1991. Microbial activity in GAC filters at the Choisy-le-Roi treatment plant. J. Amer. Water Works Assoc. 83:62–68.

Shair, S. 1975. Iron bacteria and red water. Industrial Water Eng. 12:16.

Skadsen, J., and Y. K. Cohen. 2006. Operational and Treatment Practices to Prevent Nitrification. Chapter 8 *In:* Fundamentals and Control of Nitrification in Chloraminated Drinking Water Distribution Systems. AWWA Manual M56. Denver, CO: AWWA.

Skjevrak, I., V. Lund, K. Ormerod, A. Due and H. Herikstad. 2004. Biofilm in water pipelines; a potential source for off-flavours in the drinking water. Water Sci. Technol. 49:211–217.

Smith, C. 2005. Nitrification: problems and solutions. AWWA Cal/Nevada Conference Presentation. Available at: www.CharlotteSmith.us.

Smith, C. 2006. Monitoring for nitrification prevention and control. Chapter 7 *In:* Fundamentals and Control of Nitrification in Chloraminated Drinking Water Distribution Systems. AWWA Manual M56. Denver, CO: AWWA.

Smith, D. B., A. F. Hess, and S. A. Hubbs. 1990. Survey of distribution system coliform occurrences in the United States. *In*: Proceedings of the Water Quality Technology Conference. Denver, CO: AWWA.

Smith, C. D., J. F. Smith, and B. Milosky. 1996. Excessive loss of chloramine residual: a case study. *In*: Proceedings of the AWWA Annual Conference. Denver, CO: AWWA.

Snead, M. C., V. P. Olivieri, K. Kawata, and C. W. Kruse. 1980. The effectiveness of chlorine residuals in inactivation of bacteria and viruses introduced by post-treatment contamination. Water Res. 14:403–408.

Speight, V. L., and P. C. Singer. 2005. Association between residual chlorine loss and HAA reduction in distribution systems. J. Amer. Water Works Assoc. 97(2):82–91.

Suffet, I. H., J. Ho, D. Chou, D. Khiari, and J. Mallevialle. 1995. Taste-and-Odor Problems Observed during Drinking Water Treatment. Pp. 1–21 *In:* Advances in Taste-and-Odor Treatment and Control. I. H. Suffet, J. Mallevialle, and E. Kawczynski (eds.). Denver, CO: AwwaRF.

Szewzyk, U., R. Szewzyk, W. Manz, and K. H. Schleifer. 2000. Microbiological safety of drinking water. Annu. Rev. Microbiol. 54:81–127.

Tallon, P., B. Magajna, C. Lofranco, and K.-T. Leung. 2005. Microbial indicators of faecal contamination in water: a current perspective. Water, Air, Soil Poll. 166:139–166.

Texas Commission on Environmental Quality (TECQ). 2005. 30 TAC290.110(b)(4) §290.110. Disinfectant Residuals.

Tuovinen, O. H., K. S. Button, A. Vuorinen, L. Carlson, D. Mair, and L. A. Yut. 1980. Bacterial, chemical, and mineralogical characteristics of tubercles in distribution pipelines. J. Amer. Water Works Assoc. 72:626–635.

Tuovinen, O. H., D. M. Mair, and J. Banovic. 1984. Chlorine demand and trihalomethane formation by tubercles from cast iron water mains. Environmental Technology Letters 5:97–108.

Uber, J. G. 2003. Maintaining Distribution System Residuals through Booster Chlorination. Denver, CO: AwwaRF.

Valentine, R. L., and S. W. Stearns. 1994. Radon Release from Water Distribution System Deposits. Environ. Sci. Technol. 28(3):534–537.

Valentine R. L., K. Ozekin, and P. R. Vikesland. 1998. Chloramine Decomposition in Distribution System and Model Waters. Denver, CO: AwwaRF.

van der Kooij, D. 1992. Assimilable organic carbon as an indicator of bacterial regrowth. J. Amer. Water Works Assoc. 84:57–63.

van der Kooij, D., and W. A. M. Hijnen. 1990. Criteria for defining the biological stability of drinking water as determined with AOC measurements. Pp. 1281–1333 *In:* Proceedings of the Water Quality Technology Conference. Denver, CO: AWWA.

van der Kooij, D., W. A. M. Hijnen, and J. C. Kruithof. 1989. The effects of ozonation, biological filtration, and distribution on the concentration of easily assimilable organic carbon in drinking water. Ozone Sci. Engrg. 11:297.

van der Wende E., W. G. Characklis, and D. B. Smith. 1989. Biofilm and bacterial water quality. Water Res. 23:1313–1322.

van der Leeden, F., F. L. Troise, and D. K. Todd. 1990. Water quality. Pp. 417–493 *In*: The Water Encyclopedia, Second Edition. Chelsea, MI: Lewis Publishers.

van Lieverloo, J. H. M., R. Buuren, G. Veenendaal, and D. van der Kooij. 1998. Controlling invertebrates in distribution systems with zero or low disinfectant residual. Water Supply 16(3/4):199–204.

Vasconcelos, J. J., P. F. Boulos, W. M. Grayman, L. Kiene, O. Wable, P. Biswas, A. Bhari, L. A. Rossman, R. M. Clark, and J. A. Goodrich. 1996. Characterization and modeling of chlorine decay in distribution systems. Denver, CO: AwwaRF.

Vasconcelos, J. J., L. A. Rossman, W. M. Grayman, P. F. Boulos, and R. M. Clark. 1997. Kinetics of chlorine decay. J. Amer. Water Works Assoc. 89(7):54–65.

Venczel, L. V., M. Arrowood, M. Hurd, and M. D. Sobsey. 1997. Inactivation of *Cryptosporidium parvum* oocysts and *Clostridium perfringens* spores by a mixed-oxidant disinfectant and by free chlorine. Appl. Environ. Microbiol. 63:1598–1601.

Victoreen, H. T. 1974. Control of water quality in transmission and distribution mains. J. Amer. Water Works Assoc. 66:369–370.

Vikesland, P. J., K. Ozekin, and R. Valentine. 2001. Monochloramine decay in model and distribution system waters. Water Res. 35(7):1766–1776.

Wagner, D., A. H. L. Chamberlain, W. R. Fischer, J. N. Wardell and C. A. C. Sequeira. 1997. Microbiologically influenced corrosion of copper in potable water installations—a European project review. Mater. Corrosion. 48:311–321.

Walker, J. T., K. Hanson, D. Caldwell, and C. W. Keevil. 1998. Scanning confocal microscopy study of biofilm induced corrosion on copper plumbing tubes. Biofouling 12:333–344.

Watson, S. 2004. Aquatic taste and odor: a primary signal of drinking-water integrity. J. Toxicol. Environ. Health. 67:1779–1795.

Watson, S. B., and J. Ridal. 2004. Periphyton: a primary source of widespread and severe taste and odour. Water Sci. Technol. 49:33–39.

Webster, B. J., S. E. Werner, and P. J. Bremer. 2000. Microbially influenced corrosion of copper in potable water systems—pH effects. Corrosion 56:942–950.

Wilczak, A. 2001. Chloramine decay rate: factors and research needs. *In*: Proceedings of the AWWA Annual Conference. Denver, CO: AWWA.

Wilczak, A., C. D. Smith, Y. K. Cohen, and P. B. Martin. 2003. Strategies for combining free ammonia and boosting chloramines in distribution systems—survey of utility practice. *In:* Proceedings of the AWWA Annual Conference and Exposition. Denver, CO: AWWA.

Wolfe, R. L., N. I. Lieu, G. Izaguirre, and E. G. Means. 1990. Ammonia oxidizing bacteria in a chloraminated distribution system: seasonal occurrence, distribution, and disinfection resistance. Appl. Environ. Microbiol. 56(2):451–462.

Wolfe, R. L., E. G. Means, M. K. Davis, and S. E. Barrett. 1988. Biological nitrification in covered reservoirs containing chloraminated water. J. Amer. Water Works Assoc. 80(9):109–114.

Woolschlager, J., B. Rittmann, P. Piriou, and B. Schwartz. 2001. Using a comprehensive model to identify the major mechanisms of chloramine decay in distribution systems. Water Sci. Technol.: Water Supply 1(4):103–110.

Zhang, L., W. A. Arnold, and R. M. Hozalski. 2004. Kinetics of haloacetic acid reactions with Fe(0). Environ. Sci. Technol. 38:6881–6889.

Zhang, X. R., and R. A. Minear. 2002. Decomposition of trihaloacetic acids and formation of the corresponding trihalomethanes in drinking water. Water Res. 36(14):3665–3673.

7
Integrating Approaches to Reducing Risk from Distribution Systems

The few regulations that govern water quality in distribution systems are the result of years of research leading to the demonstration of a risk to the water-consuming public from specific contaminants. The development of regulations is a complex process that includes cost analysis (EPA, 2003) and, more recently, stakeholder input as described in the Federal Advisory Committee Act. Many state regulatory agencies are either reluctant to or prohibited by statute to require measures to protect drinking water beyond those mandated by federal statute. However, drinking water utilities may independently choose to conform to industry standards to design and operate their systems beyond regulatory requirements.

Standards are useful to water suppliers that have adopted such a precautionary stance. *Recommended Standards for Water Works: Ten State Standards* (The Great Lakes-Upper Mississippi River Board of State Public Health and Environmental Managers, 2003), NSF International, and the American National Standards Institute (ANSI) are third party producers of standards that are widely used in the drinking water industry. Voluntary adoption of standards by a utility requires reallocation of resources. Nevertheless adoption of certain standards is almost universal for community water systems, such as ANSI/NSF 60 governing components that come in contact with drinking water, ANSI/NSF 61 governing additives to water, and many American Water Works Association (AWWA) standards related to design of infrastructure such as *D100-96—Welded Steel Tanks for Water Storage*. Other widely used AWWA standards related to distribution system integrity include the *C651—Disinfecting Water Mains, C652—Disinfection of Water-Storage Facilities,* and *D101-53 (R86)—Inspecting and Repairing Water Tanks, Standpipes, Reservoirs, and Elevated Tanks for Water Storage.* In addition to industry standards, AWWA "Manuals of Water Supply Practices," such as *M6 Water Audits and Leak Detection*, are commonly used by drinking water utilities to enhance their operations and service to the public.

In 1999 a technical workgroup was organized to develop a Drinking Water Distribution System Assessment Workbook, which began the process that culminated in the G200 Standard. The purpose of the G200 standard is to "define the critical requirements for the operation and management of water distribution systems, including maintenance of facilities" (AWWA/ANSI, 2004). Several components of the G200 standard relate directly to issues highlighted in the U.S. Environmental Protection Agency (EPA) Distribution System White Papers (see Chapter 1) and characterized as high priority by this committee (see Appendix

A). These include Section 4.1.1: Compliance with regulations, 4.1.3: Disinfectant residual maintenance, 4.2.1 System pressure monitoring and requirements, 4.2.2 Backflow prevention, and 4.3.1 Storage facilities. As listed in Table 7-1, G200 includes requirements related to water quality, distribution system management, and facility operation and maintenance. The standard references several existing standards such as those cited above.

TABLE 7-1 G200 Requirements

Section	Title	Requirement
4.1	***Water Quality***	
4.1.1	**Compliance with regulatory requirements**	Meet or exceed regulatory requirements.
4.1.2	**Monitoring and control**	
4.1.2.1	*Sampling plan*	Establish plan, review annually, analyze/trend data, have action plan to respond to changes.
4.1.2.2	*Sample sites*	Include all types of locations including dead ends and storage. Past problem areas require more sampling.
4.1.2.3	*Sample collection*	Use *Standard Methods*, standardized labels and chain of custody forms.
4.1.2.4	*Sample taps*	Protect from contamination. Inspect annually.
4.1.3	**Disinfectant residual maintenance**	
4.1.3.1	*Disinfectant residual*	Maintain detectable or HPC $\leq$ 500 CFU/mL.
4.1.3.2	*Nitrification control*	Monitor free ammonia, control chlorine-to-ammonia ratio.
4.1.3.2.2	Nitrification monitoring	Monitor nitrification indicator parameters.
4.1.3.3	*Booster disinfection*	
4.1.3.3.1		Document residual goals. Monitor compliance with goals.
4.1.3.3.2		Maintain operating procedures that take into account seasonal variation, quality, flow, and system operations.
4.1.3.3.3		Written Plan showing response to variation between goals and observed values.
4.1.3.4	*Disinfection byproduct monitoring and control*	
4.1.3.4.1		Monitor and control DBPs. Set goals for DBPs at critical points.
4.1.3.4.2		Have action plan to respond to levels that exceed goals.
4.1.4	**Requirements for utilities not utilizing a disinfectant residual**	Monitor and record HPC.
4.1.4.1	*Response program*	Have action plan to respond when HPC levels are above goals.

continues

TABLE 7-1 Continued

Section	Title	Requirement
4.1.5	**Internal corrosion monitoring and control**	
4.1.5.1	*Prevention and response program*	Have action plan to respond to internal corrosion and deposition.
4.1.6	**Aesthetic water quality parameters**	
4.1.6.1	*Color and staining*	Have action plan to address color and staining.
4.1.6.2	*Taste and odor*	Have action plan to address taste and odor.
4.1.7	**Customer relations**	
4.1.7.1	*Customer inquiries*	Have system to document customer inquires.
4.1.7.2	*Service interruptions*	Have system to document planned and unplanned service interruptions.
4.1.8	**System flushing**	Develop and implement a systematic flushing program.
4.2	***Distribution System Management Programs***	
4.2.1	**System pressure**	
4.2.1.1	*Minimum residual pressure*	Minimum pressure > 20 psi.
4.2.1.2	*Pressure monitoring*	Monitor pressure. Pressure alarms may be used.
4.2.2	**Backflow prevention**	Have program at least as stringent as AWWA M14.
4.2.3	**Permeation prevention**	Address in utility operation plan.
4.2.4	**Water losses**	
4.2.4.1	*Water loss*	Have goal for the amount of water loss. Document calculation.
4.2.4.2	*Response program*	Have action plan to respond if goal is not met.
4.2.4.3	*Leakage*	Quantify leakage on annual basis.
4.2.5	**Valve exercising and replacement**	
4.2.5.1	*Valve exercising program*	Have valve exercising program.
4.2.6	**Fire hydrant maintenance and testing**	
4.2.6.1	*Maintenance and testing*	Comply with AWWA M17.
4.2.7	**Materials in contact with potable water**	
4.2.7.1	*Approved coatings or linings*	Specify in accordance to AWWA standards, NSF 61, or other.
4.2.8	**Metering**	
4.2.8.1	*Metering requirements*	Determine daily peak flows and maximum day peak flows.
4.1.8.2	*Metering devices*	Meters shall meet AWWA requirements or other applicable standard.
4.2.8.3	*Testing*	Test as recommended in AWWA M6.
4.2.8.4	*Repair and replacement programs*	Have program that includes records to verify conformance with AWWA M6.

continues

TABLE 7-1 Continued

Section	Title	Requirement
4.2.9	**Flow**	
4.2.9.1	*Flow requirements*	Be capable of delivering maximum day demand and fire flow.
4.2.10	**External corrosion**	
4.2.10.1	*Leaks/breaks*	Have a standardized system for recording and reporting leaks and breaks.
4.2.10.2	*Monitoring program*	Have external corrosion monitoring plan.
4.2.11	**Design review for water quality**	
4.2.11.1	*Policies and procedures*	Have standardized design procedures that review construction projects to reduce potential for water quality degradation.
4.2.11.2	*Records*	Prepare as-built drawings.
4.2.12	**Energy management**	
4.2.12.1	*Energy management program*	Review and optimize electrical energy usage.
4.3	***Facility Operation and Maintenance***	
4.3.1	**Treated water storage facilities**	
4.3.1.1	*Storage capacity*	Establish minimum operating levels in storage facilities.
4.3.1.2	*Operating procedures*	Write Standard Operating Procedures for turning over facilities and minimizing water age.
4.3.1.3	*Inspections*	Write Standard Operating Procedures for facility inspection.
4.3.1.4	*Maintenance*	Have a maintenance program for facilities.
4.3.1.5	*Disinfection*	Facilities shall be disinfected according to ANSI/AWWA C652.
4.3.1.6	*Additional requirements*	All facilities shall be covered.
4.3.2	**Pump station operation and maintenance**	
4.3.2.1	*Operating procedures*	Write Standard Operating Procedures describing the operation of each pump station.
4.3.2.2	*Maintenance program*	Write Standard Operating Procedures describing the maintenance of the equipment in each pump station.
4.3.3	**Pipeline rehabilitation and replacement**	
4.3.3.1	*Rehabilitation and replacement program*	Have a program for evaluating and upgrading the distribution system.
4.3.4	**Disinfection of new or repaired pipes**	
4.3.4.1	*Disinfection of new or repaired pipes*	Disinfect according to ANSI/AWWA C651 requirements.
4.3.4.2	*Bacteriological testing*	Testing shall be performed according to ANSI/AWWA C651.
4.3.4.3	*Disposal of chlorinated water*	Disposal shall follow local, state, and federal regulations.

continues

TABLE 7-1 Continued

Section	Title	Requirement
5.1	*Documentation Required*	
5.1.1	General	Include statements of policy and quality objectives, standard operating procedures etc.
5.1.2	Examples of documentation	Document to include requirements of Section 4.
5.1.3	Control of documents	Establish procedures to review and approve and maintain documents.
5.1.4	Control of records	Maintain evidence of conformity to requirements of this standard.
5.2	*Human Resources*	
5.2.1	General	Personnel performing work on the DS will be competent on the basis of appropriate education, training, skills, test requirements, and experience.
5.2.2	Competence, awareness, and training	The utility shall provide training and determine competence.

SOURCE: Excerpted, with permission, from AWWA/ANSI G200 (2004). © 2004 by American Water Works Association.

As discussed in Chapter 2, the use of the standards such as ANSI/NSF 60, ANSI/NSF 61, and AWWA G200 and Manuals of Practice have advantages over programs such as Hazard Analysis and Critical Control Points (HACCP) in that they are more easily adapted to the dynamic nature of drinking water distribution systems. Use of a standard such as G200 that is intended to assess whether the system can be managed under all conditions is appropriate for utilities that desire to operate beyond regulatory requirements. To minimize the public health risks of distribution systems, it is recommended that drinking water utilities adopt G200 or an equivalent program in order to develop distribution system management plans that combine their regulatory requirements and available voluntary standards.

The purpose of this chapter is to discuss certain elements of G200 that deserve more thoughtful consideration because emerging science and technology are altering whether and how these elements are implemented by a typical water utility. Much of the current scientific thrust is in the development of new monitoring methods, models, and methods to integrate data, all to better inform decision making.

MONITORING

Drinking water of "acceptable quality" is defined by the Safe Drinking Water Act (SDWA) and its amendments and is framed in terms of the Maximum Contaminant Levels (MCLs), treatment techniques, rules, and regulations promulgated under the Act. The regulations contain significant monitoring requirements that prescribe the sampling frequency (minimum monitoring frequencies), sampling locations, testing procedures, record keeping, and the water quality parameters to be monitored, and are classified according to system size and vulnerability. The regulations also cover specific reporting procedures to be followed if a contaminant exceeds an MCL. Failure to have the proper water quality analyses performed or to report the results to the state primacy agency can result in the water system having to provide public notification.

Under the SDWA, monitoring or treatment techniques are required for all contaminants regulated under the Act, both at the entry point to a water distribution system and, in some cases, at various locations within the system. Rules and regulations that explicitly require monitoring in the distribution system include the Total Coliform Rule (TCR), the Surface Water Treatment Rule (SWTR) and Long-Term Enhanced Surface Water Treatment Rule (LTESWTR), Lead and Copper Rule (LCR), and the Stage 2 Disinfectants/Disinfection By-Products Rule (Stage 2 D/DBPR). These requirements are summarized in Table 7-2. Routine compliance monitoring is a useful tool for detecting and assessing some common water quality problems throughout a system if the event is large enough and long enough in duration to be detected (Byer and Carlson, 2005). Note that pressure monitoring is not required by any of the existing rules, which is unfortunate.

The compliance monitoring required by the SDWA is limited in its ability to protect public health because the end-point or customer tap monitoring required under the regulations is typically (1) not sufficient to provide early warning of contamination, (2) not indicative of what could have gone wrong between the treatment plant and the consumer's tap so as to effectively guide remediation, and (3) too limited across space (too few sampling locations) and time (discrete small volume samples are collected too infrequently) to provide information that applies to every potential user. The realities of financial and personnel resources in most cases preclude expanding monitoring programs to cover vastly larger areas and periods of time.

Rather, it is more useful for utilities to consider how to control the processes taking place within the distribution system, as well as activities to maintain the processes, such that the risk of the customer being exposed to contaminated drinking water is minimized. This concept hinges on viewing a water distribution system as a linkage of *processes* working together to maintain flow, pressure, and water quality. These processes include pumping, valving, metering, transmission, distribution, service, storage, and corrosion control, to name a few. Though each individual distribution system is a unique linkage of processes, the processes have common characteristics that allow generalizations to be made about their control. For example, the number of storage tanks from one system to another may be different, but there are common problems with hydraulic retention time and chlorine loss in all storage tanks. The variety of pipes used (materials and sizes) will differ from one system to another, but cast iron displays a common corrosion problem in all systems.

TABLE 7-2 Federal Distribution System Water Quality Monitoring Requirements

Regulation	Monitoring Requirement
Total Coliform Rule	• Samples must be collected at sites that are representative of the water throughout the distribution system based on a sample siting plan that is subject to review by the primacy regulatory agency. • The minimum number of samples that must be collected per month depends on the population served by the system. • For each positive total coliform sample, there are various repeat sampling requirements.
Surface Water Treatment Rule (SWTR) and multiple Long-Term Enhanced Surface Water Treatment Rules (LTESWTRs)	• Disinfectant residuals must be measured at TCR monitoring sites. • Disinfectant residual must be monitored at the entry to the distribution system. Larger systems (> 3,300 population) must provide continuous monitoring. Systems serving less than 3,300 population can take grab samples.
Lead and Copper Rule (LCR)	• All systems serving a population > 50,000 people must do water quality parameter (WQP) monitoring. • Samples must be collected for Pb/Cu at Tier I sites. The number of sample sites for Pb/Cu and water quality monitoring is based on system size.
Stage 2 Disinfectants/ Disinfection By-Products Rule (DBPR)	• Standard Monitoring Program requires one year of data on THMs and HAAs. Number of sampling locations based on utility size and source characteristics. Modeling can reduce sampling requirement.

SOURCE: Owens (2001) and Lansey and Boulos (2005).

In addition to being a linkage of processes, the distribution system is also a reactor, in that treated drinking water begins to change physically (e.g., iron and manganese particles settle out), chemically (e.g., chlorine begins to decompose) and biologically (e.g., bacterial cells begin to adhere to pipe surfaces and form biofilms) as soon as water leaves the treatment plant. Each of the processes display common tendencies to promote these changes irrespective of how they are linked within a distribution system.

Real-time feedback on whether a utility's distribution system processes are in or out of control goes beyond the regulatory requirements for water quality monitoring mentioned above. The following sections discuss monitoring for process control; they are intended to build upon discussions of detection methods and tools, such as such Geographic Information System (GIS) and hydraulic modeling, found earlier in the report. A systematic strategy for distribution system monitoring to detect water quality alterations is comprised of the following actions: (1) develop a list of parameters to be monitored, (2) assess appropriate temporal and spatial scales for monitoring, (3) develop a response plan for monitored parameters, and (4) implement. Each of these activities is discussed below, focusing on recent scientific developments that should lead to improvement in how the activities are conducted.

Parameters to be Monitored

The parameters that are useful for monitoring distribution system processes may include those that are required from a regulatory point of view (e.g., turbidity, chlorine residual), but likely would include others. The key requirements are that the monitoring parameters can be measured relatively quickly, inexpensively, and (ideally) continuously at multiple locations in the system. The parameters should be selected with consideration for the potential mechanisms that may induce adverse changes in water quality. For example, in corrosive waters passing through ductile iron pipe, conductivity, pH, and oxidation reduction potential (ORP) may be useful. In waters passing through polymeric pipe or vulnerable to intrusion in contaminated overlying soils, UV254 or TOC may be useful.

Table 7-3 lists sentinel parameters that could be used to indicate changes in distribution system integrity. These parameters include indicators of physical deterioration (pressure changes, main breaks, water loss, or corrosion), hydraulic failure (turbidity, complaints of low flow or pressure) or a water quality failure

TABLE 7-3 Sentinel Parameters for Distribution System Integrity

Parameter	Physical	Hydraulic	Water Quality
Routine (Primary)			
Pressure	X	X	
Turbidity	X	X (flow reversals)	X
Disinfectant residual		X (water age)	X
Main breaks	X		
Water loss	X		
Color	X (corrosion)		X
Coliforms	X (sanitary, main break)		X (biofilms)
Flow velocity and direction		X (pipes, tanks)	
pH, Temperature			X
Chemical parameters	X	X	X
Secondary			
TOC			X
UV Adsorption			X
T&O	X (permeation)	X (water age)	X (biofilms)
Metals	X (corrosion)		X
Nitrite			X (nitrification)
HPC			X (biofilms??)
Tank level/volume		X	

Note: **Bold** entries indicate those parameters for which on-line real-time sensors are available.

(particulates, tastes, odors, or color). For some of the parameters (listed in bold), on-line monitoring equipment is available to provide real-time control of distribution system operations. Most methods for monitoring physical characteristics of water (e.g., flow, velocity, water level in a storage tank) tend to be relatively inexpensive, quite durable, and able to generate continuous, real-time, on-line data (Grayman et al., 2004; Panguluri et al., 2005a). Less available in on-line, real-time versions are methods for detecting inorganic chemicals, synthetic organic chemicals, volatile organic chemicals, and radionuclides. The direct real time detection of biological changes within distribution systems remains beyond current technology (Bernosky, 2005).

Pressure. One of the most important parameters for utilities to consider monitoring for is transient pressure change using high-speed, electronic pressure data loggers. Recent research has documented the frequency and magnitude of pressure transient events (Friedman et al., 2004; Gullick et al., 2005). High-speed data loggers are required for monitoring distribution system pressure transients because such transients may last for only seconds and may not be observed by conventional pressure monitoring. High-speed pressure data loggers can measure pressures at a rate of up to 20 samples per second, allowing measurement of sudden changes in pressure. The units can be programmed with preset alarm levels to notify operators when specific thresholds have been exceeded. Additionally, some units can be programmed to capture and store specific data surrounding a pressure transient event, permitting the episode to be analyzed and corrective actions to be determined.

Turbidity. Turbidity in distribution systems, which can be can be caused by suspended sediments, oxidized iron or manganese, or other corrosion products, is another critical parameter for which on-line, real-time methods are available. Various models exist but in the finished water distribution system, turbidity probes need to be sensitive at low ranges (i.e., < 1 NTU). Measurement accuracy may be improved further by employing wiper or shutter mechanisms that are activated immediately prior to measurements to avoid interferences from particulates or air bubbles. In general, turbidity units from different manufacturers behave similarly, and calibration frequencies vary from weekly up to three monthly intervals, but require a good level of operator skill. On-line turbidimeters are being used successfully under the Partnership for Safe Water Program to monitor low level (< 0.3 NTU) turbidity, and therefore should prove valuable for low level turbidity in distribution system monitoring.

Disinfectant Residual. Disinfectant residual monitors can measure free chlorine, chloramines, or ORP. The principle of detection for residual on-line sensors relies on either polarographic, voltametric, or colorimetric methods which can influence their sensitivity, calibration, and interferences from other water quality parameters. Operation of an ORP sensor is similar to that of the pH sensor where a two-electrode system is used to make potentiometric measurements. The calibration frequency for these monitors is usually on a monthly basis. Typical data from continuous monitoring of total chlorine residual is shown in Figure 7-1.

Flow. In-line meters are available to measure flow in the distribution system but are typically used only to monitor flows into distribution system sub-districts. Monitoring flows by sub-district can be compared to customer meter data to indicate the amount of leakage in specific areas of the distribution system. Flows can be influenced by pumping regimes, storage tank operations, and manipulations of hydrants or blow-off valves. Use of a well-calibrated distribution system hydraulic model along with pump, tank, and flow data is required to generate detailed descriptions of distribution system water velocities and flow reversals.

pH. Measurements of pH are made with a pH meter using a glass indicator electrode. These measurements are reliable, but the meter requires regular calibration to avoid drift.

Temperature. Temperature thermistors typically work over a relatively small temperature range and can be very accurate within that range. The measurements are very reliable and typically do not require routine calibrations.

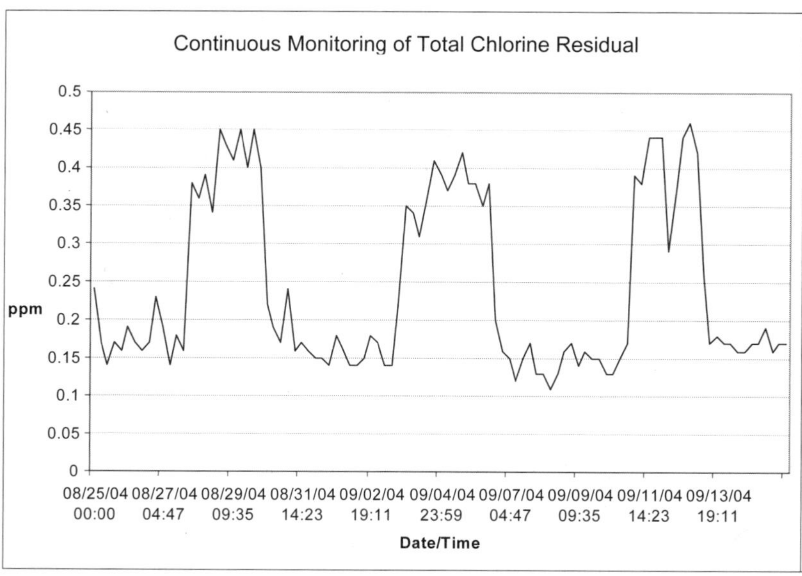

FIGURE 7-1 Data from a continuous, on-line chlorine analyzer, showing how a total chlorine residual can vary through a day and the need to relate this to system operations. SOURCE: Data from Philadelphia Water Department, Bureau of Laboratory Services.

Chemical Parameters. For chemical parameters, EPA has been examining the reliability of on-line sensors under the Environmental Technology Verification (ETV) Program (EPA, 2004a,b). Currently this program has examined 40 monitoring and treatment technologies and plans to conduct additional testing under the newly formed Technology Testing and Evaluation Program (TTEP)—an off shoot of the ETV program, which is not dependant upon voluntary vendor involvement. This independent testing is providing a valuable database on the reliability of on-line monitors (*http://www.epa.gov/etv/*). The sensors being developed are not specific for the chemical contaminants themselves. Rather, the premise of the research is that a chemical contaminant in a distribution system would elicit a pattern of changes in other, primary parameters that can be easily measured in real time, such that changes in their detection would indicate the presence of the contaminant. Table 7-4 shows the responsiveness of various water quality parameters to a range of contaminants in controlled experimental tests.

It should be noted that the actual ability of the on-line sensors shown in Table 7-4 to detect a target contaminant in field situations has not been ascertained. Hence, whether a particular pattern of shifts in a battery of on-line analysis results can be reliably associated with a particular type of contaminant (e.g., malathion) is uncertain. Another National Research Council committee is in the process of examining research needs in the area of drinking water homeland security, and further discussion of this issue may be found in its report.

TABLE 7-4 Responsiveness of parameters that can be easily measured on-line to various contaminants

Contaminant Compound	**Water Quality Parameter**								
	Free/total chlorine	**ORP**	**TOC**	**SC**	**Turbidity**	**NH_3**	**N_2**	**NO_3^-**	**Cl^-**
Ferricyanide	NC (F+ w/ DPD test)	+	++[a]	+	+	F-		F-	F-
Malathion (pesticide)	+	+	+	NC	+				+
Glyophosphate (herbicide)	+	+	+	NC	NC				+
Nicotine (organic)	++		++	NC					
Arsenic trioxide	++	++	NC		+	++	++		
Aldicarb	++		++						
Groundwater	+	+	NC	+	NC				+
Wastewater	+	+	+	+	+				+

Key: ++ = very responsive, + = responsive, F+ = false positive, F- = false negative, NC = no change
Abbreviations: ORP, oxidation/reduction potential; TOC, total organic carbon; SC, specific conductance.
[a]May be due to bound carbon in the cyanide complex, from Hall et al. (2005).
Note: For the pesticides and herbicides, commercial products were used that had different concentrations of organic compounds.
SOURCE: Adapted from Hall et al. (2005).

Secondary Parameters. In addition to the primary on-line monitoring tools, a range of other monitoring approaches can be used to enhance the measurement of distribution system integrity. Both ultraviolet absorbance and transmittance monitors operate on the optical principle where light of known wavelength (typically 254 nm) and intensity is passed through a sample cell of a known path length. A photo-detector on the opposite side to the light source measures the degree of light attenuation by the sample. The percent of UV light passing through the water determines the UV transmittance (UVT) or alternatively can be translated into absorbance. Double bonds and ring structures strongly absorb light at 254 nm and therefore absorbance (or transmittance) provides useful relative measures of the amount of organic matter, which can contribute to color in water. Because the detection procedure utilizes optics, small particles or other materials (such as dissolved iron) that can deposit on the cell windows can lead to interference.

Measurement of specific conductance, color, TOC, chemical ions, and metals can be used to measure changes in water quality baseline values. Conductivity is directly affected by the number of dissolved ions, and when adjusted for a given temperature (usually 25°C), it is referred to as specific conductance (Siemens per cm) and can be used for approximating the total dissolved solids content. Ion selective electrodes for Cl^-, NO_3^-, NH_4^+, and others analytes are available but are not entirely ion-specific and can lead to problems of ionic interference. Several on-line TOC monitors are commercially available but the routine maintenance and calibration are cumbersome and require oxidants, carrier gas, and UV lamp replacements. Some units have simplified this process by using pre-packaged chemical packs that are easily replaced. Some units use high temperature catalysis for the oxidation step, and thus eliminate the need for oxidant chemicals.

***

Various manufacturers have either single or multi-parameter sensors that can monitor distribution system water quality and directly communicate the data to the Supervisory Control and Data Acquisition (SCADA) system, as shown in Table 7-5. There are a multitude of options currently available that have been recently reviewed (Hasan, 2005). Multi-parameter sensors are available in a panel format that uses a side-stream to draw a sub-sample of water from the distribution system or alternatively exist as sondes that can be installed directly or indirectly within distribution system pipes. The former are large conspicuous units, and, to avoid being tampered with, their use in the distribution system would have to be limited to secure locations. Also, the side-stream of water drawn from the distribution system into the multi-parameter sensors requires appropriate disposal (i.e., into a sanitary sewer) under normal operation. In contrast, the in-line sondes have the potential to be installed discretely within the distribution system, which reduces their vulnerability to tampering but may complicate routine calibration.

It should be noted that the on-line monitoring technologies discussed above are currently cost-prohibitive and too complex for premise plumbing and service lines. However, this may change with technological advances.

Consumer Complaints

A final type of monitoring that utilities may want to consider is consumer complaints monitoring. Consumers can detect off odors, changes in taste or flavor, color, turbidity, and particulates resulting from system failures (e.g., water main breaks, cross connections) as well as from system operations (e.g., hydrant flushing, valve operations). While these untrained assessors are subjective and unreliable from a laboratory testing point of view, they are everywhere at all times in a distribution system and thereby serve as valuable sources of information on potential water quality problems (Burlingame, 1999a,b; Laurer, 2005).

Collection and mapping of customer complaints should be done in the context of a GIS-linked database to monitor conditions in the distribution system, track operational issues, and determine the boundary of water quality events. Critical to the functionality of using customer data as a monitoring tool is the seamless integration and transfer of complaint data into operational databases so that all functional departments (production, network, water quality, management, communications, etc.) are instantly aware of any disturbances. Use of mobile computers can effectively communicate and coordinate workforce resources in the field to respond to and mitigate any events. Unfortunately, a survey in North America showed that while 84 percent of the responding utilities have formal procedures in place for investigating customer complaints, only 61 percent had a customer complaint database (Deb et al., 2000). This was the case even though these utilities often relied on customer notification for early detection of problems.

A three-year study demonstrated the benefits of certain practices at reducing customer complaints related to water quality from a distribution system in Southern California (Wen et al., 2005). Problems with manganese, old cast iron pipes, and rusty water were addressed by keeping good records of customer complaints and developing a database to sort and track the complaints. These data were then used to show the improvements made by the chosen controls.

One drawback to customer complaints is that they are end-user in origin and so cannot distinguish between contamination originating within the customer's premise, the service line and local water main, a regional storage facility, or all the way back to the treatment plant and the source water (Burlingame, 1999b). Nonetheless, customer complaints may be the first line of detection of water quality problem short of having an exhaustive and expensive monitoring plan.

TABLE 7-5 Performance Specifications of Commercially Available Sensors

Manufacturer	***Dascore***	***YSI***	***Hydrolab-Hach***	***Analytical Technology, Inc***	
Model	Sixcense	6-series	(DS5X, DS5, MS5)	Series Q45	Model A 15/B-2-1
Brief description	Multi-parameter, on-line, free chlorine or chloramines	Multi-parameter, on-line	Multi-parameter, on-line	Multi-parameter, customizable on-line	Free chlorine or chloramines
Cost (US $)	9,700	15,000	15,000	< 10,000	3,000
Free chlorine (mg per L)	0–5			0–2, 0–20 or 0–200	0–2, 0–20 or 0–200
Chloramines (mg per L)	0–20			0–2, 0–20 or 0–200	0–2, 0–20 or 0–200
TOC (mg per L)					
ORP (Volts)	-1.4–1.4	-0.999–0.999	-0.999–0.999	-0.999–2.0	
SC (mS per cm)	0.1–10	0–100	0–100	0–0.2 and 0-40	
DO (%)	0–200	0–500	0–200*	0–40 ppm	
Turbidity (NTU)			0–3000	0.001–4 plus other wider ranges	
PH	2–12	0–14	0–14	0–14	
Temp (°C)	0–50	-5–45	-5–50		

*LDO: luminescent DO measurements. For the bottom nine rows, the values given are the detection ranges of the sensors. SOURCE: Reprinted, with permission, from Bukhari and LeChevallier (2006). © 2006 by American Water.

Wallace & Tiernon	***Emerson***	***Hach***		***ProMinent***	***In-Situ***
Depolox 3 plus	Model 1055 Comp II Analyzer	WDM Panel or Pipe Sonde™	TOC Process Analyzer	D1C &D2C	Troll 9000
Free or total chlorine (or chlorine diox-ide or ozone)	Multi-parameter, customizable, on-line	Multi-parameter, on-line	Used with WDM	Multi-parameter, on-line	Multi-parameter, on-line
3,500	15,885	12,000	18,000	7,000	11,200
0–20	0–10	0–4 (DPD)		0–0.5, 0–2, 0–10	
	0–15	0–5 (total)		0–0.5, 0–2, 0–10	
			< 5–20,000		
	-1.4–1.4	-1.5–1.5		-1.0–1.0	-1.4–1.4
	0–200	0–100		0–200	0–200
	0–20 ppm	0–20 ppm		0.1–10 or 0.1–20 ppm	0–20 ppm
	0.001–200	1–100			1–2000
	0–14	0–14		0–12	0–12
		-5–50		0–100	-5–50

Temporal and Spatial Scales for Monitoring

Two important aspects of distribution system monitoring need to be reconsidered by many utilities. The first involves the timing of sampling, which is dictated largely by the method by which samples are taken. Currently most routine water quality monitoring in a distribution system is carried out through manual grab samples followed by analysis in the field or in the laboratory. A grab sample is a single water sample collected at a specific point in time. Essentially all distribution system regulatory monitoring uses this method. For example, samples required under the SWTR are manually collected at sites within the distribution system and manually tested for disinfectant levels in the field. Samples taken to satisfy the requirements of the TCR are also manually collected in the field and subsequently analyzed in the laboratory. Manual sampling is labor intensive, and the number of samples that can be collected is limited by personnel and analysis costs. In addition, grab sampling can only show the water characteristics at the time the samples were taken. Important events (e.g., night-time events) that occur between samples are lost or unusual results may be dismissed (Premazzi and Hargesheimer, 2002). Thus, grab sampling is of limited use as an alert system to warn against potential contaminants that might pose a threat to public health.

Moving from grab-sampling to real-time, on-line monitoring is essential for more expeditious and accurate water quality assessment. This trend is being reinforced because of increased emphasis among water utilities in consolidating and automating data processing and control functions (Premazzi and Hargesheimer, 2002). On-line monitoring has the benefit of providing—in real time—early warning of intentional or accidental contamination, and when fully develop and deployed it could help water utilities take the appropriate actions to safeguard public health. On-line monitoring requires a mechanism for moving the sample water from the distribution system to an instrument, instrumentation for analyzing the water, a mechanism for communicating the results, and a means of assessing the results of the monitoring. As discussed above, relatively inexpensive on-line water quality monitoring instruments are becoming more prevalent (Byer and Carlson, 2005). Additionally, the instrumentation must be periodically calibrated and maintained for quality control/quality assurance to guarantee the reliability of generated data (i.e., minimize false positives and negatives). The issue of minimizing false positives cannot be too strongly emphasized. Unless the individual analyzer false positive rate is kept extremely low (e.g., < 1/1000 analyses), when a large number of analyzers are deployed in a single system there is a high likelihood that most of the "hits" will be the result of false positives (if the occurrence of actual true "hits" is rare). Furthermore, for any given analyzer there is likely to be a trade-off between its specificity (ability to detect specific contaminants) and its sensitivity (ability to detect lower levels of contaminants). The statistics of deploying systems of analyzers with specific false positive and negative rates must be considered during the design of the monitoring program.

A second critical consideration is the location of sampling points. The TCR requires the development of an approved distribution system sampling plan. Unfortunately, there is a requirement for routine access for sample collection to both the primary monitoring site and locations within five service connections up-stream and down-stream of the primary site. This results in decreased monitoring in residential areas of the system where access can be limited. Although use of dedicated sampling stations can be used to overcome accessibility problems in these areas, installation of sampling stations require extra cost and can be prone to vandalism, freezing, and contamination.

Current practices for on-line monitors typically locate these devices on utility-owned property where power, sewer, and telemetry to the SCADA system are available so that the results can be instantly communicated to a central operations office for improved system management. These requirements typically restrict monitoring locations to pump stations, storage tanks, well stations, and perhaps government-owned buildings (all of which, coincidentally, may be high priority areas for monitoring for security purposes because they represent points of easy access). Rather than relying on such "convenience monitoring," utilities should consider employing more "risk-based monitoring" where sensors are strategically located based on hydraulic flow, the population at risk, and sensitive locations (e.g., hospitals, government installations, etc.). In particular, water utilities should strive to sample areas where water quality may be more prone to intensive deterioration. These areas may include (depending on the system), areas of low flow, areas subject to frequent flow reversal, areas achieving variable blends of waters from different plants, and areas of old and/or deteriorating pipe. Utilities need the technology, regulatory support, and public understanding to customize their routine water quality monitoring programs to accomplish these more risk-based goals.

Recent research has examined algorithms for placing sensors based on population exposed and time and flow for contaminant detection (Berry et al., 2004). Often, there are trade-offs for one approach versus another, such that optimization programs are needed to choose a best overall strategy. EPA's Threat Ensemble Vulnerability Assessment (TEVA) program is developing an add-on tool in EPANET to allow water utilities to select the optimal number of sensors and identify strategic locations for installation of on-line sensors to maximize public health protection (*http://www.epa.gov/NHSRC/news/news111505b.htm*). Conceptually, given a spectrum of potential threats or vulnerabilities to a distribution system (from either unintentional or intentional events) it is possible to determine the optimal locations to site a given number of detectors such that the likelihood of detecting such events is maximized. The computational framework uses Monte Carlo simulations to vary parameters, such as the quantity or concentration of contaminant, location of injection, duration (or rate) of injection and the probability of ingesting an infectious or toxic dose of these selected contaminants, to generate threat ensembles (collections of many threat scenarios). These threat ensembles are collectively analyzed to estimate health impact statistics, including mean infections or mean fatalities. The public health benefits of no sensors in the distribution system have been compared with both utility convenience monitoring and TEVA designs (Table 7-6).

These computer simulations support the use of on-line sensors in providing an early indication of drinking water contamination events. However, achieving the maximum benefits from on-line installation of sensors requires optimization of the number of sensors in the distribution system. Clearly this number is likely to be system specific and will vary depending upon the distribution system network, the number of service connections, the type of service connections (i.e., primarily residential or commercial), the size of the population being served, and the length of the distribution system pipes. Modeling tools like those being developed by the TEVA program and others (Lee et al., 1991; Murray et al., 2004; Ostfeld, 2004; Ostfeld and Salomons, 2004; Uber et al., 2004a,b) are still under evaluation and will likely undergo significant refinement and validation before finalization. However, until these have been adequately tested, sensor deployment at locations serving the highest population densities may be an appropriate initial strategy. Of course, practical considerations, such as access to power, communication lines, waste disposal (from samplers), and equipment security may limit where sampling can be located. Nonetheless, TEVA analyses could be used to delineate which locations, amongst those identified as feasible, would offer the highest protection.

Data Analysis and Reporting

Although on-line monitors can provide a continuous stream of information, the data needs to be analyzed, reported, and stored at some prescribed frequency. For example, a single on-line multi-parameter sensor measuring six water quality parameters every 15 minutes on a 24-hour basis will lead to the

TABLE 7-6 Public health benefits provided by various sensor location strategies

Sensor Design	**Health Impacts (Fatalities) Biological Attack**			**Health Impacts (Fatalities) Chemical Attack**		
	Median	Mean	Max	Median	Mean	Max
No sensors	980	1,544	22,287	158	139	284
Utility convenience monitoring	671 (31.5%)	1,015 (34.3%)	5,107 (77.1%)	110 (30.0%)	113 (18.70%)	284 (0%)
TEVA	227 (76.8%)	350 (77.3%)	2,730 (87.6%)	67 (57.6%)	78 (43.0%)	229 (19.0%)

*Values in parenthesis are % public health protection relative to the system with no sensors.
SOURCE: Based on analysis by R. Janke and R. Murray, EPA National Homeland Security Research Center, as cited in Bukhari and LeChevallier (2006). Reprinted, with permission, from Bukhari and LeChevallier (2006). © 2006 by American Water.

generation of 4,032 data points per week. Thus, timely management and interpretation of large quantities of data are imperative to the efficient utilization of an on-line monitoring system. It can quickly become an onerous task without the aid of interpretative software. It would be desirable for an automated data analysis package to be capable of not only capturing data from on-line monitoring devices, but also able to (1) perform automated trend analysis that would compare real-time data with baseline historic data to define and characterize anomalies and (2) allow user-defined and programmable triggers with automated notification by means of alarms (on cell phones, pagers, or via e-mail). Presently there are only a limited number of options for predicative data management tools, making this an area ripe for research and innovation.

Advanced Monitoring for Contaminant Identification

As discussed above, real-time monitoring is currently not useful for identifying specific contaminants in distribution systems; rather, they determine baseline water quality conditions and look for deviations from historical trends (Hrudey and Rizak, 2004; Watson et al., 2004). On-line sensors that could detect a range of chemical or biological parameters are a number of years away from commercial development or utility utilization. Additionally, there is a need to advance the technology for parameters that cannot be measured in real-time and on-line. Advances in microfluidics, robotics, and miniaturized components are lowering costs and may have the potential to perform analyses for chemical and microbiological contaminants that just a few years ago required sophisticated and expensive laboratory equipment. Ultimately, a multi-tiered monitoring system is envisioned where on-line water quality monitoring sensors would detect a deviation in baseline water quality and draw a side-stream sample that would be automatically analyzed using an advanced "lab-on-a-chip" that can detect multiple contaminants. On-site microprocessors would analyze the results and send an alarm to the centralized SCADA system. While a mobile analyst is dispatched to verify the on-line monitoring results, the centralized event management software is checking other on-line monitors, customer service, and operational databases for any other anomalies to determine potential causes and a range of corrective actions.

Although the above description of distribution system monitoring is years away from implementation, some water utilities have begun developing elements of what will likely evolve into the envisioned comprehensive monitoring program. For example, the Arizona Department of Environmental Quality has partnered with the Tucson Water Department, the University of Arizona, and several Pima County agencies, businesses, and organizations to provide citizens with on-line information about drinking water quality. The effort was made possible by an Environmental Monitoring for Public Access and Community Tracking (EMPACT) grant from EPA. System monitoring focuses on three components: water quality parameters that are common to all water systems;

specific water quality parameters that focus on public health in water and wastewater treatment; and the volume of water flowing through the cycle. The monitored parameters include pH, conductivity, temperature, hardness, sodium, and total dissolved solids. Additional parameters important for public health include coliform bacteria, disinfectant residuals, total trihalomethanes, fluoride, and nitrate. The overall objectives of the project are (*http://www.ci.tucson.az.us/water/water_quality.htm*):

- Increase the amount of water quality testing by continuous on-line sampling
- Improve the access to water quality data in the potable distribution system
- Provide information for customers by identifying specific constituencies and methods to individualize data by location
- Create a context for understanding water resources data, thus removing misperceptions
- Serve as a source of reliable, authoritative information on fast-breaking water quality issues

The automatic monitoring stations are currently running and continuously updating water quality data on a map-based website; this program is a model for other utilities.

How to Interpret Data and Respond to Monitoring Data

Given a stream of data from a monitoring program, a critical task is to determine whether the results indicate an "event" or "problem" and if so, how utilities should respond. The occurrence of alterations may be ascertained by formal statistical tests (Ortiz-Estarelles et al., 2001), of which there are several types. The presence of outliers (from historical past behavior) can be ascertained using statistical quality control methods (Egan and Morgan, 1998; Lalor and Zhang, 2001). There can be tests of trends to determine if a systematic drift in water quality has occurred. The underlying concept is to assess water quality using a statistical process control concept. To do this, a utility needs to assess what the "normal" water quality, and its fluctuations, might be. Furthermore, water quality data streams require site-specific "tuning" of software in order to detect unusual deviations, and there has been reluctance on the part of some vendors to disclose their tuning and detection algorithms.

There are other fields in which similar problems to the one outlined above have been experienced; it is possible that the methods used in their solution could be adapted. These include the following:

- Identification of financial enterprises that are on the verge of difficulty (Booth et al., 1989)

- Assessment of machine malfunctioning (Javadpoour and Knapp, 2003)
- Detection of outliers in chemical (Egan and Morgan, 1998) and geochemical (Lalor and Zhang, 2001) data
- Structural health monitoring (Omenzetter et al., 2004)
- Detection of computer intrusion or other unwarranted use of computer resources (Lazarevic et al., 2003)

There are several broad approaches to the problem of identifying unusual observations in time series of multivariate data that have been outlined in the literature. These include (1) distance and generalized distance approaches, (2) regression approaches, and (3) neural network approaches. In distance and generalized distance approaches, an observation is regarded as unusual if it is away from the typical population of observations. There are a number of design alternatives for this strategy including the following:

- Euclidean versus other distance scales
- Transformation of variables prior to evaluation (including rank transformations and extraction of principal components)
- Use of direct versus cross-validation distance
- Choice of criteria to call an observation unusual (false positive and negative rates that are deemed acceptable)
- Use of "de-trending" or other preprocessing steps to eliminate non-stationary components of the data.

These methods have been reviewed and discussed by Egan and Morgan (1998) in the context of interpreting analytical chemistry data. These authors discuss use of trimming, distance measurements, use of sub-sampling and test various approaches against sample data from the literature, and specifically recommend *against* the use of ordinary distance measurements and Mahalanobis distance measurements (i.e., distance scaled by the sample variance/covariance matrix) in detecting outliers. Future research is needed to assess the applicability, sensitivity, and selectivity of the various numerical approaches applied to various combinations of measurements which might be taken in a distribution system.

Utilities deploying advanced monitoring in their distribution systems should do so with a specific response plan developed in advance. There are limited options for responses at the disposal of utilities, including boil-water, do-not-consume, or do-not-use notices, that can be applied for particular sections of a system or system-wide. The response to a detection event from a monitoring network carries risks associated with both false positives and false negatives. If an alarm signal is triggered when an actual system deterioration has not occurred (a false positive) and results in an action such as a "do-not-use" or a "boil-water" notice, there may be consequences associated with unavailability of supply for fire fighting and economic impacts on individuals and businesses. If an alarm signal fails to result when an actual deterioration has occurred (a false negative), then there is a failure to detect and respond to an event and thereby reduce its impact. The adverse consequences from such false negatives and false positives need to be considered when determining the action levels at which management will respond to monitoring data.

Implementation of an Enhanced, Process-Oriented Monitoring Program

The monitoring program discussed above, which includes pressure and chlorine residual monitoring as well as other parameters as needed, all continuously monitored and with deviations followed up on in a timely manner, represents a considerable step up from what many water utilities may already be doing. This will require an increase in training, supervision, maintenance, documentation, and management. Training will be needed for using new on-line technology and for its data management and interpretation; for using GIS, hydraulic modeling, and other data integration tools; and for identifying deviations in monitoring data and in responding to the deviations to determine if the associated processes are being adequately controlled. New tools, such as on-line water quality analyzers, will need quality assurance and quality control (calibration, maintenance, data approval) in order to provide reliable data. Software will be needed for the data management and integration. Documentation is critical to providing feedback for the whole program; not only must data be recorded but associated information is needed (on maintenance, accuracy, and quality control) to provide an appropriate level of certainty. The monitoring program outlined above will require comprehensive management, with lines of responsibility clearly outlined and funding and staffing adequately provided for. Furthermore, the monitoring program will have to evolve as the distribution system is adjusted and expanded to meet changing demands over time.

At present, a program of real-time monitoring and the use of the advanced technologies discussed in this section is likely to be feasible only for the largest and most sophisticated utilities and not to smaller (and even non-community) systems, which unfortunately are where a large fraction of disease outbreaks are reported. This mismatch highlights the need for further technical development of sensors alongside alternative strategies for protecting water quality (discussed extensively in Chapter 4, 5, and 6) among smaller utilities.

DISTRIBUTION SYSTEM MODELING

Water distribution network (mathematical) models have become increasingly accepted within the water industry as a viable mechanism for simulating the behavior of water distribution systems. They are intended to replicate the behavior of an actual or proposed system under various demand loading and operating conditions. Their purpose is to support the decision-making processes in various utility management applications including planning, design, operation, and water quality improvement of water distribution systems:

- Planning applications include capital investment decisions to identify and prioritize capital improvements to meet projected growth or to replace aging infrastructure; development of water system master plans to schedule, stage, locate, and size new facilities to support projected growth as well as to analyze the interconnection of separate systems for emergencies; infrastructure rehabilitation and replacement to identify and prioritize water mains that need to cleaned, lined, paralleled (duplicated), or replaced; and water conservation studies to maximize the use of existing supply sources and evaluate sound conservation measures to reduce overall water consumption and capital improvement costs.
- Design applications include estimation of fire protection capacity (e.g., available flow at 20 psi) to verify compliance with fire protection standards; pressure zone management to keep supply pressures within acceptable ranges in regions with significant differences in elevation; determination of the location and size (or capacity) of new water mains, storage facilities, and pump stations to keep pace with projected growth; and hydraulic transient analysis to identify weak spots and select the optimal combination of surge protection or suppression devices to ensure safe system operation.
- Systems operations include energy management applications to optimize storage-pumping trade-off and minimize energy costs; emergency planning to develop an effective emergency response program to reduce or eliminate the damage or impact of unplanned outages at wells, pump stations, pipes, storage tanks, and treatment plants; and daily operational and management decisions to optimize use of existing facilities and train system operators.
- Water quality improvement applications include calculation of water retention time for tanks, travel time in pipes, and the spatial and temporal distribution of water quality throughout the system to predict locations of poor water quality and evaluate improvement measures such as installation of rechlorination facilities and improving reservoir turnover; locating permanent water quality monitoring stations for compliance with federal regulations; and design and implementation of unidirectional flushing programs. Other applications include area isolation during repairs, water loss calculation, leakage minimization, statistical and probabilistic analyses, and, more recently, water security assessment.

Early models simulated hydraulic behavior only and were steady state (static) in nature. But with the advent of more powerful computers and numerical algorithms, extended period simulation (dynamic) models were developed (e.g., Wood, 1980) to simulate behavior under time varying demand and operational conditions, which is necessary because system demands and consequently the flows in the network vary over the course of a day. These models have become ubiquitous within the water industry and are an integral part of most water system design, master planning, and fire flow analyses. In the early 1980s investigators began introducing the concept of water quality modeling (Clark and Males, 1986; Grayman, et al., 1988; Clark and Coyle, 1990), and now most water distribution system models routinely incorporate sophisticated water quality simulation capability. In addition to hydraulic and water quality simulation, many distribution network models are capable of analyzing water hammer (surge/transient) and tank and reservoir mixing characteristics. Currently available water distribution network models have become very sophisticated and many incorporate Computer Aided Drafting and Design (CADD) and GIS capability as well as interfacing with SCADA and Asset Management Systems (AMS).

This section discusses the basic principles underlying routine hydraulic and water quality modeling in drinking water distribution systems and presents new developments. In addition, integration of network modeling and optimization with a range of information management systems into an effective decision support and utility management and protection system is presented.

Hydraulic Modeling

Hydraulic models simulate flows and pressures throughout the water distribution system and can be divided into four broad categories (Wood et al., 2005a):

(1) Steady State Theory: The basic network hydraulic approach, applicable to time-invariant conditions, solves the conservation of mass (at each node) and energy (around each loop) equilibrium expressions using an iterative scheme (e.g., Newton-Raphson) based on known (static) demand loading and operating conditions.

(2) Extended Period Simulation (EPS): The second approach, applicable to very slow transients, is called *extended period simulation* (EPS) or *quasi-steady* theory, and involves solving a sequence of steady-state solutions linked by an integration scheme for the differential equation describing the storage tank dynamics. Both inertial and elastic effects are neglected. These models have become ubiquitous within the water industry and are an integral part of most water system design, master planning, and fire-flow assessment studies. They also provide flow information used in distribution system water quality models.

(3) Rigid Water Column Theory: Another category of unsteady flow is suitable for faster (but still relatively slow) transients and is called *rigid water column* theory (*lumped parameter* approach). It considers gradually varied flow and slow moving transients under the assumption that water acts as a rigid-column and elastic properties of the pipe walls are of no consequence. In this approach, the inertia of the fluid in a particular pipe is treated as lumped instead of continuously distributed.

(4) Waterhammer (Surge) Theory: The last category of unsteady flow applicable to rapid transients is called *elastic* or *waterhammer* theory (*distrib-*

uted parameter approach) and takes into account the elasticity of both the fluid and the pipe walls in the calculations. It represents situations with more rapid and sudden changes in flow velocity (e.g., rapid valve closure, pump trip) that require consideration of liquid compressibility and pipe wall elasticity.

The last three hydraulic modeling categories are known as *unsteady* (or *dynamic*) flow analysis. These models can be effectively used to estimate intrusion potential, identify susceptible regions in the distribution system that are of greatest concern for vulnerability to objectionable (low or negative) pressure surges, and evaluate how they may be avoided and/or controlled (Boulos et al., 2005).

Rigorous optimization approaches have been developed and applied to a full range of problems associated with water distribution systems (Boulos et al., 2006). Applications include optimizing network model calibration, satellite treatment (booster disinfection station location and operation), data collection and sampling/monitoring, as well as pump and storage tank operations to minimize energy cost, and valve operation for pressure management and leakage reduction. The optimization methods applied are common between problems and include linear, nonlinear and dynamic programming, and stochastic search procedures.

Water Quality Modeling

Water quality models utilize the flow and velocity information generated by the hydraulic models to predict the temporal and spatial variability of water quality within the distribution system. They can be used to simulate water quality concentrations and water age, and they can perform source tracing through the distribution system. As with hydraulic models, water quality modeling has evolved from the initial development of steady-state models (Wood, 1980; Males et al., 1985; Clark and Males, 1986; Males et al., 1988) to more dynamic models (Liou and Kroon, 1986; Hart et al., 1986; Clark et al., 1988; Grayman et al., 1988). Dynamic water quality models are predicated on extended period simulation quasi-steady network hydraulics, and they solve the equations for nodal mixing and advective transport in pipes to compute the spatial and temporal variation in water quality parameters. Solution methods for dynamic models can be classified as either *Eulerian* or *Lagrangian* (Rossman and Boulos, 1996; Clark and Grayman, 1998; Panguluri et al., 2005b). Eulerian methods consider fixed grids or cells and move water to the grid locations or through the cells to represent the movement of a constituent in a pipe. Chemical reactions are included during transport. Lagrangian methods track locations of discrete changes in water quality known as fronts. Front locations are updated at a fixed time step or when a front reaches a junction node. Longitudinal dispersion is neglected and complete mixing at the junction nodes is assumed.

In a distribution system water quality analysis, the constituents of concern may be conservative (e.g., inert) or non-conservative (e.g., reactive). Conservative substances are useful as tracers that are purposely injected, monitored, and modeled to improve model calibration with measured travel times and dilution effects. Non-conservative constituents, such as bacteria, disinfection byproducts, or chlorine, undergo reactions that are dealt with in most water quality models via simple reaction kinetics (Panguluri et al., 2005b).

During the 1990s, advancements focused more on making existing models user friendly and less on improving predictive capabilities. For example, the EPA public domain EPANET model (Rossman et al., 1994; Rossman, 2000) greatly facilitated the easy application of existing water quality models to municipal drinking water distribution systems. In early 2000s, research initiated by Zierolf et al. (1998) and Shang et al. (2002) focused on development and application of control theory-based methods. While distribution system water quality models are forward methods (i.e., they begin at a source and track forward in time and space to determine where a constituent is going), control theory begins at a location of interest and tracks backward in time and space to identify the constituent source. This approach has at least two important uses. The first is developing injection policies; if the disinfection level is unacceptable at a certain location, control theory will determine the relative contributions from alternative booster stations in one analysis rather than performing a series of forward tracer runs. Another application is identifying the potential sources of contamination detected at a downstream monitor. A contaminant that is detected at a given location may be supplied by a number of inlet points and times, and control theory can identify the range of locations and the pipes and nodes that contribute to flow at the monitor. Because of uncertainty due to changing demands and pump and tank operations, these analyses generate a significant amount of data. Fortunately, control theory model analysis can be done on-line, which facilitates data collection and information storage. After the potential sources have been identified, a forward model can be applied to determine the extent of contamination for containment purposes and the need for flushing.

With the recent concern over water distribution systems as part of the nation's critical infrastructure, research into water quality modeling has become more active, with the intent of developing greater predictive capabilities. For example, EPA has extended EPANET to allow general multi-species reactions. However, the code lacks a user interface at present, it is intended for research purposes, and it is in the beta-testing phase (*http://www.epa.gov/nhsrc/pubs/tbEPANet051106.pdf*). Other recent developments include the application of transient analysis software (Boulos et al., 2005) and optimization tools for calibration, design, and operational purposes (Berry et al., 2004; Uber et al., 2004a; Murray et al., 2004; Ostfeld, 2004; Ostfeld and Salomons, 2004). An emerging area of research is the incorporation of stochastic analysis to water quality modeling (Buchberger et al., 2003). More complex kinetics can be used to describe multi-component interactions (relating transformation rates to the concentration of other constituents) and are known as multi-component or multi-species models. Clark (1998), Clark et al. (2001), and Clark and Sivaganesan (1998, 2002) have furthered the development of multi-species and competitive reaction models that could be included in general purpose algorithms such as EPANET. Examples of two promising areas for future research are illustrated by Uber et al. (2004a), which has extended EPANET to allow for modeling the fate and transport of multiple interacting chemical and biological components, and Uber et al. (2004b), which has developed algorithms for optimizing the location of water quality sensors in drinking water distribution systems.

Flushing Models

The last two years have seen the introduction of computerized unidirectional flushing models of water distribution systems (Boulos et al., 2006). Unidirectional flushing models utilize the flow and velocity information generated by the hydraulic models and make use of graph-theoretical algorithms to determine the sequences of fire hydrants and water main valves that should be manipulated to create a one way flow in the water mains while avoiding excessive pressure drops (e.g., below 20 psi) and maintaining the desired level of hydraulic performance in the distribution system. These models also compute the minimum flushing time, total flushing volume and pipe length, and the flushing velocity of every pipe in the sequence.

Modeling of Storage Facilities

The hydraulics and mixing of waters within storage facilities must be properly understood to accurately represent the constituent reactions and the effect of tanks on system water quality. In addition, understanding the mixing characteristics in storage facilities is useful in assessing the likely impacts of an injected contaminant. Tank models simulate both aging and mixing phenomena within distribution system tanks and reservoirs. The most complex is computational fluid dynamic or hydrodynamic modeling that includes a detailed physical tank description and divides a tank into a mesh of small discrete volumes known as finite elements (Grayman and Arnold, 2003). The basic governing laws of conservation of mass, energy, and momentum are written in partial differential form for each element to describe the flow patterns and the distribution of substances through the tank. The remaining approaches are known as systems models. Systems models are classified by the spatial representation in the tank. The simplest model, a continuous stirred tank reactor, considers the tank as a single unit and assumes complete mixing of water within the tank. The next level of detail represents the water in the tank in layers assuming plug flow. The tank is partitioned into several compartments to represent the flow patterns and mixing zones. These are known as multi-compartment tank models (Mau et al., 1995; Clark et al., 1996).

***

All network models are approximate representations of distribution systems, and many sources of error exist that can hinder their ability to accurately simulate actual system behavior. Sources of error can range from measurement and typographical errors to errors derived from system maps or introduced by the skeletal representation of the network as well as uncertainties in some system parameters and boundary (e.g., loading and operating) conditions. Network models should be properly calibrated and validated so that a level of confidence in their predictive capabilities can be established. Box 7-1 discusses these activities in greater detail.

BOX 7-1
Calibrating and Validating Network Models

A water distribution network model must be properly calibrated before it can be used to support planning, design, operation, or water quality improvement decisions. Calibration establishes the accuracy and credibility of the network model so that its predictions can be interpreted with confidence. It is the process of fine-tuning (adjusting) network model parameters so that the simulated hydraulic and water quality results sufficiently mirror field observations. If the field data and model results are reasonably close, the model is considered calibrated. The objective is to reduce the uncertainty in the model parameters to accurately reproduce actual "real-world" system behavior.

To be calibrated, the network model must accurately simulate pressure, flow, tank level, and chlorine residual values within an acceptable tolerance for a range of specified time horizons. Hydraulic parameters that are typically adjusted include pipe roughness factors, minor (local) loss coefficients, isolation valve status, control valve settings, pump curves, base demands, and demand patterns. For water quality models, the parameters include reaction rate coefficients, source quality, and initial conditions. The calibration tolerance refers to the difference between model simulated and actual field values. The smaller the tolerance the greater the accuracy of model predictions. Calibration can be performed to a single time frame such as maximum hour or dynamically such as maximum day for an extended period simulation (EPS). The more calibration time frames, the more accurate the model predictions will be. Common practice is to calibrate the network model first for maximum-hour and minimum-hour static conditions and then in an EPS mode for maximum day. The network model is first calibrated for hydraulic parameters and the water quality parameters are subsequently adjusted. Thus, if the hydraulic model is not properly calibrated, resulting in inaccurate flow and velocity estimates, the water quality model will not perform correctly. Water quality simulations require a dynamically calibrated (EPS) model. Network model parameters can be adjusted manually using an iterative trial-and-evaluation approach or automatically using optimization techniques until the desired degree of accuracy is attained (Panguluri et al., 2005b; Boulos et al., 2006).

Although automated calibration methods are becoming more readily available, manual calibration still remains the predominant methodology. However, since there is a vast number of combinations of parameter values that can be considered for adjustment, manual evaluation of all options through trial-and-error is unlikely to be practically feasible or manageable, and even knowledgeable modelers often fail to obtain good results. As a result, model calibration has generally been neglected or done haphazardly.

continues

BOX 7-1 Continued

There are currently no universally accepted standards for calibrating water distribution network models. The extent of calibration will normally depend on the intended use of the model. A greater degree of calibration will be required for models that are used for detailed analyses, such as design, operations, and water quality modeling, than for models used for more general planning purposes (e.g., master panning). The AWWA Engineering Computer Applications Committee (AWWA ECAC, 1999) has proposed a draft set of calibration guidelines for modeling based on intended use. These performance criteria were not intended as true calibration standards, but can serve as a good starting point for illustrating the extent of calibration needed for various modeling applications. These calibration criteria are summarized in Table 7-7.

Network model validation follows the calibration process and makes use of an independent field data set for use in verifying that the model is well calibrated. The model must first be calibrated using one or more sets of field data and then validated with an independent set of field data. The degree of confidence in the model increases with the number of independent data sets with which it is validated. Tracer studies can also be used to validate network models. These studies consist basically of measuring the concentration of a tracer over time (e.g., using on-line monitors and grab samples) at various locations throughout the distribution system and comparing observed values with model predictions. The most commonly used tracers are fluoride, calcium chloride, and sodium chloride. The use of tracer studies greatly enhances the ability of network models to accurately estimate water age and travel times in the system.

TABLE 7-7 Draft Calibration Criteria for Modeling

Intended Use	Level of Detail	Type of Simulation	Number of Pressure Readings	Accuracy of Pressure Readings	Number of Flow Readings	Accuracy of Flow Readings
Planning	Low	Steady or EPS	10% of Nodes	±5 psi for 100% Readings	1% of Pipes	± 10%
Design	Moderate to High	Steady or EPS	5% – 2% of Nodes	±2 psi for 90% Readings	3% of Pipes	± 5%
Operations	Low to High	Steady or EPS	10% –2% of Nodes	±2 psi for 90% Readings	2% of Pipes	± 5%
Water Quality	High	EPS	2% of Nodes	±3 psi for 70% Readings	5% of Pipes	± 2%

The efficacy of calibration and calibration techniques is highly dependent on the quality of the calibration data available and the quality of the constructed network model (e.g., skeletal representation of the network, node elevation, geometric anomalies). Poorly collected field data (e.g., from poorly calibrated measuring equipment) and poorly defined network models will result in inadequate calibrations and unreliable model predictions, and would defeat the whole purpose of the calibration process.

DATA INTEGRATION

Data and information on the additional parameters of distribution system integrity discussed above and in previous chapters—water pressure and flow, valve operations, main breaks, customer complaints, condition assessment, inventory, and new water quality data—must be integrated in order to make informed decisions, preferably using models and GIS tools. For example, if a contamination event were to occur in a water distribution system, depending on the hydraulic design and operational conditions, much of the distribution system could be impacted, potentially affecting a large percentage of the population served. It would be very difficult to track the spread of the contaminant, find its originating source, and understand its impact based on monitoring information alone. Rather, an understanding of the system's hydraulic behavior as well as a proper visualization of all system facilities (e.g., pumps, tanks, isolation valves) and rapid access to customer data and SCADA information are required. GIS and modeling can be used to predict the movement of contaminated water in the system, locate the appropriate facilities that need to be closed manually or via the SCADA system for event containment, identify populations at risk and report customer notification information, compute affected water volumes that need to be purged, and help develop an effective flushing program (e.g., which sequence of hydrants to open and how long to keep them open).

In the past few years, advances in infrastructure management technology have been occurring at an accelerated pace, with potential significant benefits for the water works industry. The development of GIS is greatly expanding the applications of water distribution network models. Because of the spatial nature of water distribution systems, many aspects of managing these systems consist of using, analyzing, and displaying geospatial data, which includes the geographic location and characteristics of various water system facilities, including pipes, pumps, storage tanks, reservoirs, and valves. Using GIS, a water valve can be identified by its geographical location in the system and its characteristics, such as valve type, size, manufacturer, pressure or flow setting, loss coefficient, year of installation, condition, maintenance records, opening direction, and location and distance to operating nut. Similarly, a pipe can be described by its route, length, diameter, material, installation date, lining, wall thickness, pressure class, service connections, ground surface type, street identifier, parallel pipe indicator, cost of installation, condition, leakage and burst records, fire service capacity, physical samples (e.g., observations of tuberculation or corrosion exhibited by the pipe samples after a breakage event), and taste and odor complaints. Indeed, a GIS can encompass all of the data collected during asset management (see Chapter 4) such as the condition of pertinent pipes, pumps, and valves as well as other water system facility characteristics that are used for model construction and emergency response (e.g., how many turns are required to close a valve and in which direction).

A GIS is able to store and maintain (keep up-to-date) these spatial data while allowing easier access to the data, flexibility in data sharing and modeling, and a substantial decrease in data storage and redundancy requirements. GIS enables the model input and output data to be developed and displayed in graphical form, the generation of accurate and detailed facility data, verification of data integrity, and the visualization and cartographic analysis provided by system maps. It can be effectively used to calculate and allocate consumption data (e.g., from land use/population data or geocoding), help identify (representative) water quality sampling sites, and quantify customer exposure to a specific event. A GIS can also assemble data on the physical characteristics of service to the customer as well as customer billing data, including customer name, street address, contact information, service line diameter, location, installation date, material, tap number, and meter number (Cesario, 1995). Customers' contact information can prove very useful for alert notification during a water quality emergency event. In addition, the GIS can store pump information and all pertinent electricity charges as well as demand and energy consumption data. This information can feed an optimization model to assist in developing improved daily pump scheduling policies that meet desired hydraulic and water quality (minimize water age) objectives while maximizing energy savings, as well as sound operational strategies for using alternative water supplies during a contamination event.

The SCADA system, which compiles real-time and historical operational data for all remote facility sites, is critical to making informed decisions during a contamination event. SCADA information is useful in defining boundary (e.g., tank and reservoir water levels, pump status, valve settings) and loading (e.g., total zone and system demands) conditions for the network model as well as real-time measurement data (e.g., pressure and flow measurements) for network model calibration and for identifying water losses during main breaks (e.g., unexplained low pressure reading, excessive pump flow). The SCADA system can effectively control the contamination spread by isolating the contaminated areas and associated facilities by shutting off all critical in-line isolation valves. The isolation can be carried out automatically either locally (i.e., at the valves) or remotely (i.e., the control room).

The Internet and the World Wide Web are also rapidly evolving to the benefit of water supply operations. Among their many improvements are easier and greater accessibility, efficient distribution, effective administration, and cross-platform flexibility (Molenaar and Songer, 2001). A web-based interface, which is becoming common in standard GIS platforms, can enable the rapid deployment of critical GIS data and modeling results over the Internet. This could facilitate the sharing of critical information with federal, state, and local emergency response and regulatory agencies during contamination events. Such data may include valve and hydrant locations for firefighting as well as information needed to isolate and flush accidentally released or intentionally introduced contaminants in the distribution system and identify all affected individuals.

These various infrastructure management systems are highly complimentary applications that, taken together with various distribution system models, constitute a decision support system for a water utility. The ability to seamlessly exchange data between these systems and models is critical to improving water distribution system operation, management, protection, and emergency planning activities. Table 7-8 lists the key components of the decision support system along with their response roles.

Managing and protecting distribution systems from contamination threats and a wide range of emergency situations will require the use of a comprehensive decision support system that integrates modeling applications with the various infrastructure management systems. The term "integration" can refer to a

TABLE 7-8 Systems and their Response Roles

System	Response Role
On-line monitoring	• Identify event (location and type) • Alarm notification
SCADA	• Alarm notification • Event isolation (operational control) • Event reporting and archiving • Resumption of normal operation after removal of threat
Hydraulic Model	• Provide system-wide flows at moment of event • Locate valves to isolate event • Compute required purge volumes • Compute available fire fighting capacity • Check whether normal operations can continue • Determine water rerouting scheme (use of alternative water supplies)
Unidirectional Flushing Model	• Determine flushing sequences (opening of hydrants and closing of valves) for proper decontamination
Water Quality Model	• Determine extent of contamination • Provide system-wide water quality at moment of event • Narrow event region • Identify grab sample locations • Track contaminant to originating source(s) • Develop re-chlorination plan • Predict future event region (as contaminant moves)
Grab Sample	• Verify event (degree of confidence) and associated emergency level
GIS	• Visualize all system facilities • Map modeling results • Coordinate response units • Identify population at risk • Report customer alert /notification information
Web Portal	• Instant access and sharing of critical information • Monitor event response progress in real-time • Provide means of central messaging exchange

wide range of capabilities related to use of network models in conjunction with various infrastructure management systems. With respect to linking a GIS to a network model, integration refers specifically to a method that combines the GIS and modeling functions into one complete seamless package, such that the network model operates within the GIS using the same spatial database. This approach is becoming more commonly used in the water and wastewater industries (www.nagcs.com), and it is facilitating the rapid development of more detailed and accurate distribution system models. Information on facilities and demands can be routinely updated. The results of a modeling application can be rapidly displayed and analyzed along with other spatial data. The potential for real-time monitoring and application of models to confirm normal system performance and assist in system operation under routine and emergency conditions can also be made possible with the additional integration of the SCADA system through the common database.

An integrated decision support system should give water utilities real-time surveillance and control on finished water quality in the distribution system. It can greatly assist water utilities in reducing infrastructure vulnerability and enhancing their ability to prepare for and respond to natural and/or man-made disasters, terrorist attacks, and other emergencies, providing the public with added security and peace of mind. The system can be used not only as an early warning system to detect potential contamination threats but also as an effective planning tool to identify viable solutions before an incident or disaster occurs (e.g., evaluating the potential impact of unforeseen facility breakdown, assessing the effect of water treatment on contaminants, as well as using surge modeling to predict and eliminate potential weak spots), or to assist in responding should it occur (e.g., increasing the chlorine dose at the treatment plant). These enhanced capabilities create significant management advantages for water utilities, including greater operational efficiency and emergency preparedness, reduced system vulnerability, improved public notification, shortened response time, more informed decision making, and stronger customer ties. In evaluating the potential implementation of an integrated decision support system, a water utility will need to balance the reduction of risks with the costs associated with implementing the system.

Although there is not currently a decision support system that satisfies all of the above requirements simultaneously, significant progress has been made to date in both GIS and modeling technologies that meet many of these needs. The availability of robust and reliable on-line, real-time sensors that can rapidly and accurately detect and report all potentially detrimental chemical and biological contaminants that will be both affordable and useful for most utilities is still years away. Water utilities should monitor the development and maturity of this technology and determine when such a system is practical for their use. Box 7-2 presents an example of integrating GIS with water quality modeling, master planning, and operational decision-making within the Las Vegas Valley Water District.

BOX 7-2
Data Integration at the Las Vegas Valley Water District

Chapter 1 mentioned an AwwaRF project that will test the concept of decentralized treatment and its application to the Las Vegas Valley Water District (LVVWD). Various technologies will be evaluated for their use in the decentralized treatment, including air stripping, granular activated carbon, and biological activated carbon. A key feature of this study is the use of distribution system modeling interfaced with a GIS. Indeed, the LVVWD is linking its master planning, operational planning, and development review functions by integrating its GIS database with distribution system modeling, SCADA, and enterprise data. Figure 7-2 presents the conceptual relationship model of these functions and potential integration benefits (Jacobsen et al., 2005).

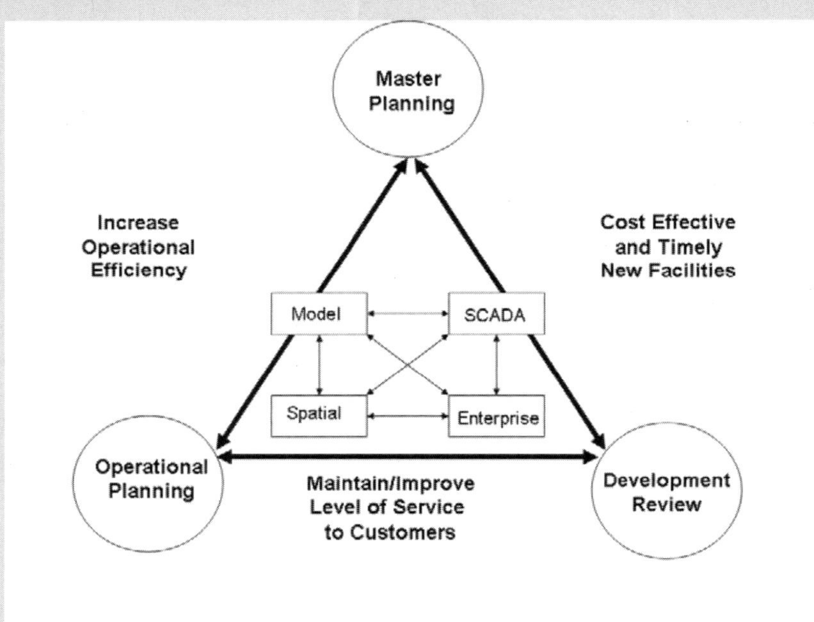

FIGURE 7-2 Conceptual relationship model for integration.

During the process of integration, LVVWD developed a one-to-one relationship between the GIS spatial data and its network model (Jacobsen and Kamojjala, 2005). The advantages of taking this approach include ease of search and retrieval with other data/applications, and ease of importation, development, and maintenance of data. For a large network model, the disadvantages include an increase in the run-time of the network model due to the addition of detailed components and relatively slow water quality simulations. To minimize this problem, LVVWD has taken an "all-pipes capable" approach where the distribution system is divided according to existing pressure zones and attached to an operational backbone network (skeletonized). Each of the zone models can be attached seamlessly to the backbone network for detailed hydraulic and water quality modeling. LVVWD uses GIS data on a day-to-day basis for pressure complaint resolution and main break analysis.

FEASIBILITY OF ADOPTING G200 FOR SMALL SYSTEMS

While it is important to design, promote, and employ voluntary improvement programs that will allow utilities to be proactive in maintaining high water quality in distribution systems, it is equally critical to scale such programs to accommodate small water systems. Water systems serving small populations have limited monetary and personnel resources, whose expertise and technical knowledge are often limited. For many small water systems, simply meeting the requirements of the existing federal and local regulations presents a continuous struggle, leaving little or no time and money for quality improvement programs described in the previous sections.

Monitoring of water quantity and quality beyond compliance requirements should be based on site-specific characteristics and priorities. For example, all small systems should seriously consider monitoring of water pressure, because it is critical to avoiding contamination via cross connections, which is likely to be a ubiquitous problem. On the other hand, monitoring for non-regulated contaminants and the maintenance of an extended water quality database could provide important information regarding changes in water quality and allow for optimization of water treatment prior to its distribution, but it may be cost-prohibitive. Additional water quality monitoring beyond compliance monitoring should (1) be targeted to contaminants of local concern and help to overcome site-specific challenges, (2) be associated with source characteristics and treatment, and (3) be conducted at critical points in the distribution system during critical times (changes in weather, flow, system maintenance, etc.).

In general, adherence to the G200 standard by small water systems, which will invariable exceed current regulatory requirements, should be implemented using the following guidelines: (1) implement new activities using a step-wise approach; (2) provide technical assistance, education, and training; and (3) develop regulatory, financial, and social incentives. Such a tiered approach to the implementation of G200 activities should allow for the prioritization of needs, for the planning of resources, and for the implementation of additional monitoring practices and maintenance activities over a long period of time. The consolidation or cooperation of small water systems may also make adoption of more advanced monitoring and modeling techniques more feasible, as discussed in NRC (1996).

Training materials, scaled for small-size systems, are essential for operators and maintenance crew. For example, the EPA provides guidance to small water systems in asset management—i.e., developing an inventory of assets and determining when repair should give way to rehabilitation or replacement (EPA, 2004c). This type of guidance would be helpful for the other important elements of a distribution system management plan mentioned above and found in G200. Technical assistance could also be provided to small systems by larger water utilities in the area. Finally, public education could result in an increased awareness and emphasis on the significance of implementing proactive voluntary efforts.

HOW TO PROVIDE INCENTIVES TO ADOPT G200

For a utility to engage in all of the activities mentioned above in addition to compliance monitoring is a considerable challenge that may require the creation of incentives, perhaps through existing regulations and associated policies. There are several instruments already in place that could be modified to better implement the monitoring, modeling, and other approaches for reducing risk from public distribution systems discussed above.

Federal Regulatory Approach

A first option is that federal regulations could require adherence to a prescribed list of activities deemed necessary for reducing the risk of contaminated distribution systems. This list could partly or fully parallel the G200 standard. Given the accreditation atmosphere in which G200 was created and its history to date, making the standard a federal requirement seems unlikely. Indeed, G200 is currently viewed as one of many available industry programs that are voluntary and include best practices, such as the Partnership for Safe Water, QualServe, and other accreditation standards. In the opinion of the committee, G200 is the most comprehensive voluntary program and should be central to a utility's development of a distribution system management plan.

MCL vs. TTR. In lieu of making G200 a federal requirement, portions of the standard could be explicitly made part of other federal requirements. For example, new maximum contaminant levels (MCLs) or treatment technique requirements (TTR) could be created that would capture essential elements of G200. There is some precedence for this line of action. For example, in the late 1980s, the turbidity requirement in the SDWA went from being an MCL to being a TTR. The regulation specified treatment techniques that would provide log removals for *Giardia* and viruses by requiring treatment plants to achieve at least 99.9 percent (3 log) and 99.99 percent (4 log) removal/inactivation, respectively. It also specified the performance criteria for treatment based on turbidity. Something similar could be done for many of the distribution system practices mentioned in this chapter. For example, with respect to water pressure monitoring, a minimum water pressure requirement could be established, not unlike the measurable chlorine residual that is now required.

Sanitary Surveys. Another mechanism to capture elements of G200 within existing federal regulations would be via the sanitary surveys conducted by the state and required for some systems under federal regulations. Under 40 CFR 142.10(b)(2), states are required to have in place a program for conducting sanitary surveys of public water systems, especially those who are out of regulatory compliance (EPA, 1999). The purpose of such surveys is to "evaluate and document the capabilities of the water system's sources, treatment, storage, distribution network, operation and maintenance, and overall management to continually provide safe drinking water and to identify any deficiencies that may adversely impact a public water system's ability to provide a safe, reliable water supply." The TCR has a requirement for periodic sanitary surveys for all small systems that collect less than five samples per month. The SWTR requires an annual on-site inspection for surface water systems that do not filter. The IESWTR now requires a survey for all surface water and groundwater-under-the-direct-influence systems and requires that each of eight elements be addressed (source; treatment; distribution system; finished water storage; pumps, pump facilities, and controls; monitoring and reporting and data verification; system management and operation; and operator compliance with state requirements) as well as the correction of significant deficiencies. These surveys are required every three to five years.

A sanitary survey might reveal an absence of training, use of standards, routine inspections, or certifications that could be predictive of a loss of distribution system integrity. The distribution system components of the sanitary survey include:

- Distribution system maps and records, field sampling and measurements, system design and maintenance
- Finished water storage location, capacity, design, painting, cleaning and maintenance, security
- Pumps and pump facilities and controls capacity, condition, pumping station
- Water system management and operation administrative records, water quality goals, water system management, staffing, operations and maintenance manuals and procedures, funding
- Operator compliance with state requirements such as certification and competency.

The Drinking Water Academy developed software for use by state sanitary inspectors in accomplishing all aspects of a sanitary survey with some level of uniformity. This software can be used during field inspections with a PDA or Tablet PC (*http://www.epa.gov/safewater/dwa/e-sansurvey.html*).

A benefit to using this mechanism for promoting G200 is that the sanitary surveys encompass a wide variety of activities and would likely capture those felt to be of highest priority for reducing risk (e.g., cross-connection control and water storage facility inspections). Indeed, the EPA's Sanitary Survey program could be reviewed and compared to G200 to see whether the former might be expanded. However, this approach is likely to succeed only if sufficient funds are provided to support more comprehensive sanitary surveys. In addition, current regulations for the sanitary survey exempt or avoid a large number of water supply systems, effectively limiting the reach of this mechanism.

GASB Accounting Requirements. Another approach for encompassing some of the elements of G200 and the committee's high priority activities under existing federal regulations is through the Government Accounting Standards Board (GASB) Statement 34 on Basic Financial Statements and Management's Discussion and Analysis for State and Local Governments (GASB, 1999). This regulation requires that all capital assets be documented and reported in financial statements by looking at the long-term health of government institutions throughout the United States, including municipally owned water utilities (Romer et al., 2004). The reporting includes the valuation of infrastructure and related disclosure of deferred maintenance costs on treatment plants, pump stations, storage facilities, and distribution systems (Donahue, 2002), and thus is well positioned to provide the asset management functions called for in G200 and in Chapter 4. Indeed, GASB 34 has encouraged the application of asset management in order to meet requirements (Cagle, 2005). With respect to the specific problems that cause water quality deterioration in distribution system, GASB 34 may "inadvertently become the regulatory mandate for corrosion control since uncontrolled asset deterioration can negatively impact financial statements and, therefore, limit or degrade the ability of a utility to raise money for capital improvement using bonds. Utilities that have good corrosion control programs will have better financial statements and bond ratings" (Romer et al., 2004).

State Regulatory Approach

In lieu of federal regulations, state regulations could require adherence to G200 or the committee's list of preferred activities for reducing risk in distribution systems. This approach is limited primarily by the fact that some states would legally be unable to make such modifications, while others could.

State and Local Building and Plumbing Codes. A logical avenue would be to consider enhancing state building and plumbing codes to cover more issues or simply to make enforcement of current codes more uniform. Tables 2-3 through 2-5 show that states vary in their enforcement of state and local codes for plumbing, health, building, real estate, etc. Clearly, more stringent details within these codes could be applied. These codes are, however, unlikely to be able to cover all activities considered to be of high priority for reducing distribution system risk.

Similarly, design and construction requirements at the state level could be modified to capture important elements of G200. Tables 2-3 through 2-5 show that the SDWA and states already have in place design and construction standards, enforced largely through the permitting process. Permits are required when a new system is built or when a significant change to an existing system is made. State building codes could be expanded, for example, to require inspections for cross connections prior to granting building permits on existing properties or prior to closing of a sale (as is the case for radon inspections in some jurisdictions). The permit process could also address the design of service lines and premise plumbing for water quality maintenance (e.g., existence of dead ends, oversized lines, compatible materials), the extent of lead and copper corrosion, hot water system maintenance, the level of disinfectant residual at taps, and the presence of scale and sediment.

State Revolving Fund. Another possibility is that to qualify for a loan from the State Revolving Fund a utility would have to demonstrate that it is adhering to G200 or an equivalent list of activities. The 1996 Amendments (Public Law 104-182) to the SDWA established the Drinking Water State Revolving Fund (DWSRF), intended to facilitate compliance with applicable national drinking water regulations or significantly further the health protection objectives of SDWA. States operate their respective DWSRF programs using annual capitalization grants from EPA and a 20 percent matching contribution from the state. Up to 30 percent of the federal grant can be used to assist public water systems serving disadvantaged communities through subsidized loans or loan forgiveness. However, under SDWA section 1452(a)(3) states are prohibited from providing DWSRF assistance to a public water supply that does not have the technical, managerial, and financial capability to ensure compliance with the requirements of the SDWA. EPA could clarify that any public water system that does not have a program for managing the distribution system such as G200 should be viewed as lacking such capability.

The SDWA does allow a public water system to receive DWSRF funding if the owner or operator of the system that lacks capacity agrees to undertake feasible and appropriate changes in operations (including ownership, management, accounting, rules, maintenance, consolidation, alternative water supply, or other procedures) that the state determines would ensure the system's technical, managerial, and financial capacity. This provision could be used to promote the use of G200 by requiring that a public water system, as a condition of receiving DWSRF funding, agree to develop a plan for the implementation of those elements of G200 that are feasible given the size, complexity and resources of the system.

It should be noted that the State Revolving Fund is generally used for capital investment, but it might be used to comply with G200 if it had to do with construction practices in some way. This might, however, dilute the objective of the Fund, which is to bring water supplies into compliance.

Bond Ratings. In addition to facilitating the acquisition of DWSRF funding by small, disadvantaged communities, implementation of G200 could also improve a drinking water utility's access to capital in other ways, particularly for municipally owned water systems. Drinking water utilities in the United States have historically depended on the municipal bond market to finance both the development of public water supplies and their expansion into surrounding areas (Cutler and Miller, 2005). The same practices that are encouraged by the implementation of G200 may also help improve a municipality's bond rating.

The two main categories of long-term bonds available to municipalities are general obligation bonds, secured by a pledge of the government's taxing power, and revenue bonds, secured by the exclusive (in most cases) pledge of a project's revenues. The following five factors are generally used by bond rating authorities for general obligation bonds: (1) general economy, (2) debt structure, (3) financial condition, (4) demographic factors, and (5) management practices of the governing body and administration. Because of this last factor (management and administrative practice), a utility that follows the elements of G200 as outlined in Table 7-1 along with a sound pricing policy should be in a position to receive a better bond rating, as well as to obtain funding under the DWSRF, than a utility that does not adhere to G200.

Federal Guidance

In lieu of a regulatory incentive for adopting G200, EPA could advocate a list of preferred activities as a way of meeting federal regulations for distribution systems. This might appear in updated versions of guidance manuals. The EPA already provides extensive guidance to help water utilities achieve and maintain compliance, including a capacity development program to assist water systems in achieving SDWA compliance (Stubbart, 2005). The program addresses managerial, technical, and financial capacities involving all aspects of the system from source water through treatment to distribution. Technical aspects include how to provide certified operators and reliable infrastructure. Especially with small systems, the program can help identify weaknesses and in turn identify avenues for support to eliminate those weaknesses. It also discusses the various support that is needed to fund and maintain an adequate distribution system maintenance and replacement program.

CONCLUSIONS AND RECOMMENDATIONS

This chapter has discussed the limitations of compliance monitoring to detect, respond to, and protect against an internal or external contamination event in the distribution system that might jeopardize public health. To affect real risk reduction from contaminated distribution systems, efforts beyond compliance monitoring are required. The AWWA G200 standard outlines voluntary activities that if implemented would provide substantial risk reduction from distribution systems. Many elements of G200 are critical to maintaining distribution system integrity, although they do not necessarily suffer from scientific or technological limitations. The reader is referred to previous chapters for conclusions and recommendations on these activities, which include cross-connection control, maintenance of storage facilities, asset management, and training and certification of system operators, inspectors, foremen, and managers. The following conclusions and recommendations pertain to those elements of G-200 for which emerging science and technology are altering whether and how these elements are implemented by a typical water utility. The committee recognizes that because of cost and personnel limitations these recommendations are probably not feasible for many medium- and small-sized utilities at the present time. Nonetheless, the monitoring and modeling activities discussed represent an endpoint toward which utilities should be striving. It is hoped that the gap between what is needed to affect water quality improvement and what utilities are capable of will shrink in the near future.

Distribution system integrity is best evaluated using on-line, real-time methods to provide warning against any potential breaches in sufficient time to effectively respond and minimize public exposure. This will require the development of new, remotely operated sensors and data collection systems for continuous public health surveillance monitoring. These types of systems should be capable of accurately (with sufficient precision) determining the nature, type, and location/origin of all potential threats to distribution system integrity. The availability, reliability, and performance of on-line monitors are improving, with tools now available for detecting pressure, turbidity, disinfectant residual, flow, pH, temperature, and certain chemical parameters. These devices have reached the point for greater full-scale implementation. Additional research is needed to optimize the placement and number of monitors.

Research is needed to better understand how to analyze data from on-line, real-time monitors in a distribution system. This should focus on algorithms that can integrate real-time hydrological conditions, water quality inputs, and operational data to evaluate and interpret on-line monitor signals, establish alarm triggers, and suggest remedial actions. A number of companies are selling (and utilities are deploying) multi-parameter analyzers. These companies, as well as EPA, are assessing numerical approaches to convert such data into a specific signal (or alarm) of a contamination event—efforts which warrant further investigation. Some of the data analysis approaches are proprietary, and there has been limited testing reported in "real world" situations. Furthermore, when multiple analyzers are installed in a given distribution system, the pattern of response of these analyzers in space provides additional information on system performance, but such spatially distributed information has not been fully utilized. To the greatest degree possible, this research should be conducted openly (and not in confidential or proprietary environments).

A rigorous standardized set of network model development and calibration protocols should be developed. While there is a general agreement in the modeling profession that the extent of development and calibration required for a water distribution network model depends largely upon its intended use, there are no universally accepted standards and there is currently no apparent movement toward establishing such standards. Poorly defined and calibrated models can lead to management decisions being made based on false or erroneous data, and recommendations that may not even work. Continued research is needed to improve network model development and calibration methodologies (including optimization techniques) and in standardization of calibration. In addition, improved monitoring technology, such as more affordable meters that can be inserted into distribution pipes and automated monitoring for use in conjunction with tracer studies, will greatly improve calibration of distribution system models.

Additional research, development, and experimental applications in data integration are needed so that distribution system models can be used in real-time operation. Real-time monitoring and modeling of water distribution systems to assist water utilities in making informed operational decisions under routine and emergency conditions requires the integration of network models with SCADA systems, which has yet to be accomplished at most utilities. The SCADA system can be used to update the boundary conditions in the network model such as tank water levels, pump on/off status, isolation valve status, control valve settings, and system demands, and the model can in turn be used to identify the "best" operational strategy for the selected facilities and pass their control logic back to the SCADA system for implementation.

The ability to integrate network models with SCADA systems offers a number of benefits to the water industry including at a minimum: confirmation of normal system performance; real-time calibration; system trouble shooting; projection of operating scenarios; evaluating "what-if" scenarios; training for and responding to emergencies; and improvement of overall operations. These benefits can only be realized if both systems can communicate quickly and properly with one another. Continued work is needed to develop data integration standards that will allow seamless data exchange for monitoring and controlling system operations and make them available to the water industry. Further development is also needed to expand the ability of GIS to enable time-series data (e.g., historical or obtained from real-time measurements) to be associated with geospatial attributes.

REFERENCES

American Water Works Association/American National Standards Institute (AWWA/ANSI). 2004. G-200: Distribution Systems Operation and Management. Denver, CO: AWWA.

AWWA Engineering Computer Applications Committee. 1999. Calibration guidelines for water distribution system modeling. *In:* Proceedings of the 1999 AWWA Information Management and Technology Conference, New Orleans, Louisiana, April 1999.

Berry, J., W. Hart, C. Phillips, and J. Uber. 2004. A General Integer-Programming-Based Framework for Sensor Placement in Municipal Water Networks. World Water & Environmental Resources Congress, EWRI-ASCE.

Bernosky, J. J. 2005. Distribution system security. Pp. 155–181 *In:* Distribution System Water Quality Challenges in the 21st Century: A Strategic Guide. M. J. MacPhee (ed.). Denver, CO: AWWA.

Booth, D. E., P. Alam, S. N. Ahkam, and B. Osyk. 1989. A robust multivariate procedure for the identification of problem savings and loan institutions. Decision Sciences Journal 20(2):320–333.

Boulos, P. F., B. W. Karney, D. J. Wood, and S. Lingireddy. 2005. Hydraulic transient guidelines for protecting water distribution systems. J. Amer Water Works Assoc. 97(5):111–124.

Boulos, P. F., K. E. Lansey, and B. W. Karney. 2006. Comprehensive Water Distribution Systems Analysis for Engineers and Planners. Pasadena, CA: MWH Soft Publisher.

Buchberger, S. G., J. T. Carter, Y. H. Lee, and T. G. Schade. 2003. Random demands, travel times and water quality in deadends. Denver, CO: AWWA.

Bukhari, Z., and M. W. LeChevallier. 2006. Early warning systems to protect distribution system water quality. A report submitted to American Water, Voorhees, NJ.

Burlingame, G. A. 1999a. Solving customers' taste and odor complaints—Part 1: the importance of the first response. Opflow 25(10):10–11.

Burlingame, G. A. 1999b. Solving customers' taste and odor complaints—Part 2: tracking odors to their source. Opflow 25(11):6–7.

Byer, D., and K. H. Carlson. 2005. Real-time detection of intentional chemical contamination in the distribution system. J. Amer Water Works Assoc. 97(7):130–133.

Cagle, R. F. 2005. Daddy, are we there yet? Underground Infrastructure Management Jan/Feb:43–46.

Cesario, L. 1995. Modeling, analysis, and design of water distribution systems. Denver, CO: AWWA.

Clark, R. M., and J. A. Coyle. 1990. Measuring and modeling variations in distribution system water quality. J. Amer. Water Works Assoc. 82(8):46–53.

Clark, R. M., and R. M. Males. 1986. Developing and applying the water supply simulation model. J. Amer. Water Works Assoc. 78(8):6l–65.

Clark, R. 1998. Chlorine demand and TTHM formation kinetics: a second-order model. J. Environmental Engineering 124(1):16–24.

Clark, R. M., and W. M. Grayman. 1998. Modeling water quality in drinking water distribution systems. Denver, CO: AWWA.

Clark, R. M., W. M. Grayman, and R. M. Males. 1988. Contaminant propagation in distribution systems. Journal of Environmental Engineering, ASCE 114(4):929–943.

Clark, R. M., and M. Sivaganesan. 1998. Predicting chlorine residuals and the formation of TTHMS in drinking water. Journal of Environmental Engineering 124(12):1203–1210.

Clark, R. M., and M. Sivaganesan. 2002. Predicting chlorine residuals in drinking water: a second order model. Journal of Water Resources Planning and Management 128(2):1–10.

Clark, R. M., R. Thurnau, M. Sivaganesan, and P. Ringhand. 2001. Predicting the formation of chlorinated and brominated by-products. Journal of Environmental Engineering 127(6):493–501.

Clark, R. M. 1998. Chlorine demand and TTHM formation kinetics: a second-order model. Journal of Environmental Engineering 124(1):16–24.

Clark, R. M., F. Abdesaken, P. F. Boulos, and R. Mau. 1996. Mixing in distribution system storage tanks: its effect on water quality. Journal of Environmental Engineering 122(9):814–821.

Cutler, D., and G. Miller. 2005. Water water everywhere: municipal finance and water supply in American cities. *In:* Proceedings of the National Bureau of Economics Corruption and Reform Conference held in Salem, MA. Grant No. T32 A00186. Washington, DC: National Bureau of Economics Research (NBER).

Deb, A. K., K. A. Momberger, Y. J. Hasit, and F. M. Grablutz. 2000. Guidance for the management of distribution system operation and maintenance. Denver, CO: AWWA and AwwaRF.

Donahue III, E. J. 2002. GASB 34 and water utilities: deferred maintenance and contributed capital. *In:* Assessing the Future: Water Utility Infrastructure Management. D. M. Hughes (ed). Denver, CO: AWWA.

Egan, W. J., and S. L. Morgan. 1998. Outlier detection in multivariate analytical chemical data. Analytical Chemistry 70:2372–2379.

Environmental Protection Agency (EPA). 1999. Guidance manual for conducting sanitary surveys of public water systems; surface water and ground water under the direct influence (GWUDI). EPA-815-R-99-016. Washington, DC: EPA.

EPA. 2003. National Primary Drinking Water Regulations; Announcement of Completion of EPA's Review of Existing Drinking Water Standards. Federal Register, 42907–42929.

EPA. 2004a. Draft Report on Evaluation of Water Quality Sensors in Distribution Systems. Washington, DC: EPA Office of Research and Development and Shaw Environmental.

EPA. 2004b. Draft Report on Water Quality Sensor Responses to Chemical and Biological Warfare Agent Simulants in Water Distribution Systems. Washington, DC: EPA Office of Research and Development and Shaw Environmental.

EPA. 2004c. Taking stock of your water system—a simple asset inventory for very small drinking water systems. EPA 816-K-03-002. Washington, DC: EPA.

Friedman, M., G. Kirmeyer, G. Pierson, S. Harrison, K. Martel, A. Sandvig, and A. Hanson. 2004. Development of distribution system water quality optimization plans. AWWA Research Foundation, Tailored Collaboration Project Final Report. Denver, CO: AwwaRF.

GASB (Government Accounting Standards Board Financial Accounting Foundation). 1999. Statement No. 34 of the Governmental Accounting Standards Board: Basic Financial Statements—and Management's Discussion and Analysis—for State and Local Governments. No. 171-A. Washington, DC: GASB.

Grayman, W. M., and C. N. Arnold. 2003. Overview of CFD Methods in Analysis of Distribution System Tanks and Reservoirs. *In:* Proceedings of the AWWA Annual Conference, Denver, CO.

Grayman, W. M., R. M. Clark, and R. M. Males. 1988. Modeling distribution system water quality: dynamic approach. Journal of Water Resources Planning and Management, ASCE 114(3):295–312.

Grayman, W. M., R. A. Deininger, R. M. Males, and R. W. Gullick. 2004. Source water early warning systems. Pp. 11.1–11.33. *In:* Water Supply Systems Security. L. W. Mays (ed.). New York: McGraw-Hill.

Gullick, R. W., M. W. LeChevallier, R. C. Svindland, and M. J. Friedman. 2004. Occurrence of transient low and negative pressures in distribution systems. J. Amer. Water Works Assoc. 96(11):52–66.

Hall, J., A. Zaffiro, R. B. Marx, P. Kefauver, R. Krishnan, R. Haught, J. G. and Herrmann. 2005. Parameters for Rapid Contamination Detection in a Water Distribution System. AWWA-Water Security Congress, Oklahoma City.

Hart, F. L., J. L. Meader, and S. N. Chiang. 1986. CLNET a simulation model for tracing chlorine residuals in a potable water distribution network. *In:* Proceedings of the AWWA Distribution System Symposium, Minneapolis, MN.

Hasan, J. 2005. Technologies and techniques for early warning systems to monitor and evaluate drinking water quality: state-of-the-art review. Available at: *http://www.epa.gov/ordnhsrc/news/news120105.htm*. Accessed May 13, 2006.

Hrudey, S. E., and S. Rizak. 2004. Discussion of rapid analytical techniques for drinking water security investigations. J. Amer. Water Works Assoc. 96(9):110–113.

Jacobsen, L. 2005. Las Vegas Valley Water District. Presented to the NRC Committee on Public Water Distribution Systems. April 18, 2005. Washington DC.

Jacobsen, L., and S. Kamojjala. 2005. Full System Models and GIS Integration. AWWA Annual Conference and Exposition, San Francisco, June 2005.

Javadpoour, R., and G. M. Knapp. 2003. A fuzzy neural network approach to machine condition monitoring. Computers and Industrial Engineering 45:323–30.

Lalor, G. C., and C. Zhang. 2001. Multivariate outlier detection and remediation in geochemical databases. The Science of the Total Environment 281:99–109.

Lansey, K. E., and P. F. Boulos. 2005. Comprehensive Handbook on Water Quality Analysis in Distribution Systems. Pasadena, CA: MWH Soft.

Lauer, W. C. 2005. Water Quality Complaint Investigator's Field Guide. Denver, CO: AWWA.

Lazarevic, A., L. Ertoz, A. Ozgur, J. Srivastava, and V. Kumar. 2003. A comparative study of anomaly detection schemes in network intrusion detection. *In:* SIAM Conference on Data Mining. San Francisco, CA. May, 2003.

Lee, B., R. Deininger, and R. Clark. 1991. Locating monitoring stations in water distribution systems. J. Amer. Water Works Assoc. 83(7):60–66.

Liou, C. P., and J. R. Kroon. 1986. Propagation and distribution of waterborne substances in Networks. *In:* Proceedings of the AWWA Distribution System Symposium, Minneapolis, MN.

Males, R. M., R. M. Clark, P. J. Wehrman, and W. E. Gates. 1985. Algorithm for mixing problems in water systems. Journal of the Hydraulics Division, ASCE III(2):206–211.

Males, R. M., W. M. Grayman, and R. M. Clark. 1988. Modeling water quality in distribution systems. Journal of Water Resources Planning and Management, ASCE 114(2):197–209.

Mau, R., P. Boulos, R. Clark, W. Grayman, R. Tekippe, and R. Trussell. 1995. Explicit mathematical models of distribution system storage water quality. J. Hydraulic Engineering 121(10):699–709.

Molenaar, K. R., and A. D. Songer. 2001. Web-based decision support systems: case study in project delivery. J. Comput. Civil Engineering 15(4):259–267.

Murray, R., R. Janke, and J. Uber. 2004. The Threat Ensemble Vulnerability Assessment Program for Drinking Water Distribution System Security. World Water & Environmental Resources Congress, EWRI-ASCE.

National Research Council (NRC). 1996. Safe Water from Every Tap: Improving Water Service to Small Communities. Washington, DC: National Academies Press.

Omenzetter, P., J. M. W. Brownjohn, and P. Moyo. 2004. Identification of unusual events in multi-channel bridge monitoring data. Mechanical Systems and Signal Processing 18:409–430.

Ortiz-Estarelles, O., Y. Martín-Bioscaa, M. J. Medina-Hernández, S. Sagrado and E. Bonet-Domingob. 2001. Multivariate data analysis of quality parameters in drinking water. The Analyst 126:91–96.

Ostfeld, A. 2004. Optimal monitoring stations allocations for water distribution system security. *In:* Water Supply Systems Security. L. Mays (ed.). New York: McGraw-Hill.

Ostfeld, A., and E. Salomons. 2004. Optimal layout of early warning detection stations for water distribution systems security. J. Wat. Res. Plan. Manag. 130(5):377–385.

Owens, J. 2001. A review of federal drinking water regulations in the U.S. Pp. 2-1–2-14 *In:* Controlling Disinfection By-products and Microbial Contaminants in Drinking Water. EPA/600/R-01/110. Washington, DC: EPA.

Panguluri, S., R. Krishnan, L. Garner, C. Patterson, Y. Lee, D. Hartman, W. Grayman, R. Clark, and Piao. 2000a. Using continuous monitors for conducting tracer studies in water distribution systems. *In:* Proceedings of the Environmental and Water Resources Institute (ASCE). Anchorage, Alaska.

Panguluri, S., W. M. Grayman, and R. M. Clark. 2005b. Distribution system water quality report: a guide to the assessment and management of drinking water quality in distribution systems. Cincinnati, OH: EPA Office of Research and Development.

Powell, J., J. Clement, M. Brandt, R. Casey, D. Holt, W. Grayman, and M. LeChevallier. 2004. Predictive Models for Water Quality in Distribution Systems. Denver, CO: AwwaRF.

Premazzi, G., and E. Hargesheimer. 2002. Introduction. Pp. 1–6 *In:* Online Monitoring for Drinking Water Utilities: Comparative Research Report. E. Hargestheimer, O. Conio, and J. Popovicova (eds.). Denver, CO: AwwaRF.

Romer, A. E., G. E. C. Bell, S. J. Duranceau, and S. Foreman. 2004. External corrosion and corrosion control of buried water mains. Denver, CO: AwwaRF.

Rossman, L A. 2000. EPANET Version 2 Users Manual. Cincinnati, OH: EPA Drinking Water Research Division.

Rossman, L. A., and P. F. Boulos. 1996. Numerical methods for modeling water quality in distribution systems: a comparison. J. Water Resources Planning and Management 122(2):137–146.

Rossman, L. A., R. M. Clark, and W. M. Grayman. 1994. Modeling chlorine residuals in drinking water distribution systems. J. Environmental Engineering 120(4):803–820.

Shang, F, J. Uber, and M. Polycarpou. 2002. A particle backtracking algorithm for water distribution system analysis. ASCE J. Environ. Eng. 128(5):441–450.

Stubbart, J. 2005. What is a Governing Board's Role? Opflow Feb.:8–9.

The Great Lakes - Upper Mississippi River Board of State Public Health and Environmental Managers. 2003. Ten-States Standards. Albany, NY: Health Education Standards.

Uber, J., R. Janke, R. Murray, and P. Meyer. 2004a. Greedy Heuristic Methods for Locating Water Quality Sensors in Distribution Systems. World Water & Environmental Resources Congress, EWRI-ASCE. Reston, VA: Environmental & Water Resources Institute of the American Society of Civil Engineers.

Uber, J., F. Shang, and L. Rossman. 2004b. Extensions to EPANET for Fate and Transport of Multiple Interacting Chemical or Biological Components. World Water &

Environmental Resources Congress, EWRI-ASCE. Reston, VA: Environmental & Water Resources Institute of the American Society of Civil Engineers.

Watson, J., H. J. Greenberg, and W. E. Hart. 2004. A Multiple-Objective Analysis of Sensor Placement Optimization in Water Networks. World Water & Environmental Resources Congress, EWRI-ASCE. Reston, VA: Environmental & Water Resources Institute of the American Society of Civil Engineers.

Wen, J., E. Busuego, II, A. Bleemers, and J. Long. 2005. Reduce customer complaints through control of water quality in distribution systems. *In*: Water Quality in the Distribution System. W. C. Lauer (ed.). Denver, CO: AWWA.

Wood, D. J. 1980. Slurry flow in pipe networks. Journal of Hydraulics, ASCE 106(1):55–70.

Wood, D. J., S. Lingireddy, and P. F. Boulos. 2005a. Pressure Wave Analysis of Transient Flow in Pipe Distribution Systems. Pasadena, CA: MWH Soft.

Wood, D. J., S. Lingireddy, P. F. Boulos, B. W. Karney, and D. L. McPherson. 2005b. Numerical methods for modeling transient flow in distribution systems. J. Amer. Water Works Assoc. 97(7):104–115.

Zierolf, M. L., M. M. Polycarpou, and J. G. Uber. 1998. Development and autocalibration of an input-output model of chlorine transport in drinking water distribution systems I.E.E.E. Trans. on Control Systems Technology 6(4):543–553.

8
Alternatives for Premise Plumbing

Premise plumbing includes that portion of the potable water distribution system associated with schools, hospitals, public and private housing, and other buildings. It is connected to the main distribution system via the service line. The quality of potable water in premise plumbing is not ensured or monitored by U.S. Environmental Protection Agency (EPA) regulation. Indeed, the only Safe Drinking Water Act (SDWA) rule in which drinking water quality is purposefully measured within premise plumbing is the Lead and Copper Rule (LCR) for which samples are collected at the tap after the water has been allowed to remain stagnant.

Virtually every problem previously identified in the main water transmission system can also occur in premise plumbing. However, unique characteristics of premise plumbing can magnify the potential public health risk relative to the main distribution system and complicate formulation of coherent strategies to deal with problems. This chapter discusses these characteristics and then considers both technical issues such as the need for monitoring of premise plumbing condition and policy alternatives for controlling public health issues related to premise plumbing.

KEY CHARACTERISTICS OF PREMISE PLUMBING

Premise plumbing systems have noteworthy differences from the main distribution system that are often under-appreciated by scientists and regulators with respect to public health goals. These are summarized in Table 8-1 and discussed more comprehensively below.

High Surface Area to Volume Ratio. Premise plumbing is characterized by relatively lengthy sections of small-diameter tubing. The total pipe length of the main distribution system has been estimated at about 1 million miles (Brongers et al., 2002; Grigg, 2005), whereas 5.3 million miles of copper tubing alone were installed in buildings between 1963 and 1999 (CDA, 2005). Premise plumbing has about ten times more surface area per unit length than in the main distribution system. One study of a distribution system in Columbia, Missouri determined that household plumbing and service connections had 82 percent of the total pipe length, 24 percent of the total surface area in the distribution system, and held just 1.6 percent of the total volume of water in the system (Brazos et al., 1985). Another 10 percent of the total distribution system volume was in premise plumbing if toilets and water heaters were considered (Brazos et al., 1985).

TABLE 8-1 Characteristics of U.S. Public and Private Transmission Systems

Characteristic	Public Infrastructure	Private Infrastructure
Approx. Pipe Surface per Volume Water*	0.26 cm^2/mL*	2.1 cm^2/mL*
Total Pipe Length (U.S.)	0.97 million miles	> 6 million miles
Replacement Value	$0.6 trillion	Much greater than $0.6 trillion
Prediction of Failure Events	Statistically predictable	Unpredictable for individual homeowner
Property Damage ($/consumer)	Relatively low	Potentially very high
Common Pipe Material	Cement, ductile iron, plastic, cast iron	Copper, plastics, galvanized iron, stainless steel, brass
Stagnation	Relatively rare except dead ends	Frequent and of variable length
Disinfectant Residual	Almost always present	Frequently absent
Regrowth Potential	Rarely realized (partly because rarely measured)	Frequently realized
Pipe Wall Thickness	> 6.6 mm	0.71–1.7 mm for copper tube
Velocity	2 to 6 ft/sec	Can be > 33 ft/sec, on/off or continuous
Infiltration	Abrupt changes in flow are relatively controllable (e.g., via scheduled flushing, proper distribution system design)	The service line can be the point of minimum pressure and experience frequent water hammer, the highest velocities, and the most leaks
Temperature	0–30° C	0–100° C (at the surface of heating elements)
Control of Water Quality	Utility treatments and operation	No control over water coming into home, but home treatment devices and selection of plumbing materials can influence water quality
Ownership	Utility	End user
Maximum Cost over 30 Years per Consumer	$500–$7,000 US	As much as $25,000 per homeowner, frequency determined by lifetime of plumbing
Financial Responsibility	Distributed burden over time and large customer base	Individual consumer
Cross Connections	Relatively rare	Widely prevalent
Frequency of Sample Collection and Evaluation of WQ Degradation	Regular sampling required by regulation and industry best standards	Often sampled only in reactive mode to consumer complaints

*Based on a 15.2-cm diameter for mains and 1.9-cm diameter for home plumbing.
SOURCE: Reprinted, with permission, from Edwards et al. (2003). © 2003 by Marc Edwards.

Water Age. Utilities and consumers have little control over water age in consumer plumbing. Water can sit stagnant in some buildings for extended periods if they are irregularly occupied, as exemplified by schools in summer months, vacation homes, or residences whose occupants have work that requires frequent prolonged travel. Even under full-time occupancy, some sections of plumbing within a given building are rarely used, and flow patterns can be highly variable dependent on water use patterns of the occupants. The upshot is that premise plumbing adds a layer of complexity to the hydraulics of distribution systems (see Chapter 5). That is, water residing in a given premise will have a wider distribution of water age than water at the entrance to the premise, resulting in greater variation in disinfectant residual levels, bacterial regrowth, and other issues than occurs in the main distribution system.

It should be noted that the negative effects of water age are exacerbated if the biological stability of the finished water is poor. Viable strategies to prevent problematic regrowth include local codes mandating premise plumbing materials that do not quickly react with disinfectants, removal of nutrients from the water to minimize regrowth potential when the disinfectant does disappear, recommendations that consumers flush unused premise plumbing lines, installation of booster stations (e.g., as is sometimes done in hospital plumbing systems) to ensure that residuals are supplied to all points of the distribution system, and use of on-demand water heaters to minimize storage volumes in premises.

Presence of Different Materials. Premise plumbing systems are comprised of a wide range of materials including copper, plastics, brass, lead, galvanized iron, and occasionally stainless steel. Many of these materials are not typically present in the main distribution system. The impact of water quality changes on the performance of materials within premises, and the effects of materials on water quality within premises, are often overlooked by water utilities. For instance, Brazos et al. (1985) show that the majority of chlorine demand in water systems often arises from pipe surfaces. Extensive work has been done investigating the reactions between chlorine and materials used in the main distribution system including polyethylene, PVC, iron, and cement (e.g., Clark et al., 1994), and routine samples collected in distribution systems reflect disinfectant loss from reaction with these materials. In general, reaction rates for chloramine with materials in the main transmission system are very low compared to free chlorine. But recent research has demonstrated that under at least some circumstances, chlorine and monochloramine decay very rapidly via reactions with copper and brass in premise plumbing (Powers, 2000; Nguyen, 2005; Nguyen and Edwards, 2005). Domestic water heaters also have very reactive aluminum and magnesium anodes that can contribute to rapid chlorine and chloramine decay in buildings.

Extreme Temperatures. Water sitting in premise plumbing is subject to greater extremes of temperature than in the main distribution system (Rushing and Edwards, 2004). In summer months, even the cold water line in premise plumbing can be 10–15° C warmer than for the mains. In addition, there is a hot-water distribution system with storage in most buildings, and often water chillers or refrigerated lines. The sampling of the main distribution system cannot capture effects of these variations on water quality in premise plumbing, particularly in relation to microbial type and concentrations. This is especially true for moderate thermophiles such as *Legionella* in water heaters.

Low or No Disinfectant Residual. Due to the high surface area to volume ratio, presence of reactive materials such as copper, long storage times, and warmer temperatures in premise plumbing, it is not possible to continuously maintain residual disinfectant throughout premise plumbing systems. Continuous contact with the water heater and copper pipe in hot water recirculation systems may be especially problematic with respect to maintaining chlorine residuals. Furthermore, water treatment devices are often installed by homeowners to remove tastes and odors—devices that also remove the disinfectant from the water. Figure 8-1 shows how the residual detected in hot water in Philadelphia residences during random sampling was well below the average disinfectant residual found in the main distribution system, even when chloramine, which is more persistent than chlorine, is used. The observed variability in disinfectant residual in homes would not be detected by a routine monitoring program for regulatory compliance; it is due to factors such as variability in water temperature, retention time in water heaters, condition of internal materials, type of heaters and pipes, etc.

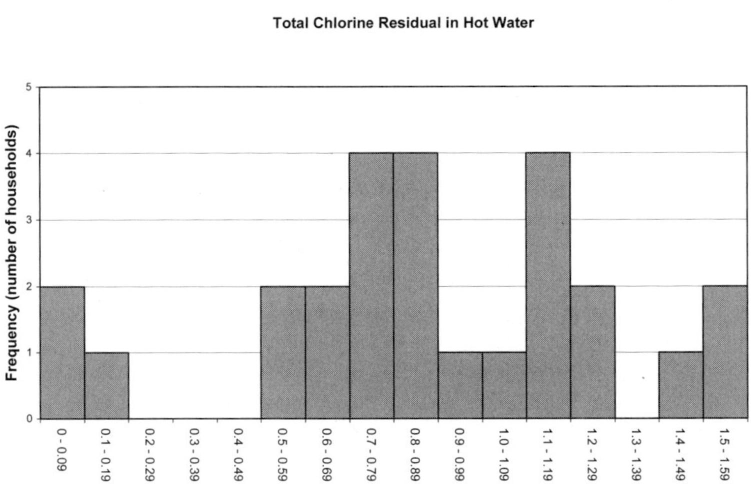

FIGURE 8-1 Water quality test results for hot water in 27 customer homes in Philadelphia. The average chloramine residual throughout the distribution system was 1.73 mg/L during December, 2003. At this time of year, chloramine decay rates in this distribution system are very low, such that the observed decreases in residual occurred primarily in premise plumbing.

Rapid chloramine decay within cold water copper pipe in consumers' homes has also been observed (Murphy et al., 1997a,b), and the resulting growth of nitrifiers can influence lead and copper leaching to water due to lowered pHs and other impacts (Garret, 1891; AWWA, 2003; Edwards and Dudi, 2004). In other instances that the committee is familiar with, chloramine has not been found to decay rapidly in premise plumbing, so additional research is needed to determine the prevalence and specifics of the problem.

Bacterial Levels and Potential for Regrowth. The lack of persistent disinfectant residuals, high surface area, long stagnation times, and warmer temperatures can make premise plumbing very suitable for microbial regrowth in at least some circumstances. Typical distribution system monitoring stipulates thoroughly flushing water through premise plumbing when sampled; consequently, problems with regrowth in premise plumbing systems can be missed. Brazos et al. (1985) noted a two- to three-order of magnitude increase in bacteria after water was held stagnant in home plumbing versus levels obtained in the same water after flushing. Using the same basic protocol in two systems experiencing difficulties with microbial control, Edwards et al. (2005) found a five-log increase in bacteria during stagnation in premise plumbing systems in Maui, Hawaii, and a three-log increase in Washington, DC. It is undoubtedly the case that high levels of bacteria in first draw samples are sometimes due to regrowth of bacteria in the faucet aerator (LeChevallier and Seidler, 1980). However, the Edwards et al. (2005) study was supplemented by bench-scale results that reproduced the problem without a faucet present. On the basis of their monitoring results, Brazos et al. (1985) recommended monitoring for bacteria in first draw samples in addition to routine monitoring of bacteria in the main transmission system.

To date there is little direct evidence that high levels of heterotrophic bacteria in premise plumbing systems have adverse health effects. However, opportunistic pathogens such as *Legionella* spp. and nontuberculous *Mycobacterium* spp. have been found in the biofilms of premise plumbing systems (Pryor et al., 2004; Tobin-D'Angelo et al., 2004; Vacrewijck et al., 2005; Flannery et al., 2006; Thomas et al., 2006; Tsitko et al., 2006). Hot-water storage tanks and showerheads may permit the amplification of these bacteria. As discussed in Chapter 3, outbreaks in healthcare facilities of Legionnaire's disease have been attributed to *Legionella pneumophila* in hot water tanks and showerheads. There is some evidence that nontuberculous mycobacteria may colonize biofilms, and the species found in treated drinking water have been linked to infections in immunocompromised individuals.

Highly Variable Velocities. Premise plumbing is characterized by start–stop flow patterns that can scour scale and biofilms from pipe surfaces. Flows up to 10 meters per second can occur. This makes premise plumbing more susceptible to the dislodgement of biofilms and associated negative health effects (see Chapter 6) than the main distribution system.

Exposure through Vapor and Bioaerosols. Stripping and formation of bioaerosols in relatively confined spaces such as home showers can be an important exposure pathway. This is relevant to waterborne disease caused by *Mycobacterium avium* and *Legionella*, as well as to overall exposure to volatile contaminants such as THMs and inhalation exposure to endotoxin (e.g., Little, 1992; Anderson et al., 2002; Mayo, 2004).

Proximity to Service Lines. As discussed in Chapter 1, service lines carry water from the distribution main to the premise plumbing in the building or property being served such that service line contamination can be a source of degraded water quality in premise plumbing. The majority of water leaks in a distribution system occur in service lines, service fittings, and connections (ferrules, corporation stops, valves, and meters) (AWWA, 2005). These locations therefore provide the greatest number of potential entry points for intrusion. The lower total chlorine residuals, lack of dilution, and short detention time before potential consumption might increase the potential health threat to individual consumers if intrusion were to occur at service lines. Little is known about the factors that might cause intrusion into service lines. Negative pressure transients could be responsible, but lower pressures and high velocities in service lines can cause a venturi effect (e.g., suction) and negative pressure waves due to water hammer that might also be significant.

Compared to the main distribution system, much less is known about the type and cause of service line failures. Possibilities include internal and external corrosion, poor installation such as improper backfilling techniques and materials, damage during handling, and improper tapping. In general, the collection of data documenting the occurrence of such failures is poor.

There is wide variation across the United States regarding ownership of service lines, which ultimately affects who takes responsibility for their maintenance. This can greatly complicate the extent to which service lines are inspected, replaced, and repaired in a timely manner when leaking. In most cases a drinking water utility, and thus most regulatory bodies, only takes responsibility for the quality of water delivered to the corporation stop, curb stop, or water meter. For that portion of the service line owned by consumers, the responsibility and cost of repairs fall on consumers, and the speed and effectiveness of repairs can therefore be even less efficient (AWWA, 2005).

Prevalence of Cross Connections. In contrast to the main transmission system, it is relatively common for untrained and unlicensed individuals to do repair work in premise plumbing. Furthermore, as discussed in Chapter 2, there is tremendous variability in state cross-connection control programs, both with respect to the breadth of the programs and the extent to which these programs are routinely enforced at the local level. As a result of these factors, premise plumbing is more likely to have cross-connections and potential backflow events than the main transmission system. For example, repairs by consumers as simple as replacing a ballcock anti-siphon valve in the toilet tank can create a direct cross connection if the line to the tank is not air-gapped. Hazardous chemicals added to the tank could then backsiphon or backflow under some circumstances. In a study in Davenport, Iowa, 9.6 percent of homes were found to have direct cross connections to a health hazard, most frequently due to failure to air gap the line in the toilet tank. Only 4.3 percent of homes investigated did not have a direct or indirect connection to a health hazard (USC, 2002).

It should be noted that for individual residences, backsiphonage is the greatest risk. However, it does not occur frequently, and when it does it would likely only affect a small population (usually only the population utilizing the building). Thus, these events are likely to be underreported. Backflow events are more likely to be reported when they occur in institutional settings, potentially affect a larger population, and are more likely to propagate back into the main distribution system.

The EPA white paper on cross connections (EPA, 2002a) makes it clear that the majority of backflow events occur in premise plumbing. As shown in Table 8-2, the portion of the distribution system controlled by the utility accounted for only 18 of 459 reported backflow events.

Responsible Party. There is lack of clarity over who is responsible for maintaining water quality in premise plumbing. Many consumers mistakenly believe that EPA regulations and their water utility guarantee that tap water is always safe to drink. Some public advertisements and educational materials reinforce the perception that EPA regulations and utility responsibility extend to the tap. Historically, however, in the United States the property line demarcating the public from the private system has not been crossed for regulatory purposes. The notable exception is the LCR, which has successfully reduced the general corrosivity of public water supplies in relation to lead and copper leaching. But ultimately individual homeowners and building supervisors bear final responsibility for protecting themselves from excessive lead or copper exposure and other degradation to water quality occurring beyond the property line.

Economic Considerations. The net present replacement value of premise plumbing and the corresponding cost of corrosion far exceed those for the main distribution system (Ryder, 1980; Edwards, 2004). Moreover, costs associated with premise plumbing failures are unpredictable and fall directly on the consumer. Leaks occurring in premises also have implications for insurance renewal and mold growth.

Leaching and Permeation. Leaching and permeation mechanisms are the same in premise plumbing as in the main distribution system. However, the higher pipe surface area to water volume ratio, very long stagnation times, and lessened potential for dilution increase the potential severity of the problem in premise plumbing. If permeation were to occur through a consumer's service

TABLE 8-2 Numbers of Documented Backflow Incidents from 1970 to 2001.

Location of cross connection	Number of reported backflow events
Homes	55
Apartments	27
Mobile homes	1
Neighborhoods	3
Public Water Supply	15
Medical buildings	27
Schools	31
Other government buildings	24
Restaurants	28
Office buildings	18
Other commercial buildings	66
Agricultural, recreational, and industrial sites	56
Unknown or other miscellaneous sites	108

SOURCE: Adapted from EPA (2002a).

line or premise plumbing, it would not be detected by routine distribution system monitoring.

Scaling/Energy. At present about eight percent of U.S. energy demand is attributable to costs of pumping, treating, and heating water, and water heating accounts for 19 percent of home energy use (EPA, 2005). Hot water systems and small diameter tubes in premises are more sensitive to build up of scale, which can increase head loss and decrease water heater efficiency. The implications and costs of scaling in buildings tend to constrain the range of feasible water chemistries that might be considered to protect public health. For example, higher pH values that might be desirable to reduce nitrification and protect public infrastructure from internal corrosion could cause unacceptable scaling.

GAPS IN RESEARCH AND MONITORING

The preceding section highlights some of the unique challenges posed by premise plumbing relative to the main water distribution system. Even more so than with the main distribution system (see Chapter 3), very few studies have been done to assess the magnitude of the public health threat posed by premise plumbing. This is partly due to a lack of water quality monitoring in premise plumbing. Normal distribution system monitoring under EPA regulations often utilizes taps located in buildings, but water is thoroughly flushed from the pipes before sampling with the exception of samples for lead and copper. Thus, if there were problems related to water quality in a given premise plumbing system it would not necessarily be detected. No drinking water maximum contaminant levels (MCLs) protect consumers against water quality degradation resulting from premise plumbing.

While solid evidence is not available, water quality degradation occurring within premise plumbing may have public health implications. For instance, recent trends in the United States to decrease water heater temperature to minimize scalding and save energy could be increasing the growth of opportunistic pathogens. Increased use of phosphate inhibitors, chloramine disinfectants, and point-of-use devices can also benefit or worsen the ultimate quality of water after it is held stagnant in premise plumbing. However, the lack of monitoring and isolated nature of problems that are discovered hinder rigorous risk analysis.

Considering the emergence of *Legionella* and *Mycobacterium* as waterborne pathogens, and recognizing the threat from these microbes arising from regrowth in premise plumbing systems, more decisive action is necessary. For instance, existing EPA regulations are likely to produce water with a low level of *Legionella* in water leaving the treatment plant, but the effective *Legionella* levels in premises may still result in adverse health effects. There are 8,000–18,000 estimated *Legionella* cases in the United States each year with a fatality rate between 10 and 15 percent (*http://www.cdc.gov/ncidod/dbmd/diseaseinfo/legionellosis_t.htm*). Drinking water was judged responsible for 12 percent of *Legionella* cases in one case study mentioned in the United Kingdom (VROM, 2005a), but the methodology and certainty of that analysis is open to question. If a similar percentage of *Legionella* cases in the United States were caused by drinking water in premise plumbing systems, the health threat from *Legionella* alone would be very high relative to all other reasonably quantified risks from waterborne disease.

Despite this relatively well established and high health risk, only a few studies have been conducted into possible broad community interventions that might reduce risk in buildings. Those studies have consistently found that chloramine was more effective than free chlorine in reducing *Legionella* levels (Kool et al., 1999a,b; Pryor et al., 2004; Stevens et al., 2004). A recent study in San Francisco demonstrated that the change from free chlorine to chloramine reduce the percentage of buildings with detectable *Legionella* from 60 percent to 4 percent, respectively (Flannery et al., 2006). However, one of the studies found higher levels of mycobacteria after chloramination (Pryor et al., 2004), and the possible impact of free ammonia as a nutrient on *Legionella* growth (if chloramine were to completely decay) has not yet been assessed. Nor have studies correlated *Legionella* occurrence and concentrations in drinking water with actual outbreaks of legionellosis.

Targeted research to improve understanding of water quality degradation within premise plumbing is recommended and must overcome several challenges. All three approaches discussed in Chapter 3 for relating distribution system contamination events to public health risk (pathogen occurrence measurements, outbreak surveillance, and epidemiology studies) have unique challenges that increase the difficulty of their execution when applied to premises. *Legionella* has only recently (since 2001) been added to the CDC outbreak surveillance system. Unfortunately, existing CDC outbreak data would rarely implicate premise plumbing because backflow and regrowth events likely would not be reported unless an institutional building with large numbers of people was affected. Furthermore, there are minimal data on exposure routes other than ingestion, and this is yet another reason why so few data exist on the health effects of *Legionella* in tap water. The CDC has recently changed it reporting requirements for the outbreak surveillance system so that outbreaks that arise from events in premise plumbing are more clearly identified (see Chapter 3). Box 8-1 presents one of the few outbreaks clearly linked to contamination of premise plumbing.

The little epidemiological research done to date has attempted to track the impacts of premise plumbing components on gastrointestinal upset (e.g., Payment et al., 1997; Colford et al., 2005), but not health problems arising from exposure to bio-aerosols as would be necessary for *Legionella* and *Mycobacteria*. The Davenport study (LeChevallier et al., 2003, 2004; Colford et al., 2005)

BOX 8-1
Waterborne Disease Outbreak Associated with Premise Plumbing Contamination: North Dakota, USA, April 1987

Ethylene glycol is a solvent with a sweet, acrid taste that is used in antifreeze solution and in heating and cooling systems in buildings. Ingestion of ethylene glycol causes acute poisoning with central nervous system depression, vomiting, hypotension, respiratory failure, coma, convulsions, and renal damage, depending on the dose. The fatal dose for adults is approximately 100 g. Several incidents of ethylene glycol ingestion have been reported to the CDC waterborne disease surveillance system. All these incidents have involved public buildings and have been linked to contamination of premise plumbing through backflow via cross-connection with an air conditioning or heating system.

In April 1987, two children in rural North Dakota were admitted to a local hospital with acute onset of somnolence, vomiting, and ataxia. Toxicologic analysis of their urine indicated the presence of ethylene glycol. Further investigation revealed that both children had been to a picnic earlier in the day at a fire hall in rural North Dakota. Approximately 400 persons had attended the picnic, and telephone interviews with about 91 percent of the attendees identified 29 additional cases of apparent ethylene glycol poisoning with 66 percent of the cases occurring in children under ten years of age. The most frequently reported symptoms were excessive fatigue and sleepiness, unsteadiness when walking, and dizziness. Data collected during the telephone interview about food and beverages consumed during the picnic indicated that one beverage, a noncarbonated soft drink, was strongly associated with illness (relative risk = 31.0). A clear dose-response was also observed among children, with no cases occurring among children who did not drink the implicated beverage, two cases among children who drank less than or equal to half a cup, five cases among children who drank one-half to one and a half cups and 12 cases among those who drank more than one and a half cups.

The implicated beverage had been prepared on-site using a powder mix and water drawn from a spigot near the fire hall heating system that used a mixture of water and antifreeze and was cross-connected to the potable water supply. There was a valve on the cross-connection but no information on whether the valve had been closed before collecting water to prepare the beverage. Other foods and beverages had been prepared in the fire hall kitchen, and the kitchen sink was about 30 feet from the spigot with the cross-connection. A water sample collected from the spigot on the evening of the picnic was determined to have an ethylene glycol concentration of 9 percent.

SOURCE: MMWR September 18, 1987/36(36):611-4.

is the only known example of an epidemiology study where premise plumbing was investigated as a source of contamination contributing to gastrointestinal upset, and no impact was observed. Payment et al. (1997) cited a lower incidence of gastrointestinal upset after water contacted premise plumbing, and speculated it was due to disinfecting properties of copper. Of course very high levels of soluble copper in water leached from premise plumbing can also cause gastrointestinal upset (Craun et al., 2001). Thus, a range of health impacts from premise plumbing issues can be expected.

With respect to pathogen occurrence measurements in premise plumbing, there is also no regulation or even voluntary standards recommending such sampling, and as a result background data are not being collected. Guidelines from the CDC (CDC, 2003, 2004) exist for *Legionella* in high risk buildings such as hospitals where an infection control officer is often responsible for monitoring and mitigating risk, but such monitoring and control measures are not routinely followed in other situations or in individual residences. Indeed, the current EPA guidelines on scalding prevention run counter to common control measures for *Legionella* (see section below under Policy Alternatives). Other opportunistic pathogens such as *Mycobacteria* are emerging concerns, for which there is a weaker link to disease and therefore even less incentive for monitoring. Monitoring samples could be collected by utilities from public buildings or from consumers' homes during Lead and Copper Rule monitoring. Box 8-2 discusses the routine monitoring conducted on tap water in Seoul, South Korea.

WHY HOME TREATMENT DEVICES ARE NOT ALWAYS THE ANSWER

Home treatment devices have become increasingly popular as a means to further treat drinking water supplied by public water systems, and they are considered to be a potential technical solution to some problems associated with premise plumbing. There are a myriad of available devices designed to remove organic and inorganic chemicals, radionuclides, and microbiological agents from tap water. Common home treatment devices include point-of-use (POU) devices that are mounted at the end of the faucet, canister type devices that are plumbed in-line under the sink, stand-alone pitchers in which water is gravity fed through a filter, and refrigerated filtered-water systems. Home treatment devices used to treat the entire flow into the premise are called point-of-entry (POE) devices. POE devices can be as simple as a water softener to more complicated devices that combine sediment filters, activated carbon filters, and ultraviolet (UV) disinfection.

Home treatment devices can range in cost from tens of dollars for a pitcher-type filter to thousands of dollars for a whole house treatment system. Most devices have components that need to be changed at a regular interval or after a specified volume of water has been treated. Membranes for reverse osmosis treatment systems are changed at a given frequency or when there is a reduction

BOX 8-2
City of Seoul Water Works' Customer Tap Water Quality Certification Program

In South Korea, the City of Seoul Water Works (SWW) has conducted a customer tap water quality certification program since November 2001. The program is part of its water quality management system to enhance reliability and meet customer satisfaction of its municipal drinking water supply. Under this program more than 50,000 drinking water taps are checked each year. A total of 344,600 taps have been covered by SWW since 2001. The targeted sites include apartment complexes, schools, households, public parks, and shopping malls.

Seoul has over 10 million inhabitants, which are provided for with six drinking water treatment plants with a total daily production capacity of 5.4 million cubic meters of finished water. The total length of distribution system pipe in the city is approximately 15,870 km. While SWW's treated water meets the national water quality standards set forth by the Korea Ministry of Environment, it is well documented that city water can deteriorate upon standing in customers' water storage systems that have been poorly maintained. Many residential and commercial facilities in the city have indoor water storage systems.

The customer tap water quality certification team consists of SWW employees and representatives from environmental or citizen groups. At each targeted site the team collects a cold water sample from a kitchen faucet and examines the conditions and integrity of the water pipes and storage systems of the customer. Each water sample is field tested for chlorine residual, turbidity, pH, iron, and copper at the site. If the water meets the national drinking water quality standards, the team issues and attaches a water quality certification on the faucet. If it does not meet the standards, the following secondary parameters are tested back in the laboratory: heterotrophic plate count, total coliforms, *E. coli*, ammonium nitrogen, zinc, and manganese. If the team finds inadequate plumbing or poor sanitary conditions that could create water quality problems, they provide guidance to the occupant to correct the problems.

SOURCE: Communicated to Gary A. Burlingame, Philadelphia Water Department via e-mail on July 20, 2005 by Jung J. Choi, Philadelphia Water Department and Dr. Lee Suwon and Lee Gyu Sub, Waterworks Research Institute, Seoul Metropolitan Government.

in total dissolved solids removal efficiency. UV disinfection systems must be inspected periodically to prevent scale from forming on the lamps, which will reduce the light intensity.

Most manufacturers of home treatment devices test their devices according to ANSI/NSF standard test protocols. However, there is no national certification program that requires such testing. Only three states (California, Iowa, and Wisconsin) mandate that, before making a health claim and selling devices in their state, home treatment devices must be tested and certified as meeting ANSI/NSF standards.

For many reasons, POU and POE devices are not a panacea to premise plumbing issues. First, although home treatment devices are effective in removing the contaminants for which they are designed, they cannot work past their point of application. For example, if a POE device is used and there are cross-connections within the home "downstream" from the device, contaminants resulting from the cross connection will not be removed. Second, the media or membranes used in POU and POE treatment devices may be susceptible to microbial colonization. Higher levels of bacteria have been found in the finished water produced by some POU and POE treatment devices, particularly those that incorporate an activated carbon element (Rollinger and Dott, 1981; Camper et al., 1985; Calderon and Mood, 1987; EPA, 2002b). Granular activated carbon in point-of-use treatment devices can accumulate nutrients and neutralize disinfectant residuals, thereby providing an ideal environment for microbial growth (Tobin et al., 1981; Geldreich et al., 1985; Reasoner et al., 1987; LeChevallier and McFeters, 1988). Several coliform bacteria (*Klebsiella, Enterobacter,* and *Citrobacter*) have been found to colonize granular activated carbon filters, regrow during warm-water periods, and discharge into the process effluent (Camper et al., 1985). The presence of a silver bacteriostatic agent did not prevent the colonization and growth of HPC bacteria in granular activated carbon filters (Tobin et al., 1981; Reasoner et al., 1987). Rogers et al. (1999) reported the growth of *Mycobacterium avium* in point-of-use filters in the presence of 1,000 μg silver/mL filter medium. The health implications of this regrowth are uncertain. Third, although POU disinfection devices are available, including UV and distillation systems, these devices are not designed to treat water used for showering and bathing. Some POE devices include UV disinfection, which can potentially be effective in reducing the levels of *Legionella* and other microorganisms entering the premise (Gilpin, 1985; EPA, 1999), but they would not stop regrowth of opportunistic pathogens in the premise plumbing system.

POLICY ALTERNATIVES

Although the magnitude of the public health threat from bacterial regrowth, cross connections, intrusion, leaching, and permeation in premise plumbing is not clearly defined, improved control should be a high priority based on existing data and best professional judgment regarding the potential for problems. It is possible to address these problems through legislation and regulation, the plumbing code, voluntary standards, and public education. Examples of each approach, including their use in other countries, are provided in the sections that follow.

Problems Addressed Through Plumbing or Building Codes: Scalding and Regrowth in Water Heaters

Several countries are addressing the complicated issue of simultaneously controlling scalding problems and preventing *Legionella* growth in water heaters. A conflict arises because the hotter temperatures that control *Legionella* also increase the likelihood of scalding (see Figure 8-2). The consumer product safety commission estimates that scalding from tap water results in 3,800 injuries

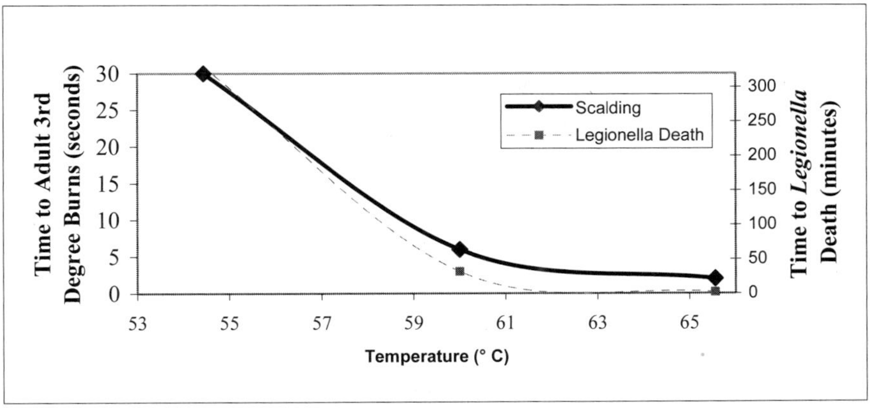

FIGURE 8-2 Higher temperature decreases the time required for *Legionella* inactivation but also decreases the time to acquire burns from scalding. SOURCES: Illustrative data derived from CPSC (2005) and Armstrong (2003).

and 34 deaths annually in homes (CPSC, 2005), and children are especially at risk from scalding (NSKC, 2004).

To ensure that water storage is hot enough to prevent microbial regrowth but that delivered water temperatures are not high enough to cause scalding, one solution is to install a plumbing device that physically limits the percentage of hot water flowing to the tap based on target delivery temperatures. For instance, Australian standards require maintenance of hot water systems at a minimum of 60° C to control *Legionella* and installation of a mixing valve at points of delivery to prevent scalding (Spinks et al., 2003). The Australian work recognized that growth of many pathogenic microbes in hot water systems other than *Legionella* would also be controlled by this change in the plumbing code.

Canada is in the process of finalizing a plumbing code that requires installation of valves that prevent water outlet temperatures exceeding 49°C at showerheads, bathtubs, or lavatories, while also requiring temperatures above 55°C in hot water recirculation or 60°C in water service heaters to prevent *Legionella* (C. R. Taraschuk, personal communication, Standing Committee on Building and Plumbing Services, final proposed wording for code, 2005). A cost/benefit analysis of the mixing valve requirement in Canada indicated a benefit of $0.7–$4.2 million in reduced scalding versus a cost of $48–$119 million per year (K. Newbert, personal communication to B. E. Clemmensen, August 26, 2005). However, the estimated benefit did not include costs of reduced *Legionella* death rates or reduced outpatient care, nor did it consider higher energy costs incurred in maintaining higher water temperatures. The code change in Canada would be relevant only for new dwellings, since retrofit cost/benefit analyses were less favorable. However, specific localities could choose to require retrofits.

The American Society of Plumbing Engineers has recommended a similar approach in the United States (George, 2001), but the official recommendation from EPA is that consumers reduce their water heater temperature to 48° C to save energy, prevent scalding, and reduce scaling (EPA, 2004). Many U.S. water utilities highlight the EPA advice to reduce water heater temperatures on their web page.

Problems Addressed By Regulation: Control of Regrowth in Premise Plumbing Systems

Consistent with the current U.S. approach, English water companies have met their obligations if a failure to meet standards at the tap can be attributed to degradation occurring in privately owned premise plumbing (Colburne, 2004; Jackson, 2004; WHO, 2005). But in public buildings, including schools, hospitals, and restaurants, water quality must meet all regulations for potable water *at the tap*. While details are still under discussion, guidelines suggest that water must be sampled at taps in 10 percent of public buildings each year. "First draw" samples for bacteria must also be collected, and disinfection of sample taps is not allowed before collecting samples (Colburne, 2004). A similar regulatory approach could be considered of U.S. utilities to detect microbially unstable water and rapid disinfectant loss in premise plumbing.

Legislation and regulation has also targeted operators of premise plumbing systems. In the Netherlands, the owners of collective water systems including hotels, camp sites, and sports facilities have been required to complete a risk analysis for microbial regrowth. The focus was mostly on *Legionella*, but a new Drinking Water Directive 98/83/EC also will eventually consider other microbial parameters at the tap (Regal et al., 2003). If a high risk is identified, the owner must indicate measures to protect against *Legionella* (VROM, 2005a,b). A recent survey of European approaches to controlling *Legionella* found that some countries directly addressed premise plumbing issues (VROM, 2005b).

Problems Addressed By Voluntary Compliance: Hong Kong

A survey in 1999 that revealed 48 percent of Hong Kong respondents rated their water quality at the tap fair to poor, but also indicated that less than 0.1 percent of the customers made complaints to the water company. Beginning in 2001 the Advisory Committee on the Quality of Water Supplies (ACQWS) began meeting to discuss strategies that would protect water to the tap. The key concerns were turbidity and discolored water from older galvanized plumbing. Various strategies were initially considered including:

1) encourage designers of new buildings to design plumbing with water quality at the tap in mind

2) educate the public to increase confidence and encourage drinking of water from taps and to maintain plumbing systems

3) encourage renovation of plumbing systems as part of routine maintenance

4) inspection programs for older buildings to determine if they need maintenance, with potential issuance of orders requiring repair

5) require building owners to inspect internal plumbing using licensed plumbers and submit a report, with possible fines for non-compliance

6) empower utilities to make repairs or remediation for consumers when problems are persistent

A staged plan was considered for implementing some of the above strategies, in which the first three years of effort would focus on education of consumers, required implementation at government or quasi-governmental buildings, and voluntary compliance. Thereafter, if progress was unsatisfactory, laws would be considered. Loans were already available to customers from the building department for maintenance of plumbing.

Consideration eventually gave rise to a Fresh Water Plumbing Quality Maintenance Recognition Scheme in buildings. The general idea is to create market forces that would make compliance desirable for participants. Voluntary successful applicants are awarded a certificate that can be used as a symbol of effective premise plumbing maintenance to the consumers' taps. To qualify, the plumbing system must (1) be inspected at least once every three months by qualified personnel, (2) have all defects quickly repaired, (3) have water tanks cleaned every three months, and (4) have water samples collected at least once a year for analysis. The program is overseen by the water supply department, and the program is confidential. Checklists are provided for tank cleaning and water quality analysis and inspection. The program was started in July 2002; 32 months later, 2,807 certificates had been issued for residential buildings, hotels, and restaurants. Logos are provided to place on taps, and the time period in which certified compliance is valid is indicated. About 34 percent of residential households were covered by the program. Other progress included issuance of a plumbing maintenance guide, which might eventually become mandatory. A system was in development to track problem premises for which frequent complaints occurred.

A survey was conducted of approaches for controlling premise plumbing problems in other Asian cities, and the results are included in Table 8-3. In general, the survey revealed that consumers in many of the Asian cities do not drink water from the tap (< 0.5 percent drink tap water directly in Hong Kong).

Problems Addressed by Public Education

Because water utilities have very limited control beyond the water meter of the customer, much of our existing information on water quality is not relevant to premises. Thus, there is a need to engage customers in identifying problems and coming up with solutions through public education.

A multifaceted and long-term approach to providing safe drinking water from the treatment plant to the water meter is already used by the drinking water industry—one that involves compliance with the Safe Drinking Water Act, use

TABLE 8-3 World-wide Perspectives on Responsible Party to Prevent Degradation of Water Within Premise Plumbing.

Country	Approach
U.S.A.	Explicit requirements for Lead and Copper only. Utility has responsibility to "Optimize" corrosion control to minimize Pb/Cu at the tap of select homes. Regulated by "action levels" for lead and copper. Lead pipe and solder banned in new construction. Guidelines for lead in schools but no regulation.
U.K.	By-laws in some instances requires draw off point for potable water directly from utility services, thereby completely avoiding home plumbing and allowing direct access to drinking water. Compliance with all regulations required at the tap in public buildings.
Hong Kong	Utility publishes free books and TV ads to encourage upgrades to plumbing and to clean storage tanks. Inspection for dirt and testing for bacteria (utility inspects based on complaints).
Singapore	Code of practice for consumers and their agents recommends that samples from various premise plumbing locations be examined periodically by water analysis. Chemical examination is beneficial in showing if corrosion is taking place, and bacterial contamination can be determined by sampling. Storage should be inspected at least once a year and cleaned. For "housing estates" and government buildings the recommendations are followed, but for "private estates" recommendations are voluntary. Reports are made to the water department. Making the recommendations into law was being considered.
Shenzen, China	At least every half year, water tanks must be cleaned and sterilized, with testing of water quality at the inlet and outlet by labs. The water company has responsibility for this task for low-rise buildings whereas the building owner has responsibility in high rises. The building management bears the cost, and a financial penalty can be given to those not complying. Reports are required to the water utility and department of health.
Taipei, Kuala Lumpur, Malaysia	Consumer generally has complete responsibility. However, Kuala Lumpur requires sufficient residual chlorine, and the desirability of regularly cleaning cisterns is publicized in newspapers and on television in Taipei.

SOURCE: Adapted from ACQWS (2005), except the entry for the United States.

of AWWA and ANSI standards for specifications and best practices, and enrollment in certification programs such as the Partnership for Safe Water and AWWA's QualServe. This broad approach involves regulations, best practices, and peer review. However, regulation might prove to be the most expensive way and least efficient way to reduce risk and achieve control when one considers the customers' premises as an integral component of the distribution system. Rather, regulation is only part of an overall approach to minimizing the risk from everyday use of tap water.

Public education is needed to spur the public to incorporate new actions into everyday life. Similar changes are needed within the water and plumbing industries, such as the sanitary handling and storage of materials that come in contact with drinking water. Concepts such as the value of water, the need to conserve water (which has already taken place in some areas of the United States), and the need for good materials in guaranteeing good water quality are basic to bringing about solutions to problems the drinking water industry is faced with. These concepts must become part of the public psyche, as natural as washing ones hands after handling raw meat. Altering public behavior with respect to water will requires a multifaceted approach, broad-based support, and long-term commitment. It will require numerous efforts directed at premise plumbing such as:

- Basic education of concepts in elementary schools and higher education
- Education of trades such as plumbing contractors and building supervisors in health effects of premise plumbing, and the need for standards in products and design
- Available, easy-to-understand information in public libraries
- Government officials, politicians, consumers, and advocacy groups who are properly educated and can represent the best interests of the public at large
- Health officials, doctors, and nurses who educate their patients and the public on how to minimize risks in practical and achievable ways.

Some progress has been made in the above areas for control of *Legionella* in institutional settings through published voluntary guidelines (ASHRAE, 2000; CDC, 2003). For the analogous problem of indoor air pollution and radon control, EPA has developed "A Guide to Indoor Air Quality" (EPA, 1995) that is easy to understand and which highlights the nature of the health threat and mitigation strategies that can be implemented to reduce the magnitude of the risk. A similar manual with an accompanying website would be highly desirable relative to premise plumbing systems. At a minimum the manual should include consideration of:

- Taste, odor, and aesthetic issues that can arise from premise plumbing
- Maintenance, including flushing of water heaters

- Issues related to energy conservation, scalding, and microbial regrowth in water heaters
- Trade-offs with different types of water heaters
- Benefits, limitations, and appropriate uses for various POU and POE devices
- The need to prevent cross connections
- Risks of untrained repair
- Recognizing obvious repairs or plumbing designs that could be problematic
- Troubleshooting premise plumbing problems, with information on who to contact for additional information, investigation, and repair.

CONCLUSIONS AND RECOMMENDATIONS

Premise plumbing should be recognized as a contributor to the loss of distribution system integrity, particularly due to microbial regrowth, backflow events, and contaminant intrusion via holes in service lines. Improper design or operation of premise plumbing systems can pose a substantial health threat to consumers, although additional research is needed to better understand its magnitude. In particular, more extensive sampling of water quality within premise plumbing by utilities or targeted sampling via research is required. The following detailed conclusions and recommendations are made.

Communities should squarely address the problem of *Legionella*, both via changes to the plumbing code and new technologies. Changes in the plumbing code such as those considered in Canada and Australia that involve mandated mixing valves would seem logical as a compromise that would prevent both scalding and microbial regrowth in premise plumbing water systems. On-demand water heating systems may have benefits worthy of consideration versus traditional large hot water storage tanks in the United States. It may be desirable for building owners to conduct risk analysis for *Legionella* on their properties as per the Netherlands, and to develop a plan to address obvious deficiencies. The possible effects of chloramination and other treatments on *Legionella* control should be quantified to a higher degree of certainty.

To better assess cross connections in the premise plumbing of privately owned buildings, inspections for cross connections and other code violations at the time of property sale could be required. Such inspection of privately owned plumbing for obvious defects could be conducted during inspection upon sale of buildings, thereby alerting future occupants to existing hazards and highlighting the need for repair. These rules, if adopted by individual states, might also provide incentives to consumers and building owners to follow code and have repairs conducted by qualified personnel, because disclosure of substandard repair could affect subsequent transfer of the property.

EPA should create a homeowner's guide and website that highlights the nature of the health threat associated with premise plumbing and mitigation strategies that can be implemented to reduce the magnitude of the risk. As part of this guide, it should be made clear that water quality is regulated only to the property line, and beyond that point responsibility falls mainly on consumers. Whether problems in service lines are considered to be the homeowner's responsibility or the water utility's varies from system to system.

Research projects are needed that specifically address potential problems arising from premise plumbing. Because no organized party has had clear responsibility for this problem, research has been under-funded. Three lines of research are needed, each of which would help to improve future understanding of the public health risks from distribution systems:

- ***Collection of data quantifying water quality degradation in representative premise plumbing systems in geographically diverse regions and climates.*** Some of the needed data include those routinely collected in the main distribution system, including water residence time, disinfectant residuals, and microbial monitoring. In addition, greater attention should be focused on understanding the role of plumbing materials. Furthermore, the role of nutrients in distributed water in controlling regrowth should be assessed for premises because their longer holding times, chronic lower disinfection residuals, warmer temperatures, and most importantly their colonization by opportunistic pathogens such as *Legionella* and *Mycobacterium avium* make the biological stability of the water even more important than in the main distribution system. Specialized sampling is needed to quantify regrowth of opportunistic pathogens such as *Legionella* and *Mycobacteria* as a function of consumer water use patterns, plumbing system layout, and water heater operation. Finally, the potential impacts of representative POU and POE devices need to be quantified.

- ***Practical insights should be developed regarding exposure routes other than ingestion, including inhalation of bioaerosols from water.*** Effects of climate, consumer behavior in bathing and showering, and the specifics of plumbing system design and operation are likely to be key contributing factors in disease transmission from premise plumbing contamination. With respect to contracting disease such as legionellosis, such information would make it possible to develop steps that might reduce risk or explain why disease is contracted in some cases and not in others.

- ***An epidemiological study to assess the health risks of contaminated premise plumbing should be undertaken in high risk communities.*** Without information from the two bullets above, it would be very difficult to identify such groups with confidence.

- ***Environmental assessments of outbreaks should begin to incorporate new insights and allow possible cause-and-effect relationships to be established.*** Such assessments have traditionally focused on documenting outcomes of waterborne disease and not on key factors related to human exposure. Chapter 3 has documented that the reporting of outbreaks is being revised to include more explicit consideration of distribution system and premise plumbing deficiencies that might contribute to waterborne disease. Much greater emphasis must also be placed on dose reconciliation in outbreaks, which would require specialized sampling techniques for bioaerosols in the case of premise plumbing, in order to develop basic practical data on dose-response relationships. It is possible to genetically link bioaerosols to microorganisms that infect humans (e.g., Angenent et al., 2005); this type of analysis must be attempted with greater frequency to establish cause-and-effect relationships in potable water.

REFERENCES

Advisory Committee on the Quality of Water Supplies (ACQWS). 2005. Papers # 7, #8, #9, #13, #15 and Minutes of Meeting April 22, 2004. Available on-line at *http://www.wsd.gov.hk/acqws/eng/home.htm*. Accessed May 10, 2006.

Angenent L. T., S. T. Kelley, A. Amand, N. R. Pace, and M. T. Hernandez. 2005. Molecular identification of potential pathogens in water and air of a hospital therapy pool. Proceedings of the National Academy of Sciences 102(13):4860–4865.

Armstrong (Armstrong International, Inc.). 2003. Controlling Legionella in domestic water systems. 2003. Available on-line at *http://www.bbriefings.com/pdf/13/Hosp031_t_Armstron.pdf*. Accessed August 10, 2005

American Water Works Association (AWWA). 2003. Nitrification. Available on-line at *http://www.epa.gov/safewater/tcr/pdf/nitrification.pdf*. Accessed May 10, 2006.

AWWA. 2005. Are Service Lines the “Achilles Heel” of Your Distribution System? Webcast August 24, 2005.

ASHRAE. 2000. Minimizing the Risk of Legionellosis Associated with Building Water Systems. Standard 12-2000. ISSN 1041-2336. Available on-line at: *http://www.lakoshvac.com/enewsimages/guide12.pdf*. Accessed May 10, 2006.

Anderson, W. B., R. M. Slawson and C. L. Mayfield. 2002. A review of drinking-water-associated endotoxin, including potential routes of human exposure. Can. J. Microbiology 48:567–587.

Brazos, B. J., J. T. O’Conner, and S. Abcouwer. 1985. Kinetics of chlorine depletion and microbial growth in household plumbing systems. Pp. 239–274 (Paper 4B-3) *In:* Proceedings of the American Water Works Association Water Quality Technology Conference. Houston, Texas.

Brongers, M. P. H. 2002. Appendix K of drinking water and sewer systems in corrosion costs and preventative strategies in the United States. Report FHWA-RD-01-156. Washington, DC: U.S. Department of Transportation Federal Highway Administration.

Calderon, R. L., and E. W. Mood. 1987. Bacteria colonizing point-of-use, granular activated carbon filters and their relationship to human health. EPA CR-811904-01-0. Washington, DC: EPA.

Camper, A. K., M. W. LeChevallier, S. C. Broadaway, and G. A. McFeters. 1985. Growth and persistence of pathogens on granular activated carbon filters. Appl. Environ. Microbiol. 50(6):1378–1382.

Centers for Disease Control and Prevention (CDC). 2003. Guidelines for preventing health care associated pneumonia 2003. MMWR 2004(RR-3):1–36.

CDC. 2004. Legionellosis: Legionnaires' Disease (LD) and Pontiac Fever. Available on-line at *http://www.cdc.gov/ncidod/dbmd/diseaseinfo/legionellosis_g.htm*. Accessed May 10, 2006.

Clark, R. M., B. W. Lykins, Jr., J. C. Block, L. J. Wymer, and D. J. Reasoner. 1994. Water quality changes in a simulated distribution system. J. Water Supply Research and Technology Aqua 43(6):263–277.

Colbourne, J. 2004. Monitoring drinking water at establishments where water is supplied to the public. Personal communication to Board Level and Day to Day Contacts of Water and Sewerage Companies in England and Wales. Available on-line at *http://www.dwi.gov.uk/regs/infolett/ 2004/info1004.shtm*. Accessed May 15, 2006.

Colford, J. M., T. J. Wade, S. K. Sandhu, C. C. Wright, S. Lee, S. Shaw, K. Fox, S. Burns, A. Benker, M. A. Brookhart, M. van der Laan, and D.A. Levy. 2005. A randomized controlled trial of in-home drinking water intervention to reduce gastrointestinal illness. American Journal of Epidemiology 161(6):472–482.

Consumer Product Safety Commission (CPSC). 2005. Tap water scalds. Document #5098. Available on-line at *http://www.cpsc.gov/cpscpub/pubs/5098.html*. Accessed August 10, 2005.

Copper Development Association (CDA). 2005. Search for nation's oldest copper plumbing continues. Available on-line at *http://www.copper.org/copperhome/ HomePlan/search_for_nations_oldest_copper_plumbing_continues.html*. Accessed October 5, 2005.

Craun, G. F., and R. L. Calderon. 2001. Waterbourne disease outbreaks caused by distribution system deficiencies. J. Amer. Water Works Assoc. 93(9):64–75.

Edwards, M., D. Bosch, G. V. Loganathan, and A. M. Dietrich. 2003. The future challenge of controlling distribution system water quality and protecting plumbing infrastructure: focusing on consumers. *In:* Proceedings of the IWA Leading Edge Conference, Noordwijk, Netherlands.

Edwards, M., and A. Dudi. 2004. Role of chlorine and chloramine in corrosion of lead-bearing plumbing materials. J. Amer. Water Works Assoc. 96(10):69–81.

Edwards, M. 2004. Corrosion control in water distribution systems: one of the grand engineering challenges for the 21st century. Water Science and Technology 49(2):1–8.

Edwards, M., B. Marshall, Y. Zhang and Y. Lee. 2005. Unintended consequences of chloramine hit home. *In:* Proceedings of the WEF Disinfection Conference. Mesa, Arizona.

Environmental Protection Agency (EPA). 1995. The Inside Story: A Guide to Indoor Air Quality. Washington, DC: U.S. Environmental Protection Agency and the U.S. Consumer Product Safety Commission, Office of Radiation and Indoor Air (6604J) EPA Document # 402-K-93-007.

EPA. 1999. Alternative Disinfectants and Oxidants Guidance Manual. EPA 815-R-99-014. Washington, DC: EPA.

EPA. 2002a. Potential contamination due to cross-connections and backflow and the associated health risks: an issues paper. Available on-line at: *http://www.epa.gov/ safewater/tcr/pdf/ccrwhite.pdf*. Accessed May 10, 2006.

EPA. 2002b. Guidance for Implementing a Point-of-Use or Point-of-Entry Treatment Strategy for Compliance with the Safe Drinking Water Act – Revised Final Draft. Washington, DC: EPA.

EPA. 2004. Energy Efficiency. Available on-line at *http://www.epa.gov/opptintr/p2home/ aboutp2/energy.htm.* Accessed May 10, 2006.

EPA. 2005. EPA promotes water efficiency in the home. Available on-line at *http://www.epa.gov/water/water_efficiency.html.* Accessed May 10, 2006.

Flannery, B., L. B. Gelling, D. J. Vugia, J. M. Weintraub, J. J. Salerno, M. J. Conroy, V. A. Stevens, C. E. Rose, M. R. Moore, B. S. Fields, and R. E. Besser. 2006. Reducing *Legionella* colonization of water systems with monochloramine. Emerg. Infect. Dis. [serial on the Internet]. Available at: *http://www.cdc.gov/ncidod/EID/vol12no04/05-1101.htm.* Accessed April 13, 2006.

Garret, J. H. 1891. The Action of Water on Lead. London, England: H. K. Lewis.

Geldreich, E. E., R. H. Taylor, J. C. Blannon, and D. J. Reasoner. 1985. Bacterial colonization of point-of-use water treatment devices. J. Amer. Water Works Assoc. 77 (2):72–80.

George, R. 2001. A scalding hot topic. Plumbing Engineer. Available on-line at: *http://www.plumbingengineer.com/pdf/pe/articles/1001PE22.PDF.* Accessed May 10, 2006.

Gilpin, R. W., S. B. Dillion, P. Keyser, A. Androkites, M. Berube, N. Carpendale, J. Skorina, J. Hurley and A. M. Kaplan. 1985. Disinfection of circulating water system by ultraviolet light and halogenation. Water Research 19(7):839–848.

Grigg, N. S. 2005. Assessment and renewal of water distribution systems. J. Amer. Water Works Assoc. 97(2):58–68.

Jackson, P. J., N. M. Williams, K. L. Rule, L. J. Davis, S. Blake, S. G. Warburton, and J. C. Ellis. 2004. Quality of drinking water in public buildings. Final Report to the Drinking Water Inspectorate No: DWI 6348.

Kool, J. L., D. Bergmire-Sweat, J. C. Butler, E. W. Brown, D. J. Peabody, D. S. Massi, J. C. Carpenter, J. M. Pruckler, R. F. Benson, and B. S. Fields. 1999a. Hospital characteristics associated with colonization of water systems by *Legionella* and risk of nosocomial legionnaires' disease: a cohort study of 15 hospitals. Infect Control Hosp Epidemiol. 20:798 –805.

Kool, J. L., J. C. Carpenter, and B. S. Fields. 1999b. Effect of monochloramine disinfection of municipal drinking water on risk of nosocomial Legionnaires' disease. Lancet 353:272–277.

LeChevallier, M. W., and R. J. Seidler. 1980. *Staphylococcus aureus* in rural drinking water. Appl. Environ. Microbiol. 39:739–742.

LeChevallier, M. W., and G .A. McFeters. 1988. Microbiology of activated carbon. Pp. 104–119 *In:* Drinking Water Microbiology, Progress and Recent Developments. G. A. McFeters (ed.). New York: Springer-Verlag.

LeChevallier, M. W., M. Karim, R. Aboytes, R. Gullick, J. Weihe, B. Earnhardt, J. Mohr, J. Starcevich, J. Case, J. S. Rosen, J. Sobrinho, J. L. Clancy, R. M. McCuin, J. E. Funk, and D. J. Wood. 2003. Profiling Water Quality Parameters: From Source Water To The Household Tap. Denver, CO: AWWA and AwwaRF.

LeChevallier, M. W., T. J. Wade, S. Shaw, D. A. Very, R. L. Calderon, and J. M. Colford. 2004. Results of the Big Wet: an epidemiology study of the microbiological quality of drinking water in Davenport, Iowa. *In:* Proceedings of the 2004 AWWA WQTC. Denver, CO: AWWA.

Little, J. C. 1992. Applying the two-resistance theory to contaminant volatilization in showers. Environ. Sci. Technol. 26:1341–1349.

Mayo (Mayo Clinic). 2004. Hot tub lung. Available on-line at *http://www.mayoclinic.com/invoke.cfm?id=AN00660.* Accessed May 10, 2006.

Murphy, B., J. T. O'Connor, and T. L. O'Connor. 1997a. Willmar, Minnesota battles copper corrosion. Part 2: nitrification, bacteria and copper corrosion in household plumbing. Public Works 128(11):44.

Murphy, B., J. T. O'Connor, and T. L. O'Connor. 1997b. Willmar, Minnesota battles copper corrosion. Part 3: results of pilot plant column and copper pipe test loop studies. Public Works 128(12):37.

Nguyen, C. K. 2005. Interactions between copper and chlorine disinfectants: chlorine decay, chloramine decay and copper pitting. M.S. Thesis. Blacksburg, VA: Virginia Tech.

Nguyen, C., and M. Edwards. 2005. Chemistry of rapid chloramine decay in water contacting copper plumbing. *In:* Proceedings of the AWWA Water Quality Technology Conference, Quebec City, Canada. Denver, CO: AWWA.

National SAFE KIDS Campaign (NSKC). 2004. Burn injury fact sheet. Washington, DC: National SAFE KIDS Campaign.

Payment, P., J. Siemiatycki, L. Richardson, G. Renaud, E. Franco, and M. Prevost. 1997. A prospective epidemiological study of gastrointestinal health effects due to the consumption of drinking water. International Journal of Environmental Health Research 7:5–31.

Powers, K. 2000. Fundamentals and Practical Implications of Cupric Hydroxide Aging. M.S. Thesis. Blacksburg, VA: Virginia Tech.

Pryor, M., S. Springthorpe, S. Riffard, T. Brooks, Y. Huo, G. Davis, and S. A. Satter. 2004. Investigation of opportunistic pathogens in municipal drinking water under different supply and treatment regimes. Water Sci. Technol. 50(1):83–90.

Reasoner, D. J., J. C. Blannon, and E. E. Gelderich. 1987. Microbiological characteristics of third-faucet point-of-use devices. J. Amer. Water Works Assoc. 79(10):60–66.

Regal, S., J. Ashworth, M. Benoliel, V. Cardoso, H. Jagt, H., J. Klinger, M. Ottaviani, E. Trickquel, E. Veschetti, I. Wagner, and E. J. Hoekstra. 2003. Assessment of effect of high level of disinfectants on products in contact with drinking water. Development of a harmonised test to be used in the European Acceptance Scheme Concerning Construction Products in Contact with Drinking Water. Available on-line at *http://cpdw.jrc.it/WP4%20FR%20main.pdf.* Accessed May 10, 2006.

Rodgers. M. R., B. J. Backstone, A. L. Reyers, and T. C. Covert. 1999. Colonisation of point-of-use water filters by silver resistant non-tuberculous mycobacteria. J. Clin. Pathol. 52(8):629.

Rollinger, Y., and W. Dott. 1981. Survival of selected bacterial species in sterilized activated carbon filters and biological activated carbon filters. Appl. Environ. Microbiol. 53(4):777–781.

Rushing, J. C., and M. Edwards. 2004. The role of temperature gradients in copper pipe corrosion. Corrosion Science 46:1883–1894.

Ryder, R. A. 1980. The costs of internal corrosion in water systems. J. Amer. Water Works Assoc. 72(5):267–279.

Spinks, A. T., R. H. Dunstan, P. Coombes, G. Kuczera. 2003. Thermal destruction analyses of water related pathogens at domestic hot water system temperatures. The Institution of Engineers. 28th International Hydrology and Water Resources Symposium. November, 2003.

Stevens, V. A., Gelling, L., Flannery, B., Conroy, M. Vugla, D., Salerno, J., Weintraub, J., Besser, R., and B. Fields. 2004. Characterization of *Legionella* and amoeba populations in a municipal water system over a one-year time period—San Francisco, California. Poster presented at ASM annual conference.

Thomas, V., K. Herrera-Rimann, D. S. Blanc, and G. Greub. 2006. Biodiversity of amoebae and amoeba-resisting bacteria in a hospital water network. Appl. Environ. Microbiol. 72(4):2428–2438.

Tobin, R. S., D. K. Smith, and J. A. Lindsay. 1981. Effects of activated carbon and bacteriostatic filters on microbiological quality of drinking water. Appl. Environ. Microbiol. 41(3):646–651.

Tobin-D'Angelo, M. J., M. A. Blass, C. del Rio, J. S. Halvosa, H. M. Blumberg, C. R. Horsburgh. 2004. Hospital water as a source of complex isolates in respiratory specimens. Journal of Infectious Diseases 189(1):98–104.

Tsitko, I., R. Rahkila, O. Priha, T. Ali-Vehmas, Z. Terefework, H. Soini, and M. S. Salkinoja-Salonen. 2006. Isolation and automated ribotyping of *Mycobacterium lentiflavum* from drinking water distribution system and clinical specimens. FEMS Microbiology Letters 256(2):236–243.

University of Southern California (USC). 2002. Prevalence of cross connections in household plumbing systems. Available on-line at: *http://www.usc.edu/dept/fcchr/epa/hhcc.report.pdf.* Los Angeles, CA: USC Foundation for Cross-Connection Control and Hydraulic Research.

Vacrewijck, M. J. M., G. Huys, J. C. Palomino, J. Swings, and F. Portaels. 2005. Mycobacteria in drinking water distribution systems: Ecology and significance for human health. FEMS Microbiology Reviews 29(5):911–934.

VROM (VROM International; Netherlands Ministry of Housing, Spatial Planning and the Environment) 2005a. Hotels do too little to prevent *Legionella* contamination. Available on-line at *http://international.vrom.nl/pagina.html?id=8739.* Accessed May 10, 2006.

VROM. 2005b. *Legionella* in Europe: problems and prevention: Summary by the Chairman. International Congress September 28–29, Amsterdam, The Netherlands. Available on-line at *http://international.vrom.nl/docs/internationaal/congres%20verslag%20Engels%2025-11-04.pdf.* Accessed May 10, 2006.

World Health Organization (WHO). 2005. Water Safety Plans in Public Buildings. University of East Anglia-Norwich, UK. Available on-line at *http://www.who.int/water_sanitation_health/hygiene/settings/watsafpubuildings.pdf.* Accessed May 10, 2006.

Acronyms

ABPA	American Backflow Prevention Association
ANSI	American National Standards Institute
AODC	acridine orange direct count
ASDWA	Association of State Drinking Water Administrators
ASTM	American Society for Testing and Materials
AWWA	American Water Works Association
BAC	biologically active carbon
BTEX	benzene, toluene, ethylbenzene, and xylenes
CCC	cross-connection control
CFD	computational fluid dynamics
AOC	assimilable organic carbon
CFU	colony forming units
CPVC	chlorinated polyvinyl chloride
CWS	community water systems
D/DBPR	Disinfectants/Disinfection By-Products Rule
DBP	disinfection byproduct
DWSRF	Drinking Water State Revolving Fund
EPS	extended period simulation
ETV	environmental technology verification
GAC	granular activated carbon
GI	gastrointestinal symptoms
GIS	geographic information system
GPS	global positioning system
HAA	haloacetic acid
HACCP	Hazard Analysis and Critical Control Points
HCGI	highly credible gastrointestinal illness
HPC	heterotrophic bacterial plate count
IAPMO	International Association of Plumbing and Mechanical Officials
ICC	International Code Council
ICR	Information Collection Rule
IDSE	Initial Distribution System Evaluation
IESWTR	Interim Enhanced Surface Water Treatment Rule
IPC	International Plumbing Code
LCR	Lead and Copper Rule
LT1ESWTR	Long Term 1 Enhanced Surface Water Treatment Rule
LT2ESWTR	Long Term 2 Enhanced Surface Water Treatment Rule

MAC	*Mycobacterium avium* complex
MCL	maximum contaminant level
MCLG	maximum contaminant level goal
MRDL	maximum residual disinfectant level
NOM	natural organic material
NSPC	National Standard Plumbing Code
POE	point-of-entry
POU	point-of-use
PVC	polyvinyl chloride
SCADA	Supervisory Control and Data Acquisition
SDWA	Safe Drinking Water Act
SWTR	Surface Water Treatment Rule
TCR	Total Coliform Rule
THM	trihalomethane
TOC	total organic carbon
UCMR	Unregulated Contaminant Monitoring Rule
UPC	Uniform Plumbing Code
VOC	volatile organic chemical
WHO	World Health Organization

Appendixes

Appendix A

Public Water Supply Distribution Systems: Assessing and Reducing Risks

FIRST REPORT

Committee on Public Water Supply Distribution Systems: Assessing and Reducing Risks

Water Science and Technology Board

Division on Earth and Life Studies

NATIONAL RESEARCH COUNCIL
OF THE NATIONAL ACADEMIES

THE NATIONAL ACADEMIES PRESS
Washington, D.C.
www.nap.edu

1
Introduction

The distribution systems of public drinking water supplies include the pipes and other conveyances that connect treatment plants to consumers' taps. They span almost 1 billion miles in the United States (Kirmeyer et al., 1994) and include an estimated 154,000 finished water storage facilities (AWWA, 2003). Public water supplies serve 273 million residential and commercial customers, although the vast majority (93 percent) of systems serves less than 10,000 people (EPA, 2004). As the U.S. population grows and communities expand, 13,200 miles of new pipes are installed each year (Kirmeyer et al., 1994).

Distribution systems constitute a significant management challenge from both an operational and public health standpoint. Furthermore, they represent the vast majority of physical infrastructure for water supplies, such that their repair and replacement represent an enormous financial liability. The U. S. Environmental Protection Agency (EPA) estimates the 20-year water transmission and distribution needs of the country to be $83.2 billion, with storage facility infrastructure needs estimated at $18.4 billion (EPA, 1999).

Most federal water quality regulations pertaining to drinking water, such as Maximum Contaminant Levels (MCLs) and treatment technique requirements for microbial and chemical contaminants, are applied before or at the point where water enters the distribution system. The major rules that specifically target water quality within the distribution system are the Lead and Copper Rule (LCR), the Surface Water Treatment Rule (SWTR), which addresses the minimum required detectable disinfectant residual and the maximum allowed heterotrophic bacterial plate count, and the Total Coliform Rule. In addition, the Disinfectants/Disinfection By-Products Rule (D/DBPR) addresses the maximum disinfectant residual and concentration of disinfection byproducts like total trihalomethanes and haloacetic acids allowed in distribution systems. Of all these rules, the Total Coliform Rule (TCR) of 1989 explicitly addresses microbial water quality in the distribution system. The TCR applies to all public water supplies, both groundwater and surface water, and established (among other things) an MCL of less than 5 percent of water samples testing positive for total coliforms in any month for systems serving more than 33,000, and that there be no more than one positive sample per month for systems serving less than 33,000 (Guilaran, 2004). Sampling of distribution systems for total coliforms varies widely, from as many as hundreds of samples per month to one sample per year, depending on the size and type of system. Most contaminants that have the potential to contaminate the distribution system are not monitored for in the distribution system. Therefore, contamination of the distribution system will typically be detected by only other means (e.g., taste and odor complaints). This and other information gathered since the rule was first promulgated suggest that the TCR maybe limited in its ability to ensure public health protection from microbial contamination of distribution systems. Monitoring required under the TCR does not include monitoring for chemical contaminants. Indeed, some epidemiological and outbreak investigations conducted in the last five years suggest that a substantial proportion of waterborne disease outbreaks, both microbial and chemical, is attributable to problems within distribution systems (Craun and Calderon, 2001; Blackburn et al., 2004). Distribution system deficiencies were pinpointed as the cause of 57 reported community outbreaks from 1991 to 1998 (EPA, 2002b). Since chemically-related waterborne illnesses typically result from long-term exposures to chemicals, waterborne outbreak surveillance systems, which focus on acute exposures to contamination, do not capture the scope of illness resulting from chemical contamination of water. Epidemiology studies on chemical exposures in drinking water are also more difficult, since a long-term study is required for long-term exposures and a variety of other sources of exposure may influence the outcomes. There is no evidence that the current regulatory program has resulted in a diminution in the proportion of outbreaks attributable to distribution system related factors.

In 2000, the Federal Advisory Committee for the Microbial/Disinfection By-products Rule (M/DBPR) recommended that EPA evaluate available data and research on aspects of distribution systems that may create risks to public health. Furthermore, in 2003 EPA committed to revising the TCR—not only to consider updating the provisions about the frequency and location of monitoring, follow-up monitoring after total coliform positive samples, and the basis of the MCL, but also to consider addressing the broader issue of whether the TCR could be revised to encompass "distribution system integrity." That is, EPA is exploring the possibility of revising the TCR to provide a comprehensive approach for addressing water quality in the distribution system environment. To aid in this process, EPA requested the input of the National Academies' Water Science and Technology Board, which was asked to conduct a study of water quality issues associated with public water supply distribution systems and their potential risks to consumers.

The expert committee formed to conduct the study will consider, but not be limited to, specific aspects of distribution systems such as cross connections and backflow, intrusion caused by pressure transients, nitrification, permeation and leaching, repair and replacement of water mains, aging infrastructure, and microbial growth. The committee's statement of task is to:

> 1—Identify trends relevant to the deterioration of drinking water in water supply distribution systems, as background and based on available information.

2—Identify and prioritize issues of greatest concern for distribution systems based on review of published material.

3—Focusing on the highest priority issues as revealed by task #2, (a) evaluate different approaches for characterization of public health risks posed by water-quality deteriorating events or conditions that may occur in public water supply distribution systems; and (b) identify and evaluate the effectiveness of relevant existing codes and regulations and identify general actions, strategies, performance measures, and policies that could be considered by water utilities and other stakeholders to reduce the risks posed by water-quality deteriorating events or conditions. Case studies, either at state or utility level, where distribution system control programs (e.g., Hazard Analysis and Critical Control Point System, cross connection control, etc.) have been successfully designed and implemented will be identified and recommendations will be presented in their context.

4—Identify advances in detection, monitoring and modeling, analytical methods, information needs and technologies, research and development opportunities, and communication strategies that will enable the water supply industry and other stakeholders to further reduce risks associated with public water supply distribution systems.

This first report relates the committee's progress on Tasks 1 and 2—that is, trends relevant to the deterioration of distribution system water quality and the issues that the committee thinks are the highest priorities for consideration during TCR revision to encompass distribution system integrity. Conclusions and recommendations related to distribution system issues that EPA may want to take into consideration are sprinkled throughout the text, and a short summary of the committee's prioritization is given at the end.

2
Trends Relevant to the Deterioration of Drinking Water in Distribution Systems

In the past two decades, a number of changes have occurred that may affect the quality of drinking water in distribution systems, consumer exposure to tap water, and the consequent risks of exposure. This section discusses trends in pipe age in water distribution systems and pipe replacement rates, waterborne disease outbreaks, host susceptibility in the U.S. population, consumer use of bottled water, and installation of home water treatment devices. This is not a comprehensive list of all the factors that may affect water quality and health risks from distributions systems. Furthermore, for many of these factors, there are limited data on recent trends such that additional research is needed to better understand current practices.

DISTRIBUTION PIPE AGE AND REPLACEMENT RATES

There is a large range in the type and age of the pipes that make up American water distribution systems, depending on the population and economic booms of the previous century. For many cities, the periods of greatest population growth and urban expansion were during the late 1800s, around World War I, during the 1920s, and post-World War II. The water pipes installed during these growth periods differ in their manufacture, materials, and life span. The oldest cast iron pipes from the late 19th century are typically described as having an average useful lifespan of about 120 years because of the pipe wall thickness (AWWA, 2001; AWWSC, 2002). In the 1920s the manufacture of iron pipes changed to improve pipe strength, but the changes also produced a thinner wall. These pipes have an average life of about 100 years. Pipe manufacturing continued to evolve in the 1950s and 1960s with the introduction of ductile iron pipe that is stronger than cast iron and more resistant to corrosion. Polyvinyl chloride (PVC) pipes were introduced in the 1970s and high-density polyethylene in the 1990s. Both of these are very resistant to corrosion but they do not have the strength of ductile iron. Post-World War II pipes tend to have an average life of 75 years (AWWA, 2001; AWWSC, 2002). Approximately 20 percent of the pipe in place in North America is lined with asbestos or cement. Furthermore, the overwhelming majority of ductile iron pipe is mortar-lined and about 40 percent of cast iron pipe in place is mortar-lined. These facts may be of great importance where the life of pipe is concerned, as linings are meant to prevent corrosion and increase pipe longevity.

In the 20th century, most of the water systems and distribution pipes were relatively new and well within their expected lifespan. However, a recent report by the American Water Works Association (AWWA, 2001) and a white paper by the American Water Works Service Company, Inc. (AWWSC, 2002) point out that these different types of pipes, installed during different time periods, will all be reaching the end of their expected life spans in the next 30 years. Analysis of main breaks at one large Midwestern water utility that kept careful records of distribution system management documented a sharp increase in the annual number of main breaks from 1970 (approximately 250 breaks per year) to 1989 (approximately 2,200 breaks per year) (AWWSC, 2002). Thus, the water industry is entering an era where it must make substantial investments in pipe repair and pipe replacement. An EPA report on water infrastructure needs (EPA, 2002c) predicted that transmission and distribution replacement rates will need to be around 0.3 percent per year in 2005 and will rise to 2.0 percent per year by 2040 in order to adequately maintain the water infrastructure (see Figure 1). Cost estimates for drinking water infrastructure range from \$4.2 to \$6.3 billion per year (AWWSC, 2002). The trends of aging pipe and increasing numbers of main breaks are of concern because of the potential relationship between waterborne disease outbreaks and main breaks (see the subsequent section on New and Repaired Water Mains).

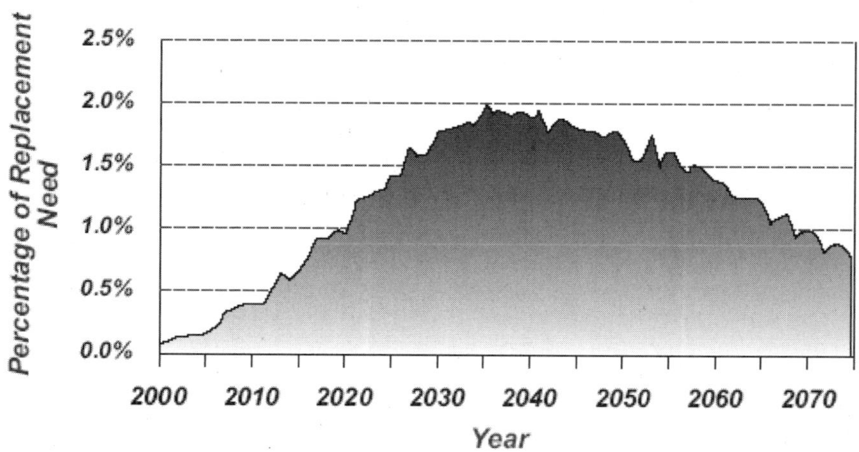

FIGURE 1 Projected annual replacement needs for transmission lines and distribution mains, 2000–2075. SOURCE: EPA (2002c).

WATERBORNE DISEASE OUTBREAKS

A voluntary, passive surveillance system for waterborne disease outbreaks in the U.S. has been maintained by the Centers for Disease Control and Prevention (CDC) in collaboration with EPA since 1971. Summary reports from this surveillance system are published every two years and describe the number of outbreaks, where they occurred, the etiologic agents, type of water systems involved, and factors that contributed to the outbreak. While the current waterborne disease surveillance summary states that the data are useful "for identifying major deficiencies in providing safe drinking water" (Blackburn et al., 2004), caution in the interpretation of these data is important, in that the proportion of outbreaks reported may vary with time, location, and the size of the water supply. With this caveat in mind, analyses of the data from this surveillance system indicate that the total number of reported waterborne disease outbreaks has decreased since 1980. However, the proportion of waterborne disease outbreaks associated with problems in the distribution system is increasing (see Figure 2). Craun and Calderon (2001) examined causes of reported waterborne outbreaks from 1971–1998 and noted that in community water systems, 30 percent of 294 outbreaks were associated with distribution system deficiencies. From 1999 to 2002, there have been 18 reported outbreaks in community

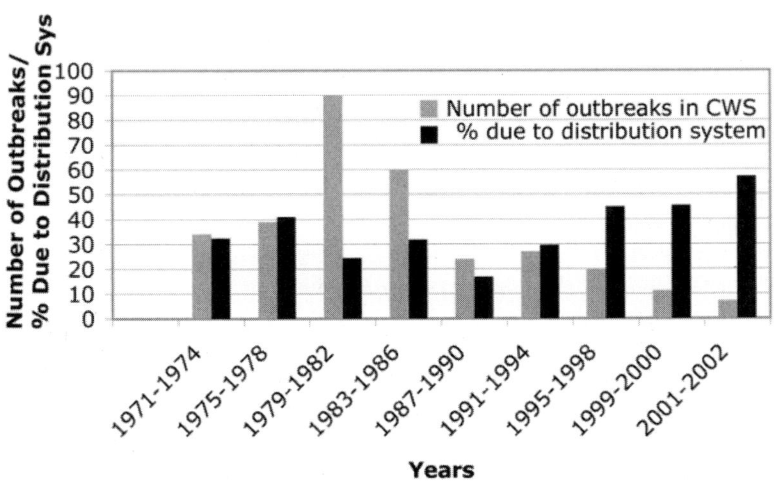

FIGURE 2 Waterborne disease outbreaks in community water systems (CWS) associated with distribution system deficiencies. Note that the majority of the reported outbreaks have been in small community systems and that the absolute numbers of outbreaks have decreased since 1982. SOURCE: Data extracted from Craun and Calderon (2001) and MMWR summary reports on waterborne disease surveillance (Lee et al., 2002 and Blackburn et al., 2004).

water systems, and nine (50 percent) of these were related to problems in the water distribution system (Lee et al., 2002, Blackburn et al., 2004). The decrease in numbers of water-borne disease outbreaks per year is important and probably attributable to improved water treatment practices and SWTR compliance that reduced the risk from waterborne protozoa (Pierson et al., 2001; Blackburn et al., 2004). The increase in the percentage of outbreaks attributable to the distribution system is probably also due to this factor (i.e., the SWTR); regulations for distribution systems have not been as extensive (other than the Lead and Copper Rule).

Most reported outbreaks associated with distribution systems occur in community water systems because of their greater size and complexity, but there have been a number of outbreaks associated with noncommunity water systems that have been attributed to deficiencies in the distribution system. Craun and Calderon (2001) reported that most distribution system-related outbreaks were linked to cross-connections and backsiphonage and most of the rest were attributed to main breaks or repair and contamination of municipal water storage tanks. The magnitude and severity of outbreaks associated with distribution systems vary, with an average of 186 illnesses per outbreak (Craun and Calderon, 2001) and a total of 13 deaths. These outbreaks have been associated with chemical (copper, chlordane, ethylene glycol and others) and microbial contaminants, including enteric protozoa (*Giardia, Cyclospora*), enteric bacteria (*Salmonella, Shigella, Campylobacter*, and *E. coli* O157:H7) and enteric viruses (noroviruses and Hepatitis A virus).

It should be noted that endemic waterborne infection and illness may be associated with contaminants entering the distribution system. If low levels of contaminants enter the system and affect small numbers of persons, it might not be recognized and investigated as an outbreak. Indeed, it has been acknowledged that a fairly sizable number of cases of cryptosporidiosis could be occurring in a large city such as New York City without detection of a possible outbreak (NRC, 1999, page 249). Thus, not only are all waterborne disease outbreaks not detected, even those that are detected and reported will not address possible endemic illness risks. It is noted that consumer confidence and legal liability may create a disincentive to report outbreaks and even water quality problems. A number of sources show that endemic risks can be greater than epidemics, for example, Frost et al., (1996) and Payment and Hunter (2001). The CDC and EPA have recently completed a series of epidemiologic studies designed to assess the magnitude of endemic waterborne illness associated with consumption of municipal drinking water; a joint report on the results of these studies is forthcoming (Blackburn et al., 2004).

CHANGES IN THE UNITED STATES POPULATION

Another cause for concern regarding the risks of waterborne disease transmission is increasing host susceptibility to infection and disease in the U.S.

population. Due to weaker immune systems, older Americans are at increased risk for morbidity and mortality from a number of infectious diseases, including influenza, pneumonia, and enteric diseases (Gerba et al., 1996). Decreased gastric acid secretion in the elderly may also result in increased susceptibility to infection from acid-sensitive enteric organisms (Morris and Potter, 1997). The U.S. population older than 74 years of age has increased from 13.1 million in 1990 to 16.6 million in 2000. The most rapid growth during this decade occurred in the size of the oldest age groups with a 38 percent increase in the population greater than 85 years of age. In 2003, more than 12 percent of the total U.S. population was 65 and older, and this proportion will increase dramatically between 2010 and 2030 as the "baby boomers" start turning 65 in 2011. By 2030, nearly 20 percent of the total U.S. population will be over 65 years of age, and the population over 85 years of age will have grown rapidly (Older Americans, 2004).

The numbers of immunocompromised persons in the U.S. due to disease and immunosuppressive therapy is also increasing. Of particular note are growing numbers of AIDS patients, cancer survivors, and organ transplant patients. The number of new AIDS cases reported in the U.S. has increased more than five-fold from 8,131 in 1985 to 44,232 in 2003. Because of more effective treatments, AIDS patients are living longer but are still at increased risk of enteric infections. Cancer patients and transplant patients often require immunosuppressive therapy that puts them at greater risk of infection during the course of their treatment. The CDC estimates that the number of persons living with cancer more than tripled from 3.0 million in 1971 (1.5 percent of the U.S. population) to 9.8 million in 2001 (3.5 percent of the population) (CDC, 2004). The number of organ transplants performed each year in the U.S. has almost doubled from 12,619 in 1988 to more than 22,554 in 2004 (Organ Procurement and Transplantation Network. Available on-line at *http://www.optn.org/latestData/rptData.asp*. Accessed February 6, 2005).

USE OF BOTTLED WATER AND HOUSEHOLD WATER TREATMENT DEVICES

There has been a dramatic increase in the proportion of the U.S. population that drinks bottled water or uses some type of water treatment device in their homes. The Natural Resources Defense Council (NRDC) reported in 1999 that more than half of all Americans drink bottled water and about one third of the population regularly drink bottled water. The sales of bottled water tripled between 1986 and 1997 and reached about $4 billion per year (NRDC, 1999). The International Bottled Water Association reported a 10.1 percent growth in sales between 1997 and 1998 (Available on-line at *http://www.bottledwater.org/public/pressrel.htm.* Accessed March 16, 2005). The cost of bottled water ranges from 240 to over 10,000 times more per gallon than tap water, and yet bottled water use is not limited to high-income households. One study reported that "Black, Asian, and Hispanic households are more likely than whites to use bottled water" despite lower household incomes (Mogelonsky, quoted in NRDC, 1999). The use of home water treatment devices has also risen steadily from 32 percent in 1997 to 38 percent in 1999 to 41 percent in 2001 (WQA, 2001). As discussed in a subsequent section, these devices can support the regrowth of microbes, such that their use is not necessarily correlated with a decrease in contaminant exposure.

Several consumer surveys and studies have attempted to determine the driving forces behind these trends and have reported that perceived safety and health, taste of tap water, and concern about some contaminants are the most frequently reported reasons people drink bottled water instead of tap water (NRDC, 1999; Anadu and Harding, 2000; WQA, 2001; Mackey et al., 2003). Although these trends are occurring, the health implications of these trends are unknown.

***

Taken together, the trends data suggest that water distribution system infrastructure in the U.S. is deteriorating and that health risks associated with distribution system water quality may be increasing. Although the proportion of the U.S. population that may be more susceptible to waterborne disease is growing, fewer Americans are drinking tap water. These trends need to be investigated to determine if they are important factors that should be taken into account when developing a distribution system rule.

3
Highest Priority Issues

The second major task of the committee was to identify the highest priority issues for consideration during TCR revision to encompass distribution system integrity. The issues considered for prioritization stem directly from nine white papers prepared by EPA, AWWA, and the AWWSCo. based on a series of expert and stakeholder workshops from 2000 to 2003. The nine white papers focused on the following events or conditions that can bring about water quality degradation in public water supply distribution systems:

1. Cross-Connections and Backflow (EPA, 2002b)
2. Intrusion of Contaminants from Pressure Transients (LeChevallier et al., 2002)
3. Nitrification (AWWA and EES, Inc., 2002e)
4. Permeation and Leaching (AWWA and EES, Inc., 2002a)
5. Microbial Growth and Biofilms (EPA, 2002d)
6. New or Repaired Water Mains (AWWA and EES, Inc., 2002e)
7. Finished Water Storage Facilities (AWWA and EES, Inc., 2002c)
8. Water Age (AWWA and EES, Inc., 2002b)
9. Deteriorating Buried Infrastructure (AWWSC, 2002)

In addition to these papers, the committee considered the summary of the Distribution System Exposure Assessment Workshop (ICF Consulting, Inc., 2004), held in Washington, DC in March, 2004, which attempted to collate all of the information gathered in the previous workshops. Additional white papers are currently being written on the following topics, but were not available to the committee in time to be considered for this first report:

- Indicators of Drinking Water Quality
- Evaluation of Hazard Analysis and Critical Control Points
- Causes of Total Coliform Positives and Contamination Events
- Inorganic Contaminant Accumulation
- Distribution System Inventory and Condition Assessment.

Some qualitative outcomes of the many workshops, as communicated by EPA officials, are that there are demonstrated adverse health effects and large potential exposure resulting from distribution system contamination. The stakeholder and industry experts who attended the workshops agreed on the need to evaluate and prioritize potential health risk.

The approach to prioritization taken by the committee was based on a careful assessment of the issues presented in the nine white papers, critical evaluation of other materials, and on the committee's assessment of the health importance of the various events. Given limited data on the specific causes of waterborne disease outbreaks, the best professional judgment of the committee was used to assess the magnitude of the health problem associated with an event, including how often the event occurs and how much contamination results when an event occurs. The use of surveillance and epidemiology data to prioritize the issues has its limitations. This approach may result in more uncertainty surrounding the interpretation of the risks for the issues which have limited surveillance or epidemiology data when compared to those issues where surveillance and epidemiology data are more abundant. In addition to prioritizing the issues presented in the nine white papers, the committee also considered whether any significant issues had been overlooked by EPA when the white papers were written.

Developing the white papers was a difficult task. A water distribution system is a complex engineering and ecological system wherein multiple adverse changes may result from the same or similar underlying causes. For example, water that has a long residence time in the system (high water age) may also have the potential to lose disinfectant residual and undergo biological nitrification. Considering any of these occurrences in the absence of the others oversimplifies the nature of the problem. However, the committee decided to follow the structure of the EPA white papers in preparing this report, with the recognition that overlaps and difficult-to-separate phenomena exist.

Of the issues presented in the nine white papers, cross connections and backflow, new or repaired water mains, and finished water storage facilities were judged by the committee to be of the highest importance based on their associated potential health risks. In addition, there are two other issues that should also be accorded high priority: premise plumbing and distribution system operator training.

CROSS CONNECTIONS AND BACKFLOW

Points in a distribution system or premise plumbing where non-potable water comes into contact with the potable water supply are called cross connections. A backflow event occurs when non-potable water flows into the drinking water supply through a cross connection, either because of low distribution system pressure (termed backsiphonage) or because of pressure on the non-potable water caused by pumpage or other factors (termed backpressure). Backflow incidents have long been recognized as significant contributors to waterborne disease. From 1981 to 1998, the CDC documented 57 waterborne outbreaks related to cross-connections, resulting in 9,734 detected and reported illnesses (Craun and Calderon, 2001). EPA compiled a total of 459 incidents resulting in 12,093 illnesses from backflow events from 1970 to 2001 (EPA, 2002b). For the period 1981 to 1998, EPA found that only 97 of 309 incidents were reported to public health authorities, demonstrating that the magnitude of the public health concern due to cross-connections is underreported. The situation may be of even greater concern because incidents involving domestic plumbing are even less recognized. In a study of 188 households, the University of Southern California's Foundation for Cross-Connection Control and Hydraulic Research reported that 9.6 percent of the homes had a direct cross connection that constituted a health hazard, and more than 95 percent had either direct or indirect cross connections (USC, 2002). Cross-connections are also of great concern where a potable system is in close proximity to a reclaimed water system (such as in dual distribution systems like that of the Irvine Ranch water district). A direct cross connection is a (not necessarily permanent) cross connection that is subject to both backpressure and siphonage whereas an indirect cross connection is subject to backsiphonage only. In most cases, the extent of the health problem caused by cross connections in homes is unknown—knowledge that could be obtained through epidemiological studies.

Although most state and primacy agencies require that utilities have a cross connection control program, the elements of such programs, their implementation, and oversight vary widely. Because of inconsistent application of these programs, cross connections and backflow events remain a significant potential cause of waterborne disease. Proven technologies and procedures are available to mitigate the impact of cross connections on potable water quality. State plumbing codes define the type of plumbing materials that are approved for use, including cross connection control devices, but whether these codes are adhered to is questionable. Regulatory options that could be considered include requiring inspections for household cross connections at the time of home sale. Furthermore, training programs such as those offered by the New England Water Works Association to train and certify backflow device installers and testers have been successful in gaining support from the plumbing community and in developing local plumbing codes that require cross connection control. Given the availability of effective technologies for preventing cross connections, opportunity exists for substantial reductions in public health risk through the implementation of more effective cross connection control programs by primacy and state agencies. At the current time, it is unknown how effective various state programs are in actually preventing cross connections—an issue that is also ripe for further investigation.

Because of the long history of recognized health risk posed by cross connections and backflow, the clear epidemiological and surveillance data, and the proven technologies to prevent cross connections, cross connection and backflow events are ranked by the committee as the highest priority. Efforts to provide implementation of a more uniform national cross connection program would have clear public health benefits.

NEW AND REPAIRED WATER MAINS

This section focuses on contamination arising from the exposure of distribution system water and pipe interior, appurtenances, and related materials to microbial and chemical contaminants in the external environment (1) during water main failures and breaks and (2) due to human activities to install new, rehabilitate old, or repair broken mains and appurtenances. When a pipe break or failure occurs, there is immediate potential for external contamination from soil, groundwater, or surface runoff (see Kirmeyer et al., 2001) to enter the distribution system or come into contact with the pipe interior in the area of the failure and be carried to downstream consumers. Furthermore, the storage, installation, rehabilitation, and repair of water mains and appurtenances provide an opportunity for microbial and chemical contamination of materials that come into direct contact with drinking water. Pierson et al. (2001) confirmed the possibility of such events by surveying distribution system inspectors and field crews.

This section does not address contamination from the external environment that enters through cracks or leaks in pipe, pipe joints, or appurtenances (even though these can exist undetected for long periods of time), as these are covered under the intrusion section. Furthermore, periodic changes to the operation of the distribution system, such as valving the local water system to shut down mains for work and then reloading the mains before their return to use, can allow for contamination of the drinking water supply from backflow through unprotected domestic and fire connection services, which is covered under the section on cross connections and backflow.

Craun and Calderon (2001), in summarizing waterborne outbreaks from 1971 to 1998, found that of the 12 largest outbreaks caused by distribution system deficiencies, one in Indiana was associated with contamination of main interiors during storage in the pipe yard or on the street prior to pipe installation. This review also recalled the well-documented 1989 waterborne disease outbreak in Cabool, Missouri, which was associated with water meter repair and two large water main breaks during the winter (see Clark et al., 1991; Geldreich et al., 1992). The loss of system pressure and ineffective system controls allowed external contamination, such as sewage, to come in contact with the water meters and pipe interior. In addition, the insufficient use of best practices such as post-repair disinfection allowed the contamination to spread through the distribution system. The EPA white paper, New and Repaired Water Mains (AWWA and EES, Inc., 2002e), states that about 5 percent of reported waterborne disease outbreaks attributed to distribution system deficiencies in the United States over a 27-year period were associated with new or repaired water mains. Over 200,000 water main breaks occur every year (over 555 breaks per day) according to Kirmeyer et al. (1994). Over 4,000 miles of pipe are replaced every year and over 13,000 miles of new pipe are installed every year. These numbers, along with the fact that disease occurrence is underreported and contamination from such activities would be highly localized and undetectable in most cases, suggest that exposure to contaminated drinking water from main breaks and installation, repair, and replacement activities is likely to be significantly greater than has been documented.

ANSI/AWWA standards, particularly C600-99 for the installation of ductile iron mains and C651-99 for the disinfection of mains, are commonly employed to prevent microbial contamination during main rehabilitation and replacement (Pierson et al., 2001). However, the actual documentation and inspection of sanitary practices varies widely. Even well-run utility operations, for example, can experience a 30 percent failure rate in the approval of new mains based on water quality testing (Burlingame and Neukrug, 1993). Haas et al. (1998) reported that interior pipe surfaces are not free of microbial contaminants even under best case conditions. Thus, when a new main is installed or a valve is repaired, it is advisable to act as if some level of contamination has occurred to both the water and the materials. In both cases testing is required, and care should be taken to address potential contamination before the affected portion of the water system is returned to use. Prevention of contamination can also be facilitated by including the existing standards and additional training on sanitary practices in distribution system operator training requirements and sanitary survey guidelines. There is more variability in practices when it comes to preventing contamination during main breaks and failures than with the installation of new mains. Haas et al. (1998) reported that while 90 percent of utilities surveyed said that new mains must meet water quality criteria before they are released back to service, only 29 percent said samples were required to be collected in response to a main break. One common practice is to simply flush hydrants on the water mains in the affected area until the water runs clear. Because water main repairs are of varying complexity and occur under a variety of environmental conditions, and due to their unplanned nature may require quick response and return to service, the application of the same level of specifications used for new water mains may not be feasible.

The chemical and microbial contamination of distribution system materials and drinking water during mains breaks and during the installation, rehabilitation, and repair of water mains and appurtenances is a high priority issue. As discussed above, there have been many documented instances of significant health impacts from drinking water contamination associated with pipe failures and maintenance activities. The improved application of best practices, and operator training and certification, can reduce this risk.

FINISHED WATER STORAGE

Treated water storage facilities, of which there are 154,000 in the United States (AWWA, 2003), are of vital importance for drinking water distribution systems. Storage facilities are traditionally designed and operated to secure system hydraulic integrity and reliability, to provide reserve capacity for fire fighting and other emergencies, to equalize system pressure, and to balance water use throughout the day. To meet these goals, large volumes of reserve storage are usually incorporated into system operation and design, resulting in long detention times. Long detention times and improper mixing within such facilities provide an opportunity for both chemical and microbial contamination of the water. One of the most important manifestations of water quality degradation during water storage is a loss of disinfectant residual, which can be further compromised by temperature increases in water storage facilities under warm weather conditions. Internal chemical contamination can also occur due to leaching from coatings used in the storage facility, or solvents, adhesives, and other chemicals used to fabricate or repair floating covers. Until recently, water quality issues associated with such facilities have usually been considered as only secondary maintenance items such as cleaning and coating.

In addition to the internal degradation of water quality that occurs over time in water storage facilities, they are also susceptible to external contamination from birds, animals, wind, rain, and algae. This is most true for uncovered storage facilities, although storage facilities with floating covers are also susceptible to bacterial contamination due to rips in the cover from ice, vandalism, or normal operation. Even with covered storage facilities, contaminants can gain access through improperly sealed access openings and hatches or faulty screening of vents and overflows. The white paper, Finished Water Storage Facilities (AWWA and EES, Inc., 2002c), identified four waterborne disease outbreaks associated with covered storage tanks. In particular, in Gideon, Missouri, a *Salmonella typhimurium* outbreak occurred due to a bird contamination of a covered municipal water storage tank (Clark et al., 1996).

Water quantity and quality requirements in distribution storage management decisions are frequently in conflict. While water quantity objectives promote excessive storage, water quality objectives are geared toward minimizing residence times and frequent exercising of treated water facilities to maximize the stored water disinfectant residual. Appropriate balancing is therefore required to ensure disinfection effectiveness and sufficient level of service (Boulos et al., 1996; Hannoun and Boulos, 1997). Numerous standards prepared by ANSI/AWWA, Ten States Standards, and the National Sanitation Foundation (NSF) are available for the design, construction, and maintenance of water storage facilities. However, if retention times are long, disinfectant residual can drop, via reaction with oxidizable material in the water, to a level that is nonprotective of microbial contamination. To minimize this potential problem, adequate turnover of the water in the facility is an essential operational parameter. It is also desirable to adequately mix (to eliminate dead zones) or prevent the short circuiting of the water entering and leaving the facility to shorten the water age in the facility.

The documented cases of waterborne disease outbreaks and the potential for contamination due to the large number of these facilities make this a high priority distribution system water quality maintenance and protection issue.

ADDITIONAL ISSUES OF CONCERN

Two distribution system issues not mentioned in the nine white papers that the committee believes are of significance to public health protection include the management of premise plumbing and the training of distribution system operators.

Premise Plumbing

Premise plumbing is that portion of the water distribution system from the main ferrule or water meter to the consumer's tap in homes, schools, hospitals, and other buildings. Virtually every problem identified in potable water transmission systems can also occur in premise plumbing. However, due to premise plumbing's higher surface area to volume ratio, longer stagnation times, and warmer temperatures (especially in the hot water system), the potential health threat can be magnified (Edwards et al., 2003). This is an important problem because it requires that individual homeowners be responsible for making decisions that will affect the safety of their drinking water. Premise plumbing is also a valuable asset, with more than 5.3 million miles of copper tube installed in buildings since 1963 (CDA, 2004). The estimated replacement value of premise plumbing in buildings is over 1 trillion dollars (Parsons et al., 2004) and the cost of a plumbing failure to an individual homeowner can exceeded $25,000. The problems of greatest concern within premise plumbing include microbial regrowth, leaching, permeation, infiltration, cross connections, leaks and the resulting indoor mold growth, scaling, and the high costs of failure.

Regrowth problems are exacerbated in premise plumbing due to very long stagnation times resulting in a loss of chlorine residual, to the presence of numerous microclimates, and to nutrient release from some pipes. Some clear links have been established between regrowth of opportunistic pathogens such as *Legionella* and *Mycobacterium* and waterborne disease amongst immunocompromised patients in hospitals (EPA, 2002b). A special concern is regrowth of pathogens within water heaters and showers, which may be exacerbated by recent efforts to reduce temperatures to prevent scalding and save energy (Spinks et al., 2003). The impact of low-flow shower fixtures on consumer exposure to airborne aerosols (a potential route for *Legionella* exposure—Blackburn et al., 2004) and endotoxins is unclear and in need of examination (see Rose et al., 1998, and Anderson et al., 2002).

A subset of the regrowth issue in homes deals with the presence of granular activated carbon in point-of-use treatment devices that can accumulate bacterial nutrients and neutralize disinfectant residuals, thus providing an ideal environment for microbial growth (Tobin et al., 1981; Geldreich et al., 1985; Reasoner et al., 1987; LeChevallier and McFeters, 1988). Several coliform bacteria (*Klebsiella, Enterobacter,* and *Citrobacter*) have been found to colonize granular activated carbon filters, regrow during warm-water periods, and discharge into the process effluent (Camper et al., 1986). The presence of a silver bacteriostatic agent did not prevent the colonization and growth of HPC bacteria in granular activated carbon filters (Tobin et al., 1981; Reasoner et al., 1987). Rogers et al. (1999) reported the growth of *Mycobacterium avium* in point-of-use filters in the presence of 1,000 μg/mL silver. While general microbial parameters such as HPC are also higher after point-of-use devices, there is currently no evidence that these microbes cause significant human health problems (WQA, 2003).

Leaching and permeation mechanisms within premise plumbing are the same as for public water supply transmission lines. However, the wide variety of materials used in building plumbing and associated treatment devices, the higher surface area to volume ratio, very long stagnation times, and lessened dilution increase the potential severity of the problem in premise plumbing. If permeation were to occur through the consumer's service line it would rarely be detected by routine monitoring of the distribution system. For new materials, levels of allowable contaminant leaching have been established through health based standards in ANSI/NSF Standard 61. However, ANSI/NSF Standard 61 is a voluntary test for premise plumbing, ANSI/NSF certification may not be required in all states for materials installed in buildings, and it is not clear whether states are applying materials tested under Standard 61 to premise plumbing. Indeed, there is reason to believe that certain existing NSF standards may not be sufficiently protective of public health in the context of lead leaching from in-line brass devices (Abhijeet, 2004).

Problems with infiltration and cross connections are also greater in premise plumbing than in the main distribution system. Premise plumbing is routinely subject to pressure and flow transients and is the site of minimum pressure in the distribution system. Moreover, the pressures measured in distribution systems during high flow events are an upper bound for pressures that occur in premise plumbing. Considering that direct or indirect cross connections are likely to be found in a majority of household plumbing systems (see previous discussion of backflow events and USC, 2002), and the known occurrence of leaks in premise service lines, there is significant potential for backflow and intrusion. Previous white papers on intrusion and cross connections (LeChevallier et al., 2002; EPA, 2002b) do not explicitly include risks in premise plumbing. Thus, additional research is needed to determine the potential impacts of building hydraulics on intrusion and contamination from cross connections.

Other problems beyond the scope of the current effort, but which are nonetheless important, include reduced flow and high energy costs associated with scaling in pipes and water heaters, leak-induced mold growth in buildings, and the high cost of water damages and re-plumbing due to material failure.

The premise plumbing issue poses unique challenges because there is no obvious single party who could assume responsibility for the problem, which might be best addressed through changes in and enforcement of plumbing codes, third party standards, and public education. Utility involvement in overseeing solutions may be appropriate where distribution system water quality directly contributes to the problem, as is currently the case with the Lead and Copper Rule.

Distribution System Operator Training

Traditionally, training of drinking water distribution system operators has focused primarily on issues related to the mechanical aspects of water delivery (pumps and valves) and safety. However, system operators also are responsible for ensuring that the operation of the system from a quantity perspective does not cause degradation of water quality (EPA, 1999), and it is important that they receive adequate training to meet this need. The training should include an understanding of constituents that affect public health, such as disinfectants, disinfection by-products (DBPs), and metals, and how distribution system operations affect their concentrations. Such training should also include guidance on meeting water quality monitoring and reporting requirements, on how to interpret monitoring results, and on actions that should be taken when "positive" hits are detected (such as increased levels of coliforms or turbidity and decreased or depleted chlorine residuals). Distribution system operators should understand the health related implications of total coliform monitoring. Most importantly, the distribution operator must be trained to make decisions regarding the proper balance of quality and quantity issues, such as in proper operation of distribution system storage facilities (Smith and Burlingame, 1994; Kirmeyer et al., 1999). The need to train operators is more pronounced in small systems where there are fewer staff members to aid operators in day-to-day decisions.

The need for the continuing and intensive training of operators of distribution systems has increased recently for three reasons. First, recent federal and state regulations (EPA, 1990) are more sophisticated and require enhanced skills for proper sample collection and preservation, as well as better understanding of aquatic chemistry and biology. Second, in many systems the new regulations (EPA, 1998) created a shift in the use of disinfectants in the distribution systems from a relatively simple application of chlorine to rather complicated application and maintenance of chloramines. Finally, with an increase in the importance of security of drinking water pipes, pumps, reservoirs, and hydrants, there is a corresponding increase in the responsibility of operators to make decisions during perceived security events.

4
Medium Priority Issues

Four of the white papers dealt with topics considered by the committee to have some potential impact on public health. These medium priority issues are biofilm growth, loss of disinfectant residual during nitrification and water aging, and intrusion. Control of these issues certainly should be considered as part of system-wide management.

BIOFILM GROWTH

Virtually every water distribution system is prone to the formation of biofilms, regardless of the purity of the water, type of pipe material, or the presence of a disinfectant. Growth of bacteria on surfaces can occur in the distribution system or in household plumbing. It is reasonably well documented that the suspended bacterial counts observed in distribution systems are the result of biofilm cell detachment rather than growth of organisms in the water (Characklis, 1988; Haudidier et al., 1988; van der Wende et al., 1989; van der Wende and Characklis, 1990). This phenomenon extends to autotrophic and heterotrophic organisms, coliforms (as noted in the TCR), and opportunistic pathogens. As a result of detachment, the biofilm can act as a continuous inoculum into finished water. The organisms can then be inhaled through bathing and showering or ingested.

The larger question is not whether biofilms are present, but whether biofilms are associated with disease. Biofilms in drinking water distribution systems are primarily composed of organisms typically found in the environment, and as such, are likely to be of limited health concern. These organisms are often enumerated by HPC, and yet epidemiological studies relating HPC counts to health effects are scarce due to their high cost and lack of funding, with only a few studies having been completed. In 1991, Calderon and Mood studied the impact of point-of-entry devices containing granular activated carbon on microbial water quality and health effects. Because granular activated carbon can enhance the growth of organisms detected by HPC, counts were elevated in the treated water. In this study, there was no correlation between elevated HPC and health effects. Other studies conducted by Payment et al. (1991a,b, 1997) in Canada suggested that there may be some association between distributed water, HPC counts, and gastroenteritis, but the findings were not clear.

Biofilms can be a reservoir for opportunistic pathogens. *M. avium* complex (*M. avium* and *M. intracellulare*; MAC) have been isolated from drinking water throughout the United States (Haas et al., 1983; duMoulin and Stottmeier, 1986; Carson et al., 1988; duMoulin et al., 1988; Fischeder et al., 1991; von Reyn et al., 1993, 1994; Glover et al., 1994). They have been implicated in infections and tuberculosis-like disease in the immunocompromised population, particularly those with AIDS (Horsburgh, 1991; Nightingale et al., 1992). The MAC are of such significant public health concern that the organisms are included on the EPA's Contaminant Candidate List (CCL). *Legionella* species have been shown to proliferate in drinking water (Wadowsky and Yee, 1983, 1985; Stout et al., 1985; Rogers et al., 1994). These bacteria can be found in water heaters, shower heads, and cooling towers, where their release can lead to respiratory disease in sensitive individuals. *Legionella* are specifically mentioned in the EPA's SWTR, with the maximum contaminant level goal set at zero. Both MAC and *Legionella* are more likely to proliferate in premise plumbing systems rather than the main distribution system. Biofilms also provide protection of pathogens (including some primary pathogens) including protection from disinfectants. *Helicobacter pylori* was noted in biofilms of drinking water mains by Park et al. (2001). Coxsackie virus B was detected in the biofilms of mains exiting a water treatment plant (Vanden Bossche and Kreitmeyer, 1995).

The elimination of biofilms is essentially impossible and their control is difficult. Attempts to eliminate them are complicated by the observation that conditions that reduce the numbers of organisms of one type may potentially select for others. In general, biofilms can be managed by removing organic matter during water treatment that would support biofilm growth, judicious use of disinfectants, good distribution hygiene practices such as flushing, minimizing the corrosion of iron pipe surfaces, and managing contamination from external sources. Biofilm management is ideally accomplished by best practices that also reduce the magnitude of other water quality problems such as disinfection byproduct concentration, corrosion, and aesthetic concerns. This is true for both the utility owned distribution system and premise plumbing.

Biofilm growth is considered to be of medium priority because the potential for public health risk from this source of exposure is of lesser importance than phenomena included in the high risk category. The risks associated with biofilms appear to be most significant with opportunistic pathogens that may cause disease in the immunocompromised population. Because coliform regrowth in biofilms can lead to TCR violations, biofilm control can assist utilities in meeting the requirements of this regulation. Therefore, mechanisms for controlling biofilms may be of benefit both to reducing coliform levels and reducing the levels of opportunistic pathogens.

LOSS OF RESIDUAL VIA WATER AGE AND NITRIFICATION

This issue is related to an effect identified in two of the nine white papers—that of loss of disinfectant residual brought about by the aging of water and nitrification. Water age, which is considered to be synonymous with "hydraulic residence time," depends on both the physical characteristics of the system (such as flow rate, pipe size, configuration, and the amount of storage) as well as its mode of operation. From the point of entry into a distribution system to an individual consumer tap, water may be in transport for days to weeks. Systems that are "looped" may have shorter maximum water ages than systems that contain long pipelines with dead end sections. However, pipe looping may reduce the ability of utilities to contain contamination. A distribution system is not generally uniform in structure but consists of a network of various elements having different physical, chemical, and biological characteristics such as differing size pipes and pipe materials, occurrence of pipe scales, and biofilms. Furthermore, some characteristics such as surface roughness may change with time, which in turn may influence the hydraulic residence time and the path water takes as it flows through the system.

Unlike specific degradative processes influencing water quality that are considered in this report, retention time is a characteristic that only indirectly affects water quality. Many degradative processes are time dependent and, therefore, more adversely affect water quality with increasing retention time. The degradative processes that are most influenced by residence time can be attributed to reactions occurring in the bulk water and at the pipe wall interface.

Biological nitrification is a process in which bacteria in distribution systems oxidize reduced nitrogen compounds (e.g., ammonia) to nitrite and then nitrate. It is an important process associated with nitrifying bacteria in distribution systems and long retention times in systems practicing chloramination. Like water age, it has a variety of direct and indirect effects on distribution system water quality.

In the committee's opinion, the most important problem exacerbated by both nitrification and by long retention times is loss of disinfectant residuals. Chlorine and chloramine loss during water aging is attributable to reactions with demand substances such as reduced iron in corrosion deposits, ammonia, and natural organic matter (NOM) both on the pipe surface and in the bulk phase. In so far as a residual in the distribution system is desirable, the microbial integrity of the system is compromised. Increased occurrence of microorganisms such as coliforms is associated with the loss of disinfectant residual (Wolfe et al., 1988, 1990). Similarly, the loss of chloramine residual driven by biological nitrification was deemed by the committee to be a significant health threat, and more important than the issue of high concentrations of nitrate and nitrite that result from nitrification. Indeed, a positive feedback loop between growth of nitrifying bacteria and chloramine loss can be established, since the loss of disinfectant residual removes one of the controls on the growth of these organisms.

The precise influence of water age on water quality is complex and clearly system specific, complicating potential control strategies. Water age, unlike other causes of distribution system water quality degradation, such as backflow, cannot be eliminated, only managed within the framework of numerous constraints. Water age is determined by flow rate and the internal volume of the distribution system network, and it can be estimated using tracer studies, mathematical models, system models, and computational fluid dynamics models. The physical aspects of pipe sizes and network layout are important considerations in minimizing water age. Research indicates that "dead ends" and low velocities should be avoided (AwwaRF, 2004). This would favor the use of small diameter pipes and careful consideration of flow paths ("looped" geometry). Current design practice, however, typically dictates a design not based on water needed at the tap but on peak flows associated with fire fighting. This tends to result in a design incorporating comparatively large pipes with resulting lower flow rates. Network operation is also an important determinant of water age at any particular point in the system. Water may be routed to avoid excessively long residence times. Periodic flushing of system elements associated with long water age may also minimize water quality degradation by removal of pipe scales and sediment associated with disinfectant consumption and release of iron into the water. Finally, in the case of systems with multiple sources of supply, hydraulic modeling can be used to assess system operations to reduce maximum water age.

A strategy to control nitrification involves periodically switching to free chlorine, which is thought to reduce the active microbial population, or wholesale replacement of chloramines with free chlorine or chlorine dioxide. Other control strategies include reducing the ratio of chlorine to free ammonia, increasing pH, reducing the residence time by managing the flow, lowering the TOC levels in the distribution system via advanced treatment, and maintaining a fairly high residual as well as a low level of free ammonia.

The loss of disinfectant residual caused by increased water age and nitrification is considered a medium priority concern because it is an indirect health impact that compromises the biological integrity of the system and promotes microbial regrowth. See two sections below for further discussion of other, lower priority effects of water age and nitrification on distribution system water quality.

LOW PRESSURE TRANSIENTS AND INTRUSION

Ensuring safe distribution of treated water to consumers' taps requires, among other measures, protection from intrusion of contaminants into the distribution system during low pressure transients. Intrusion refers to the flow of non-potable water into drinking water mains through leaks, cracks, submerged air valves, faulty seals, and other openings resulting from low or negative pressures. Transient pressure regimes are inevitable; all systems will, at some time, be started up, switched off, or undergo rapid flow changes such as those caused by hydrant flushing, and they will likely experience the effects of human errors, equipment breakdowns, earthquakes, or other risky disturbances (Boulos et al., 2004; Wood et al., 2005). Transient events can have significant water quality and health implications. These events can generate high intensities of fluid shear and may cause resuspension of settled particles as well as biofilm detachment. Moreover, a low-pressure transient event, arising for example from a power failure or intermittent/interrupted supply, has the potential to cause the intrusion of contaminated groundwater or surface water into pipes with leaky joints or cracks. This is especially important in systems with pipes below the water table. Dissolved air (gas) can also be released from the water whenever the local pressure drops considerably, and this may promote the corrosion of steel and iron sections with subsequent rust formation and pipe damage. Even some common transient protection strategies, such as air relief valves or air chambers, if not properly designed and maintained may permit pathogens or other contaminants to find a "back door" route into the potable water distribution system. In the event of a large intrusion of pathogens, the chlorine residual normally sustained in drinking water distribution systems may be insufficient to disinfect contaminated water, which can lead to adverse health effects.

Low water pressure in distribution systems is a well-known risk factor for outbreaks of waterborne disease (Hunter, 1997), although there are limited field data (suggesting that additional field studies are needed). In 1997, a massive epidemic of multidrug-resistant typhoid fever (8,901 cases of typhoid fever and 95 deaths) was reported in the city of Dushanbe, Tajikistan, which affected about 1 percent of the city's population. Low water pressure and frequent water outages had contributed to widespread increases in contamination within the distribution system (Hermin et al., 1999). More recently (April 2002), a *Giardia* outbreak was reported at a trailer park in New York State causing six residents to become seriously ill (Blackburn et al., 2004). Contamination was attributed to a power outage, which created a negative pressure transient in the distribution system. This allowed contaminated water to enter the system through either a cross-connection inside a mobile home or through a leaking underground pipe that was near sewer crossings. During the same period (February 2001 to May 2002), a large-scale case-control study conducted in England of the risk factors for sporadic cryptosporidiosis suggested a strong association between self-reported diarrhea and reported low water pressure events (Hunter et al., 2005).

Intrusion events can be controlled or prevented by developing and implementing best distribution system operational practices such as the requirement for maintaining a sufficient water pressure and an adequate level of disinfectant residual throughout the distribution system, leak detection and control, replacement and rehabilitation of nearby sewer lines, proper hydrant and valve operations, redesign of air relief venting (above grade), routine monitoring program (a sudden drop in the chlorine residual could provide a critical indication to water system operators that there is a source of contamination in the system), use of transient modeling to predict and eliminate potential weak spots in the distribution system, and more rigorous applications of existing engineering standards.

Although there are currently insufficient data in the literature to indicate whether intrusion from pressure transients is a substantial source of risk to treated distribution system water quality, nevertheless intrusion is inherently a subset of backflow events, has health risks and is, therefore, an important distribution system water quality maintenance and protection issue.

5
Lower Priority Issues

Of the issues discussed in the white papers, four were felt to be of lower priority in terms of their potential for adverse health effects. Nonetheless, these issues, together with two additional issues, are important for maintaining a well-managed system, upholding high aesthetic quality, minimizing the energy required for distribution, and providing adequate quantities of water.

OTHER EFFECTS OF WATER AGE

As discussed above, water age has an indirect effect on water quality, with the most important being the reduction in disinfectant residual over time. However, there are a number of other alterations that may occur as water ages that merit discussion. First, with increasing age there can be increasing formation of DBPs (e.g., trihalomethanes and haloacetic acids). In-system production of some DBPs may be prevalent, for example, where pipe sediments contain significant organic matter and/or booster chlorination is practiced. There may also be increasing potential for nitrification with increasing water age, especially at higher temperatures. These latter effects of water age may be reduced by reducing the concentration of byproduct precursors (e.g., total organic carbon) and ammonia entering the distribution system.

The presence of high concentrations of corrosion products is frequently associated with long water age. Corrosion in distribution systems, as well as household plumbing, is a complex process still not adequately understood despite much research into the causes. A number of relevant water quality parameters such as disinfectant residual, redox potential, and pH are affected by water age, and these are believed to play an important role in the corrosion of pipe materials and the release of iron, copper, and lead from pipe scales, especially in low alkalinity waters. Control strategies are sometimes utilized such as changing the pH or adding phosphates to reduce lead and copper corrosion and the release of iron from pipe scales, but these measures are more effective if water age and the amount of stagnation are minimized. Other problems associated with water age include the development of objectionable taste and odors, water discoloration, and sediment accumulation.

Of these lower priority concerns, DBP formation and corrosion are the most important because of obvious health risks. Even so, the health risk of

DBPs within a given system may be low compared to contaminants that have an acute health effect, and DBPs are covered by the Disinfectants/Disinfection By-Products Rule. While corroded metals such as lead may be found at very high levels at the tap in some instances, the relationship of their concentrations to water age is not yet adequately understood.

OTHER EFFECTS OF NITRIFICATION

As discussed above, nitrification is a process carried out by ammonia-oxidizing bacteria in the environment that produces nitrite and nitrate, and thus occurs whenever the substrate (ammonium) is present in the waters. There exist abundant data on the impact of nitrate and nitrite on public health, especially on methemoglobinemia (blue baby syndrome, an acute response to nitrite that results in a blockage of oxygen transport—Bouchard et al., 1992). It affects primarily infants below six months of age, but it may occur in adults of certain ethnic groups (Navajos, Eskimos) and those suffering from a genetic deficiency of certain enzymes (Bitton, 1994). Pregnant women may also be at a higher risk of methemoglobinemia than the general population (Bouchard et al., 1992).

Nitrate levels may be important under certain conditions, although the relative source contribution from drinking water is expected to be a maximum of about 1−2 mg/L as nitrogen and typically would be much less that this. Numerous papers have focused on the impact of nitrate nitrogen (nitrate plus nitrite) in drinking waters (Sandor et al., 2001; Gulis et al., 2002; Kumar et al., 2002; De Roos et al., 2003; Coss et al., 2004; Fewtrell, 2004). However, the concentration at which nitrate nitrogen in drinking waters presents a health risk is unclear (Fewtrell, 2004). Nitrate may be reduced to nitrite in the low pH environment of the stomach, reacting with amines and amides to form N-nitroso compounds (Bouchard et al., 1992; De Roos et al., 2003). Nitrosamines and nitrosamides have been linked to different types of cancer, but the intake of nitrate from drinking water and its causal relation to the risk of cancer is still a matter of debate (Bouchard et al., 1992). A study by Gulis et al. (2002) in Slovakia related increased colorectal cancer and non-Hodgkins lymphoma to medium (10.1–20 mg/l) and high (20.1–50 mg/l) concentrations of nitrate nitrogen in drinking waters. Similarly, Sandor et al. (2001) showed a correlation between the consumption of waters containing greater than 88 mg/l nitrate nitrogen and gastric cancer.

Current nitrite and nitrate MCLs, which are regulated at the entry point to the distribution system, have been set at 1 and 10 mg/l as nitrogen, respectively, in the United States and Canada. The World Health Organization recommends 11.3 mg/l nitrate nitrogen as a guideline value. van der Leeden et al. (1990) presented data up to 1962 in which 93 percent of all U.S. water supplies contain less than 5 mg/l nitrate (it was not specified if the concentrations were nitrate nitrogen or nitrate). However, this may be changing as a result of the increased use of nitrate-containing fertilizers. It has also been shown that chloramination, which is on the increase as an alternative disinfectant, may result in elevated levels of nitrate in waters because of partial nitrification (Bryant et al., 1992), but the increment in nitrate plus nitrite nitrogen from this source would typically be less than 1 mg/L. EPA predicts that about 50% of surface water plants in the United States may use chloramination as a result of the Stage 1 Disinfectants/Disinfection Byproducts Rule (EPA, 2003). This may have the unintended result of a possible increase in the final concentration of nitrate in drinking water. In most cases, the current MCL seems to be well below the concentrations at which risk has been observed. However, some special populations (pregnant women, infants) as well as some ethnic groups may more susceptible to adverse health effects as a result of elevated nitrate concentrations in drinking waters (Bitton, 1994; De Roos et al., 2003).

Lesser effects are that nitrification in low alkalinity waters can reduce alkalinity and decrease the pH. This may cause the pH to decrease to the point that corrosion of lead or copper becomes a problem.

The formation of nitrate and nitrite is considered a relatively low priority concern for distribution systems compared to the other concerns mentioned in this report, primarily because the amount of nitrate generated would likely be less than 10 percent of the MCL. Furthermore, except in very special situations drinking water is not a major source of these substances in the average diet. For example, nitrate is especially abundant in many leafy green vegetables.

PERMEATION

Permeation in water distribution systems occurs when contaminants external to the pipe materials and non-metallic joints pass through these materials into the drinking water. Permeation is generally associated with plastic non-metallic pipes (Holsen et al., 1991). The contaminants that are most commonly found to permeate plastic pipes are organic chemicals that are lipophilic and non-polar such as highly volatile hydrocarbons and organic solvents (Holsen et al., 1991; Burlingame and Anselme, 1995). These chemicals can readily diffuse through the plastic pipe matrix, alter the plastic material, and migrate into the water within the pipe.

The most common example of permeation of water mains and fittings is associated with soil contamination of the area within which the pipe was placed (Glaza and Park, 1992). The majority of permeation incidents appear to be associated with gasoline related organic chemicals. These incidents have occurred at high-risk sites, such as industrial sites and near underground chemical storage tanks, as well as at lower risk residential sites (Holsen et al., 1991). In some cases the integrity of the pipe has been irreversibly compromised, requiring the complete replacement of the contaminated section.

Although there is the potential for water quality degradation as a result of the permeation of plastic pipe, especially in the water's taste and odor aspects, the health impacts associated with such permeation are not expected to be significant. In some permeation incidents, the concentrations of certain chemicals have been shown to reach levels in the low parts per million, which are well above their respective MCLs (AWWA and EES, Inc., 2002a.). However, these MCLs are based on long-term exposure, and the short-term risk levels for these chemicals are generally much higher. In the case of permeation by gasoline components, the taste or odor thresholds of the majority of these chemicals are below the levels that would pose a short-term risk (EPA, 2002e,f,g,h). Since the taste and odor thresholds for some contaminants may be above the MCLs, consumers could be unaware of contamination without monitoring in the distribution system, and some long-term exposures could result, unless otherwise detected. Therefore, the magnitude of long-term exposures resulting from permeation is unknown. In addition, these high concentrations would be expected to occur during worst case situations where water has been in contact with the affected pipe for a considerable length of time. During periods of normal water use these concentrations would expected to be much lower.

Appropriate measures can be taken to minimize the occurrence of permeation, such as issuing regulations or guidelines that define the conditions under which plastic pipe should be used. For example, the State of California precludes the use of plastic pipe in areas subject to contamination by petroleum distillates (California Code of Regulations, Title 22, Division 4, Chapter 16, Article 5, Section 64624f).

After assessing the potential health impacts associated with permeation, the committee has concluded that the potential health impacts are low and that distribution systems can best be protected through measures that minimize the conditions under which permeation can occur.

LEACHING

All materials in the water distribution system undergo reactions that introduce substances to the water via a process known as "leaching." Pipes, fittings, linings, and other materials used in joining or sealing pipes leach at least some substances to water through corrosion, dissolution, diffusion, or detachment. Internal coatings in water storage facilities can also leach substances. Most known substances leaching to water from materials in the distribution system do not appear to pose a public health threat due the fact they are non-toxic, present only at trace levels, or are in a form unlikely to cause health problems. Taste and odor complaints are possible, however (see Choi et al., 1994, and Khiari et al., 2002, for examples).

Under some circumstances, leaching of toxic contaminants occurs at levels that pose a substantial health threat. PVC pipes manufactured before about 1977 are known to leach carcinogenic vinyl chloride into water at levels above the MCL (AWWA and EES, Inc., 2002a). It should be noted that the MCL is based on a measurement of samples at the treatment plant and not within the distribution system. To protect against a health problem from this source, sampling in the distribution system would have to be required after installation of new PVC pipe. Cement materials have, under unusual circumstances, leached aluminum to drinking water at concentrations that caused death in hemodialysis and other susceptible patients (Berend et al., 2001). Because levels of aluminum normally present in drinking water can also threaten this population, the FDA has issued guidance for water purification pre-treatments in the U.S. for dialysis and other patients (Available on-line at *http://www.gewater.com/library/tp/1111_Water_The.jsp*). Finally, excessive leaching of organic substances from linings, joining and sealing materials have occasionally been noted in the literature, and asbestos fibers in water are regulated with an MCL. Potential problems with lead and copper leaching to water are managed via the LCR and, thus, are not considered further here.

Problems from older distribution system materials can be managed by monitoring of contaminant leaching in the distribution system, adjustments to water chemistry, or by costly replacement of the material. Lead leaching to water from old lead pipe is managed in this way via the LCR. For new materials, the NSF establishes levels of allowable contaminant leaching through ANSI/NSF Standard 61. It should be noted that ANSI/NSF Standard 61, which establishes minimum health effect requirements for chemical contaminants and impurities, does not establish performance, taste and odor, or microbial growth support requirements for distribution system components. There is uncertainty regarding what the states require and systems use in terms of the application and testing of ANSI/NSF Standard 61, particularly for cement materials. Therefore, further investigation is needed to determine the potential public health implications of cement materials used in distribution systems. Research has shown that distribution system components can significantly impact the microbial quality of drinking water via leaching. For example, pipe gaskets and elastic sealants (containing polyamide and silicone) can be a source of nutrients for bacterial proliferation. Colbourne et al. (1984) reported that *Legionella* were associated with certain rubber gaskets. Organisms associated with joint-packing materials include populations of *Pseudomonas aeruginosa, Chromobacter spp., Enterobacter aerogenes,* and *Klebsiella pneumoniae* (Schoenen, 1986; Geldreich and LeChevallier, 1999). Coating compounds for storage reservoirs and standpipes can contribute organic polymers and solvents that may support regrowth of heterotrophic bacteria (Schoenen, 1986; Thofern et al., 1987). Liner materials may contain bitumen, chlorinated rubber, epoxy resin, or tar-epoxy resin combinations that can support bacterial regrowth (Schoenen, 1986). PVC pipes and coating materials may leach stabilizers that can result in bacterial growth. Studies performed in the United Kingdom reported that coliform isolations were four times higher when samples were collected from plastic taps than from metallic faucets (cited in Geldreich and LeChevallier, 1999). Although procedures are available to evaluate growth stimulation potential of different materials (Bellen et al., 1993), these tests are not applied in the United States by ANSI/NSF. Standards or third-party certification that establishes performance, taste and odor, or microbial growth support requirements for distribution system components could be considered. In spite of these limitations and occasional problems, it is currently believed that leaching is a relatively low priority relative to other distribution system problems.

ADDITIONAL ISSUES OF CONCERN

Control of Post Precipitation

Control of post precipitation in distribution systems is an important part of any program to control the quality of water in distribution systems. Post precipitation can result from introduction of water to distribution systems that is super-saturated with calcium carbonate, from introduction of a phosphate corrosion inhibitor into the filter effluent of an alum coagulation plant creating an aluminum phosphate precipitate, from water that is supersaturated with aluminum hydroxide, from water that is supersaturated with selected silicate minerals, as well as other causes. Post-precipitation causes an increase in pipe roughness and a decrease in effective pipe diameter, resulting in loss of hydraulic capacity accompanied by an increase in energy required to distribute water, in production of biofilms, and in deterioration of the aesthetic quality of tap water. If the material is loosely attached to the pipe wall, such as some aluminum precipitates, hydraulic surges can result in substantial increases in the turbidity of tap water. Treatment of water to avoid excessive post-precipitation thus is an important asset management issue. It is not amenable to regulation, but it is an important part of the guidance that should accompany distribution system regulations.

6
Summary

The purpose of this report was to review published material in order to identify trends relevant to the deterioration of drinking water quality in water supply distribution systems and to identify and prioritize issues of greatest concern. The trends relevant to the deterioration of drinking water quality in distribution systems include:

- **Aging distribution systems**. Increasing numbers of main breaks and pipe replacement activities are a possibility as systems age, depending on the pipe materials and linings used, the water quality, and system operation and maintenance practices.
- **Decreasing numbers of waterborne outbreaks reported per year since 1982, but an increasing percentage attributable to distribution system issues**. This trend is probably is related to better treatment of surface water.
- **Increasing host susceptibility to infection and disease in the U.S. population**. This trend is caused by aging of the U. S. population, the increase in the incidence of AIDS, and the increasing use of immunosuppressive therapy.
- **Increasing use of bottled water and point of use treatment devices.** This trend suggests that exposure to tap water on a per capita basis may be decreasing. However, it should be kept in mind that point-of-use devices can support microbial regrowth.

The issues from the nine white papers have been prioritized using categories of highest, medium, and lower priority. The committee also identified a number of additional issues not addressed in previous reports. The highest priority issues are those that have a recognized health risk based on clear epidemiological and surveillance data.

Cross connections and backflow. Cross connections and backflow events are ranked as the highest priority because of the long history of recognized health risks posed by cross connections, the clear epidemiological and surveillance data implicating these events with outbreaks or sporadic cases of waterborne disease, and the availability of proven technologies to prevent cross connections.

Contamination during installation, rehabilitation, and repair of water mains and appurtenances. Chemical and microbial contamination of distribution system materials and drinking water during mains breaks and during the installation, rehabilitation, and repair of water mains and appurtenances is a high priority issue because there have been many documented instances of significant health impacts from drinking water contamination associated with pipe failures and maintenance activities.

Improperly maintained and operated distribution system storage facilities. Several documented waterborne disease outbreaks and the potential for contamination due to the large number of these facilities makes this a high priority distribution system water quality maintenance and protection issue.

Control of water quality in premise plumbing. Virtually every problem identified in potable water transmission systems can also occur in premise plumbing, and some are magnified because of premise plumbing characteristics and the way in which water is used in residences. Health risks associated with premise plumbing are hard to assess because the majority of health problems are likely to be sporadic, unreported cases of waterborne disease that affect single households. Waterborne disease outbreaks due to premise plumbing failures in residential buildings have been documented.

Distribution system operator training. Training of drinking water distribution system operators traditionally has focused on issues related to the mechanical aspects of water delivery (pumps and valves) and safety. System operators are also responsible for ensuring that conveyance of the water does not allow degradation of water quality, and it is important that they receive adequate training to meet this responsibility.

Medium priority issues are those where existing data suggest that the health risks are low or limited to sensitive populations. Issues where there were insufficient data to determine the magnitude of the health risk were also classified as medium priority.

Biofilm Growth. Although biofilms are widespread in distribution systems, the public health risk from this source of exposure appears to be limited to opportunistic pathogens that may cause disease in the immunocompromised population. Some data suggest that biofilms may protect microbial pathogens from disinfection, but there are few studies directly linking health effects to biofilms.

Loss of Disinfectant Residual. The loss of disinfectant residual caused by increased water age and nitrification is considered a medium priority concern because it is an indirect health impact that compromises the biological integrity of the system and promotes microbial regrowth.

Intrusion. Intrusion from pressure transients is a subset of the cross-connection and backflow issue. It has associated health risks, and is therefore an important distribution system water quality maintenance and protection issue. There are insufficient data to indicate whether it is a substantial health risk, however.

Lower priority issues are those that are already covered by current regulations, well-managed in the majority of water distribution systems, or unlikely to pose a health risk.

Other Effects of Water Age. The quality of distributed water, in particular water age, also has indirect effects such as (1) DBP formation in distribution systems with increasing water age that might cause the MCLs for these substances to be exceeded and (2) enhanced corrosion and the release of metals from corrosion scales. DBPs and common corrosion products are covered by the D/DBPR and the LCR, respectively.

Nitrification. Nitrification that results in (1) the formation of nitrite and nitrate in quantities that cause the MCLs for these substances to be exceeded or (2) the release of excessive concentrations of metal ions should be avoided. However, the formation of nitrate and nitrite is considered a relatively low risk for distribution systems compared to the other concerns mentioned in this report.

Permeation. Permeation of chemicals through plastic pipe can occur, but the potential health impacts are low and distribution systems can best be protected through measures that minimize the conditions under which permeation can occur.

Leaching. Excessive leaching of organic substances from pipe materials, linings, joining and sealing materials, coatings, and cement mortar pipe have occasionally been noted in the literature. Leaching is a relatively low priority relative to other distribution system problems and can be controlled by regulating the materials that are used in distribution and premise plumbing systems, by specifying the water chemistry that must be used if certain materials are to be employed, and by appropriate monitoring requirements.

Post-precipitation. An additional issue of lower priority is the control of post-precipitation, which causes an increase in pipe roughness and a decrease in effective pipe diameter, resulting in loss of hydraulic capacity accompanied by an increase in the energy required to distribute water, in the production of biofilms, and in the deterioration of tap water's aesthetic quality.

Deteriorating infrastructure was not included as one of the issues that the committee prioritized because it is the ultimate cause of many of the other events that are discussed in this report, such as:

- water main breaks and contamination that results during their repair,
- contamination from decaying storage structures and their inadequate maintenance,
- intrusion and water loss,
- occurrence of excessive biofilms and nitrification,
- system design and operation practices that cause the water quality to degrade, and
- excessive deposits from corrosion and post-precipitation.

Solutions to problems caused by deteriorating infrastructure are thus expected to be applicable to most of the problems already discussed in this report. It should be noted that the rate of degradation of distribution system materials will vary from system to system depending on water quality and system operation and maintenance practices, such that the relationship between the age of a given system, its state of deterioration, and risk cannot be easily predicted. Confronting deteriorating infrastructure requires good asset management, including procedures to monitor and assess the condition of the distribution system and water quality changes that occur during distribution. Furthermore, appropriate maintenance, repair, and replacement should be carried out as needed, and operating and capital budgets should be available to finance this work.

References

Abhijeet, D. 2004. Reconsidering lead corrosion in drinking water: product testing, direct chloramine attack and galvanic corrosion. Virginia Tech MS Thesis.

American Water Works Association (AWWA). 2001. Reinvesting in drinking water structure: dawn of the replacement era. Denver, CO: AWWA.

AWWA. 2003. Water Stats 2002 Distribution Survey CD-ROM. Denver, CO: AWWA.

AWWA and EES, Inc. 2002a. Permeation and leaching. Available on-line at *http://www.epa.gov/safewater/tcr/pdf/permleach.pdf.* Accessed March 16, 2005.

AWWA and EES, Inc. 2002b. Effects of water age on distribution system water quality. *http://www.epa.gov/safewater/tcr/pdf/waterage.pdf.* Accessed March 16, 2005.

AWWA and EES, Inc. 2002c. Finished water storage facilities. Available on-line at *http://www.epa.gov/safewater/tcr/pdf/storage.pdf.* Accessed March 16, 2005.

AWWA and EES, Inc. 2002d. Nitrification. Available on-line at *http://www.epa.gov/safewater/tcr/pdf/nitrification.pdf.* Accessed March 16, 2005.

AWWA and EES, Inc. 2002e. New or repaired water mains. Available on-line at *http://www.epa.gov/safewater/tcr/pdf/maincontam.pdf.* Accessed March 16, 2005.

American Water Works Association Research Foundation (AwwaRF). 2004. Managing distribution retention time to improve water quality: phase I. Report no. 91006F (RFP 2769). Denver, CO: Binnie Black and Veatch and AwwaRF.

American Water Works Service Co., Inc. (AWWSC). 2002. Deteriorating buried infrastructure management challenges and strategies. Available on-line at *http://www.epa.gov/safewater/tcr/pdf/infrastructure.pdf.* Accessed March 16, 2005.

Anadu, E. C., and A. K. Harding. 2000. Risk perception and bottled water use. J. Amer. Water Works Assoc. 92 (11):82–92.

Anderson, W. B., R. M. Slawson, and C. I. Mayfield. 2002. A review of drinking-water-associated endotoxin, including potential routes of human exposure. Can. J. Microbiol. 48:567–587.

Bellen, G. E., S. H. Abrishami, P. M. Colucci, and C. Tremel. 1993. Methods for assessing the biological growth support potential of water contact materials. Denver, CO: AwwaRF.

Berend, K., G. Van Der Voet, and W. H. Boer. 2001. Acute aluminum encephalopathy in a dialysis center caused by a cement mortar water distribution pipe. Kidney International 59(2):746–753.

Bitton, G. 1994. Role of microorganisms in biogeochemical cycles. Pp. 51–73 *In:* Wastewater Microbiology. New York: John Wiley & Sons, Inc.

Blackburn, B. G., G. F. Craun, J. S. Yoder, V. Hill, R. L. Calderon, N. Chen, S. H. Lee, D. A. Levy, and M. J. Beach. 2004. Surveillance for waterborne-disease outbreaks associated with drinking water—United States, 2001–2002. MMWR 53(SS-8):23–45.

Bouchard, D. C., M. K. Williams, and R. Y. Surampalli. 1992. Nitrate contamination of groundwater: sources and potential health effects. J. Amer. Water Works Assoc. 84(9):84–90.

Boulos, P. F., W. M. Grayman, R. W. Bowcock, J. W. Clapp, L. A. Rossman, R. M. Clark, R. A. Deininger, and A. K. Dhingra. 1996. Hydraulic mixing and free chlorine residual in reservoirs. J. Amer. Water Works Assoc. 88(7):48–59.

Boulos, P. F., K. E. Lansey, and B. W. Karney. 2004. Comprehensive Water Distribution Systems Analysis Handbook for Engineers and Planners. Pasadena, CA: MWH Soft Pub.

Bryant, E. A., G. P. Fulton, and G. C. Budd. 1992. Chloramination. Pp. 128–170 *In:* Disinfection Alternatives for Safe Drinking Water. New York: Van Nostrand Reinhold.

Burlingame, G. A., and H. M. Neukrug. 1993. Developing proper sanitation requirements and procedures for water main disinfection. *In:* Proceedings of AWWA Annual Conference. Denver, CO: AWWA.

Burlingame, G. A., and C. Anselme. 1995. Distribution system tastes and odors. *In:* Advances in Taste-and-Odor Treatment and Control. AWWA Research Foundation Cooperative Research Report. Denver, CO: AwwaRF.

Calderon, R. L., and E. W. Mood. 1991. Bacteria colonizing point-of-entry, granular activated carbon filters and their relationship to human health. CR-813978-01-0. Washington, DC: EPA.

Camper, A. K., M. W. LeChevallier, S. C. Broadway, and G. A. McFeters. 1986. Bacteria associated with granular activated carbon particles in drinking water. Appl. Environ. Microbiol. 52:434–438.

Carson, L. A., L. A. Bland, L. B. Cusick, M. S. Favero, G. A. Bolan, A. L. Reingold, and R. A. Good. 1988. Prevalence of nontuberculous mycobacteria in water supplies of hemodialysis centers. Appl. Environ. Microbiol. 54:3122–3125.

CDA. 2004. Copper facts. Available on-line at *http://www.copper.org/education/c-facts/ c-plumbing.html.* Accessed February 11, 2005.

Centers for Disease Control and Prevention. 2004. Cancer survivorship—United States, 1971–2001. MMWR 53(24):526–529.

Characklis, W. G. 1988. Bacterial regrowth in distribution systems. Denver, CO: AwwaRF.

Choi, J., M. Fadel, L. Gammie, J. Rahman, and J. Paran. 1994. Sniff new mains...before customers complain. Opflow 20(10):3.

Clark, R. M., W. M. Grayman, J. A. Good rich, R. A. Deininger, and A. F. Hess. 1991. Field testing distribution water quality models. J. Amer. Water Works Assoc. 83(7):67–75.

Clark, R. M., E. E. Geldreich, K. R. Fox, E. W. Rice, C. H.Johnson, J. A. Goodrich, J. A. Barnick, and F. Abdesaken. 1996. Tracking a *Salmonella* serovar *typhimurium* outbreak in Gideon, Missouri: role of contaminant propagation modelling. Journal of Water Supply Research and Technology-Aqua. 45:171–183.

Colbourne, J. S., D. J. Pratt, M. G. Smith, S. P. Fisher-Hoch, and D. Harper. 1984. Water fittings as sources of *Legionella pneumophila* in hospital plumbing system. Lancet i:210–213.

Craun, G. F., and R. L. Calderon. 2001. Waterborne disease outbreaks caused by distribution system deficiencies. J. Amer. Water Works Assoc. 93(9):64–75.

Coss, A., K. P. Cantor, J. S. Reif, C. F. Lynch, and M. H. Ward. 2004. Pancreatic cancer and drinking water and dietary sources of nitrate and nitrite. Amer. J. Epidemiol. 159(7):693–701.

De Roos, A. J., M. H. Ward, C. F. Lynch, and K. P. Cantor. 2003. Nitrate in public water supplies and the risk of colon and rectum cancers. Epidemiology 14(6):640–649.

duMoulin, G. C., K. D. Stottmeier, P. A. Pelletier, T. A. Tsang, and J. Hedley-Whyte. 1988. Concentration of *Mycobacterium avium* by hospital hot water systems. J. Amer. Medical Assoc. 260:1599–1601.

duMoulin, G. C., and K. D. Stottmeir. 1986. Waterborne mycobacteria: an increasing threat to health. American Society for Microbiology News 52:525–529.

Edwards, M., D. Bosch, G. V. Loganathan, and A. M. Dietrich. 2003. The Future Challenge of Controlling Distribution System Water Quality and Protecting Plumbing Infrastructure: Focusing on Consumers. Presented at the IWA Leading Edge Conference in Noordwijk, Netherlands. May 2003.

Environmental Protection Agency (EPA). 1990. Fact sheet: drinking water regulations under the Safe Drinking Water Act. Washington, DC: EPA Office of Drinking Water Criteria and Standards Division.

EPA. 1998. National Primary Drinking Water Regulations: disinfectants and disinfection byproducts; Final Rule. Federal Register 63(241).

EPA. 1999. Guidelines for the certification and recertification of the operators of community and nontransient noncommunity public water systems. Federal Register 64(24).

EPA. 2002a. 2000 Community water system survey. EPA 815-R-02-005A. Washington, DC: EPA Office of Water.

EPA. 2002b. Potential contamination due to cross-connections and backflow and the associated health risks: an issues paper. Available on-line at *http://www.epa.gov/safewater/tcr/pdf/ ccrwhite.pdf.* Accessed March 16, 2005.

EPA. 2002c. The clean water and drinking water infrastructure gap analysis. Washington, DC: EPA.

EPA. 2002d. Health risks from microbial growth and biofilms in drinking water distribution systems. Available on-line at *http://www.epa.gov/safewater/tcr/pdf/biofilms.pdf.* Accessed on March 16, 2005. Washington, DC: EPA.

EPA. 2002e. Technical fact sheet on: Benzene. Available on-line at *http://www.epa.gov/ OGWDW/ dwh/t-voc/benzene.html.* Accessed on March 16, 2005.

EPA. 2002f. Technical fact sheet on: xylenes. Available on-line at *http://www.epa.gov/ OGWDW/dwh/t-voc/xylenes.html.* Accessed on March 16, 2005.

EPA. 2002g. Technical fact sheet on: Toluene. Available on-line at *http://www.epa.gov/ OGWDW/dwh/t-voc/toluene.html.* Accessed on March 16, 2005.

EPA. 2002h. Technical fact sheet on: Ethylbenzene. Available on-line at *http://www.epa.gov/ OGWDW/dwh/t-voc/ethylben.html.* Accessed on March 16, 2005.

EPA. 2003. Economic Analysis for Proposed Stage 2 DBPR. EPA 815-D-03-001. Washington, DC: EPA.

EPA. 2004. Factoids: Drinking water and ground water statistics for 2003. EPA 816-K-03-001. Washington, DC: EPA Office of Water.

Fewtrell, L. 2004. Drinking-water nitrate, methemoglobinemia, and global burden of disease: as discussion. Environmental Health Perspectives 112(14):1371−1374.

Fischeder, R. R., R. Schulze-Robbecke, and A. Weber. 1991. Occurrence of mycobacteria in drinking water samples. Zbl. Hygiene 192:154−158.

Frost, F. J., G. F. Craun, and R. L. Calderon. 1996. Waterborne disease surveillance. J. Amer. Water Works Assoc. 88(9):66−75.

Geldreich, E. E., R. H. Taylor, J. C. Blannon, and D. J. Reasoner. 1985. Bacterial colonization of point-of-use water treatment devices. J. Amer. Water Works Assoc. 77(2):72−80.

Geldreich, E. E., K. R. Fox, J. A. Goodrich, E. W. Rice, R. M. Clark, D. L. and Swerdlow. 1992. Searching for a water supply connection in the Cabool, Missouri disease outbreak of *Escherichia coli* 0157:H7. Water Research 26(8):1127−1137.

Geldreich, E. E., and M. W. LeChevallier. 1999. Microbial water quality in distribution systems. Pp. 18.1−18.49 *In:* Water Quality and Treatment, 5th edition. R. D. Letterman (ed.). New York: McGraw-Hill.

Gerba, C. P., J. B. Rose, et al. 1996. Sensitive populations: Who is at the greatest risk? International Journal of Food Microbiology 30(1−2):113−123.

Glaza, E. C., and J. K. Park. 1992. Permeation of organic contaminants through gasketed pipe joints. J. Amer. Water Works Assoc. 84(7):92−100.

Glover, N. A., N. Holtzman, T. Aronson, S. Froman, O. G. W. Berlin, P. Dominguez, K. A. Kunkel, G. Overturf, G. Stelma, Jr., C. Smith, and M. Yakrus. 1994. The isolation and identification of *Mycobacterium avium* complex (MAC) recovered from Los Angeles potable water, a possible source of infection in AIDS patients. International J. Environ. Health Res. 4:63−72.

Guilaran, Y.-T. 2004. EPA Presentation to the NRC Committee of Public Water Supply Distribution Systems on October 27, 2004.

Gulis, G., M. Czompolyova, and J. R. Cerhan. 2002. An ecologic study of nitrate in municipal drinking water and cancer incidence in Trnava District, Slovakia. Environmental Research 88(3):182−187.

Haas, C. N., M. A. Meyer, and M. E. Paller. 1983. The ecology of acid-fast organisms in water supply, treatment, and distribution systems. J. Amer. Water Works Assoc. 75:39−144.

Haas, C. N., R. B. Chitluru, M. Gupta, W. O. Pipes, and G. A. Burlingame. 1998. Development of disinfection guidelines for the installation and replacement of water mains. Denver, CO: AwwaRF.

Hannoun, I. A., and P. F. Boulos. 1997. Optimizing distribution storage water quality: a hydrodynamic approach. Journal of Applied Mathematical Modeling 21(8):495−502.

Hasit, Y. J., A. J. DeNadai, H. M. Gorrill, R. S. Raucher, and J. Witcomb. 2004. Cost and benefit analysis of flushing. Denver, CO: AwwaRF.

Haudidier, K, J. L. Paquin, T. Francais, P. Hartemann, G. Grapin, F. Colin, M. J. Jourdain, J. C. Block, J. Cheron, O. Pascal, Y. Levi, and J. Miazga. 1988. Biofilm growth in drinking water network: a preliminary industrial pilot plant experiment. Water Sci. Technol. 20:109.

Hermin, J. H. R. Villar, J. Carpenter, L. Roberts, A. Samaridden, L. Gasanova, S. Lomakina, C. Bopp, L. Hutwagner, P. Mead, B. Ross, and E. D. Mintz. 1999. A massive epidemic of multidrug-resistant typhoid fever in Tajikistan associated with consumption of municipal water. Journal of Infectious Diseases 179:1416–1422.

Holsen, T. M., J. K. Park, D. Jenkins, and R. E. Selleck. 1991. Contamination of potable water by permeation of plastic pipe. J. Amer. Water Works Assoc. 83(8):53–56.

Horsburgh, C. R. 1991. *Mycobacterium avium* complex infection in the acquired immunodeficiency syndrome. New England Journal of Medicine 324:1332–1338.

Hunter, P. R. 1997. Waterborne disease: epidemiology an ecology. Chichester, UK: Wiley.

Hunter, P. R., R. M. Chalmers, S. Hughes, and Q. Syed. 2005. Self-reported diarrhea in a control group: a strong association with reporting of low-pressure events in tap water. Clinical Infectious Diseases 40:e32–34.

ICF Consulting, Inc. 2004. Exposure assessment of pathogens and toxic chemicals in drinking water distribution systems workshop. Washington, DC: EPA.

Khiari, D., S. Barrett, R. Chinn, A. Bruchet, P. Piriou, L. Matia, F. Ventura, I. Suffet, T. Gittelman, and P. Luitweiler. 2002. Distribution generated taste-and-odor phenomena. Denver, CO: AWWARF.

Kirmeyer, G., W. Richards, and C. D. Smith. 1994. An assessment of water distribution systems and associated research needs. Denver, CO: AWWARF.

Kirmeyer, G. J., L. Kirby, B. M. Murphy, P. F. Noran, K. D. Martel, T. W. Lund, J. L. Anderson, and R. Medhurst. 1999. Maintaining and operating finished water storage facilities. Denver, CO: AWWARF.

Kirmeyer, G. K., M. Freidman, K. Martel, D. Howie, M. LeChevallier, M. Abbaszadegan, M. Karim, J. Funk, and J. Harbour. 2001. Pathogen intrusion into the distribution system. Denver, CO: AWWARF.

Kumar, S., A. B. Gupta, and S. Gupta. 2002. Need for revision of nitrates standards for drinking water: a case study of Rajasthan. Indian Journal of Environmental Health 44(2):168–172.

LeChevallier, M. W., and G. A. McFeters. 1988. Microbiology of activated carbon. Pp. 104–119 In: Drinking Water Microbiology, Progress and Recent Developments. G. A. McFeters (ed.). New York: Springer-Verlag.

LeChevallier, M., R. Gullick, and M. Karim. 2002. The potential for health risks from intrusion of contaminants into the distribution system from pressure transients. Draft Distribution System White Paper. Washington, DC: EPA

Lee, S. H., D. A. Levy, G. F. Craun, M. J. Beach, and R. L. Calderon. 2002. Surveillance for waterborne-disease outbreaks in the United States, 1999–2000. MMWR 51(No. SS-8):1–49.

Mackey, E. D., J. Davis, L. Boulos, J. C. Brown, and G. F. Crozes. 2003. Consumer perceptions of tap water, bottled water, and filtration devices. Denver, CO: AWWARF.

Morris J. G., and M. Potter. 1997. Emergence of new pathogens as a function of changes in host susceptibility. Emerg. Inf. Dis. 3(4):435–441

National Research Council (NRC). 1999. Watershed Management for Potable Water Supply: the New York City Strategy. Washington, DC: The National Academies Press.

National Resources Defense Council (NRDC). 1999. Bottled water: Pure drink or pure hype? Executive Summary, Chapter 2. Available on-line at *http://www.nrdc.org/water/drinking/bw/exesum.asp*. Accessed February 6, 2005.

Nightingale, S. D., L. T. Byrd, P. M. Southern, J. D. Jockusch, S. X. Cal, and B. A. Wynne. 1992. Mycobacterium avium-intracellulare complex bacteremia in human immunodeficiency virus positive patients. Journal of Infectious Disease 165:1082–1085.

Older Americans. 2004. Key indicators of well-being. Federal Interagency Forum on Aging-Related Statistics. Available on-line at *http://www.agingstats.gov/chartbook2004/population.html*. Accessed February 6, 2005.

Park, S. R., W. G. Mackay, and D. C. Reid. 2001. Helicobacter sp. recovered from drinking water biofilm sampled from a water distribution system. Wat. Res. 35(6):1624–1626.

Parsons, S., R. Stuetz, B. Jefferson, and M. Edwards. 2004. Corrosion control in water distribution systems: One of the grand engineering challenges for the 21st century. Water Science and Technology 49(2):1–8.

Payment, P. L., E. Franco, L. Richardson, and J. Siemiatycki. 1991. Gastrointestinal health effects associated with the consumption of drinking water produced by point-of-use domestic reverse-osmosis filtration units. Appl. Environ. Microbiol. 57:945–948.

Payment, P. L., L. Richardson, J. Siemiatycki, R. Dewar, M. Edwards, and E. Franco. 1991. A randomized trial to evaluate the risk of gastrointestinal disease due to consumption of drinking water meeting current microbiological standards. American Journal of Public Health 81:703–708.

Payment, P. L., J. Siemiatycki, L. Richardson, G. Renaud, E. Franco, and M. Prevost. 1997. A prospective epidemiological study of gastrointestinal health effects due to the consumption of drinking water. International Journal of Environmental Health Research 7:5–31.

Payment, P., and P. R. Hunter. 2001. Endemic and epidemic infectious intestinal disease and its relationship to drinking water. Pp. 62–88 *In:* Water Quality: Guidelines, Standards and Health: Assessment of Risk and Risk Management for Water-Related Infectious Disease. L. Fewtrell and J. Bartram (eds.). London: IWA Publishing.

Pierson, G., K. Martel, A. Hill, G. Burlingame, and A. Godfree. 2001. Methods to prevent microbiological contamination associated with main rehabilitation and replacement. Denver, CO: AwwaRF.

Reasoner, D. J., J. C. Blannon, and E. E. Geldreich. 1987. Microbiological characteristics of third-faucet point-of-use devices. J. Amer. Water Works Assoc. 79(10):60–66.

Rogers. M. R., B. J. Backstone, A. L. Reyers, and T. C. Covert. 1999. Colonisation of point-of-use water filters by silver resistant non-tuberculous mycobacteria. J. Clin. Pathol. 52(8):629.

Rogers, J., A. B. Dowsett, P. J. Dennis, J. V. Lee, and C. W. Keevil. 1994. Influence of materials on biofilm formation and growth of *Legionella pneumophila* in potable water systems. Appl. Environ. Microbiol. 60:1842–1851.

Rose, C. S., J. W. Martyny, L. S. Newman, D. K. Milton, T. E. King ,Jr., J. L. Beebe, J. B. McCammon, R. E. Hoffman, and K. Kreiss. 1998. Lifeguard lung: endemic granulomatous pneumonitis in an indoor swimming pool. American Journal of Public Health 88(12):1795–1800.

Sandor, J., I. Kiss, O. Farkas, and I. Ember. 2001. Association between gastric cancer mortality and nitrate content of drinking water: ecological study on small area inequalities. European Journal of Epidemiology 17(5):443–447.

Schoenen, D. 1986. Microbial growth due to materials used in drinking water systems. In: Biotechnology, Vol. 8. H. J. Rehm and G. Reed (eds.). Weinheim: VCH Verlagsgesellschaft.

Smith, C. D., and G. Burlingame. 1994. Microbial problems in treated water storage tanks. Proceedings of the 1994 Annual AWWA Conference. Denver, CO: AWWA.

Spinks, A. T., R. H. Dunstan, P. Coombes, and G. Kuczera. 2003. Thermal destruction analyses of water related pathogens at domestic hot water system temperatures. The Institution of Engineers. 28th International Hydrology and Water Resources Symposium.

Stout, J. E., V. L. Yu, and M. G. Best. 1985. Ecology of *Legionella pneumophila* within water distribution systems. Appl. Environ. Microbiol. 49:221–228.

Thofern, E., D. Schoenen, and G. J. Tuschewitzki. 1987. Microbial surface colonization and disinfection problems. Off Gesundh.-wes. 49:Suppl:14–20.

Tobin, R. S., D. K. Smith, and J. A. Lindsay. 1981. Effects of activated carbon and bacteriostatic filters on microbiological quality of drinking water. Appl. Environ. Microbiol. 41(3):646–651.

University of Southern California (USC). 2002. Prevalence of cross connections in household plumbing systems. www.usc.edu/dept/fcchr/epa/hhcc.report.pdf. Los Angeles, CA: USC Foundation for Cross-Connection Control and Hydraulic Research.

Vanden Bossche, G. and Kreitemeyer. 1995. Detergent conditioning of biofilm samples: a most sensitive method for the detection of enterovirus infectivity. Paper presented to the IAWQ health-related water microbiology symposium, Budapest.

Van der Leeden, F., F. L. Troise, and D. K. Todd. 1990. Water quality. Pp. 417–493 *In:* The Water Encyclopedia, Second Edition. Michigan: Lewis Publishers.

van der Wende, E., and W. G. Characklis. 1990. Biofilms in potable water distribution systems. Chapter 12 *In:* Drinking Water Microbiology. G. A. McFeters (ed.). New York: Springer-Verlag.

van der Wende, E., W. G. Characklis, and D. B. Smith. 1989. Biofilms and bacterial drinking water quality. Water Research 23:1313.

Vikesland, P., K. Ozekin, and R. L. Valentine. 2001. Monochloramine decay in model and distribution system water. Water Research 35(7):1766–1776.

von Reyn, C. F., J. N. Maslow, T. S. Barber, J. O. Falkinham, III, and R. D. Arbeit. 1994. Persistent colonisation of potable water as a source of *Mycobacterium avium* infection in AIDS. Lancet 343:1137–1141.

von Reyn, C. F., R. D. Waddell, T. Eaton, R. D. Arbeit, J. N. Maslow, T. W. Barber, R. J. Brindle, C. F. Gilks, J. Lumio, J. Lahdevirta, A. Ranki, D. Dawson, and J. O. Falkinham, III. 1993. Isolation of *Mycobacterium avium* complex from water in the United States, Finland, Zaire, and Kenya. Journal of Clinical Microbiology 31:3227–3230.

Wadowsky, R. M., and R. B. Yee. 1983. Satellite growth of *Legionella pneumophila* with an environmental isolate of *Flavobacterium breve*. Appl. Environ. Microbiol. 46:1447–1449.

Wadowsky, R. M., and R. B. Yee. 1985. Effect of non-legionellaceae bacteria on the multiplication of *Legionella pneumophila* in potable water. Appl. Environ. Microbiol. 49:1206–1210.

Water Quality Association (WQA). 2001. National consumer water quality survey fact sheet. April 23, 2001. Lisle, IL: WQA.

WQA. 2003. Heterotrophic bacteria in drinking water from POU & POE devices. Lisle, IL: WQA.

Wolfe, R. L., E. G. Means, M. K. Davis, and S. E. Barrett. 1988. Biological nitrification in covered reservoirs containing chloraminated water. J. Amer. Water Works Assoc. 80(9):109–114.

Wolfe, R. L., N. I. Lieu, G. Izaguirre, and E. G. Means. 1990. Ammonia oxidizing bacteria in a chloraminated distribution system: seasonal occurrence, distribution, and disinfection resistance. Appl. Environ. Microbiol. 56(2):451–462.

Wood, D. J., S. Lingireddy, and P. F. Boulos. 2005. Pressure Wave Analysis of Transient Flow in Pipe Distribution Systems. Pasadena, CA: MWH Soft Pub.

Appendix B
Committee Biographical Information

Vernon L. Snoeyink (NAE), Chair, is the Ivan Racheff Professor of Environmental Engineering at the Department of Civil and Environmental Engineering, University of Illinois at Urbana-Champaign. Dr. Snoeyink's research interests involve the physical and chemical processes of drinking water purification, including the removal of trace organic compounds by adsorption onto activated carbon and the development of integrated adsorption-low pressure membrane systems for removing particles and trace contaminants from water. His work has also focused on precipitation of solids in water distribution systems and the chemistry of the formation of colored water in corroded iron pipes. He has been a trustee of the American Water Works Association Research Foundation, president of the Association of Environmental Engineering Professors, a member of the editorial advisory board of the Journal of the American Water Works Association, and vice-chair of the Drinking Water Committee of the EPA's Science Advisory Board. Dr. Snoeyink is a member of the National Academy of Engineering and has participated in several NRC committees as either the chair or a member. He has a B.S. in civil engineering, an M.S. in sanitary engineering, and a Ph.D. in water resources engineering from the University of Michigan.

Charles N. Haas is the L. D. Betz Professor of Environmental Engineering, and Chair of the Department of Civil, Architectural and Environmental Engineering at Drexel University. His areas of research involve microbial and chemical risk assessment, chemical fate and transport, hazardous waste processing and disposal practices, industrial wastewater treatment, and water and wastewater disinfection processes. He is currently conducting research on disinfection processes, water microbiology, and microbial risk assessment. Dr. Haas has co-authored 14 books or major works on water and wastewater treatment and/or microbial risk assessment. He is currently a member of the Water Science and Technology Board and a fellow of the American Academy of Microbiology, the Society for Risk Analysis, and the American Association for the Advancement of Science. Dr. Haas received a B.S. in biology and an M.S. in environmental engineering from the Illinois Institute of Technology and a Ph.D. in environmental engineering from the University of Illinois.

Paul F. Boulos is president and COO of MWH Soft. Dr. Boulos is a recognized world-leading expert on water distribution engineering. He oversees worldwide operations and strategic directions from MWH Soft Corporate headquarters with more than 18 years of industry experience, primarily master planning, hydraulic modeling, and regional water quality studies to large IT integration projects. Dr. Boulos has written over 100 technical papers and co-authored three books on water distribution systems analysis. He has also served on technical review committees for several municipal drinking water projects throughout the United States and is currently a consultant to the EPA Science Advisory Board Drinking Water Committee (Stage 2 Disinfection/Disinfectant Byproduct Rule and Long-Term 2 Enhanced Water Treatment Rule). Dr. Boulos received his B.S., M.S., and Ph.D. in civil engineering from the University of Kentucky and his MBA from the Harvard Business School.

Gary A. Burlingame is the administrative scientist of the Philadelphia Water Department's organic chemistry and aquatic biology laboratories at the Bureau of Laboratory Services, where since 1983 he has been involved in a wide range of operational and distribution system activities. He oversees monitoring of drinking water, source water, wastewater, sediment, sludge, and related media for disinfection byproducts, natural organic matter, PCBs, emerging chemicals, VOCs and SOCs, algae, coliform bacteria, and *Giardia* and *Cryptosporidium*. Best known for his contributions to taste and odor control of drinking water and odor control at wastewater treatment plants, his current research addresses the issues of watershed monitoring and distribution system quality control. He has participated in the review of the Total Coliform Rule that is currently under development and in the expert workshop on Exposure Assessment of Contamination of Distribution Systems. Mr. Burlingame also chaired The Unsolicited Proposal Review Committee for AWWA Research Foundation. He received his B.S. and M.S. degrees in environmental science from Drexel University.

Anne K. Camper is a professor of civil engineering, adjunct associate professor of microbiology, and faculty member of the Center for Biofilm Engineering at Montana State University. She is also Associate Dean for Research and Graduate Studies in the College of Engineering. Her primary research interests are in biofilm formation in low nutrient environments, including microbial physiology and ecology, as well as the persistence of pathogenic bacteria in biofilms. She also specializes in biological treatment of drinking water and microbial regrowth in drinking water distribution systems. She recently participated in the EPA workshop on Exposure Assessment of Pathogens and Toxic Chemicals in Drinking Water Distribution Systems, the outcomes of which are to be coupled with revisions to the Total Coliform Rule. Dr. Camper is presently on the editorial boards of both Microbial Ecology and Biofilms. She received her B.S. and M.S. in microbiology and her Ph.D. in civil and environmental engineering, all from Montana State University.

Robert M. Clark is an environmental and engineering and public health consultant. He served as a consultant to Shaw Environmental and Infrastructure (SE&I) working on homeland security issues and to the University of Cincinnati working on risk assessment methodology for water system vulnerability. He spent over 40 years in government, first for the U.S. Public Health Service Commissioned Corps, and then for EPA where he directed the Water Supply and Water Resources Division for 14 years. Among other things, his research interests have focused on modeling water quality in drinking water distribution systems, including understanding the many factors that can cause the quality of distribution system water to deteriorate such as the chemical and biological quality of source water, the effectiveness and efficiency of treatment processes, the adequacy of the treatment facility and storage facilities, distribution system age and design, the maintenance of the distribution system, and the quality of treated water. He received the 2004 Lifetime Achievement Award from the American Society of Civil Engineers' Environmental & Water Resources Institute. Dr. Clark holds a B.S. in civil engineering from Oregon State University; a B.S. in mathematics from Portland State University; a M.S. in mathematics from Xavier University; an M.S. in water resources and environmental planning from Cornell University; and a Ph.D. in environmental engineering from the University of Cincinnati.

Marc A. Edwards is the Charles Lunsford Professor of Civil Engineering at Virginia Polytechnic and State University. Prior to joining the faculty at Virginia Tech, he was an assistant professor at the University of Colorado in Boulder and a senior engineer with Montgomery Engineers. Dr. Edwards is the current president of the Association of Environmental and Engineering Science Professors. His research interests are internal corrosion processes in home plumbing, water treatment, scaling, arsenic removal, and applied aquatic chemistry. The White House honored him in 1996 with a National Science Foundation Presidential Faculty Fellowship. In 2003 he was awarded the Walter Huber Research Prize from the American Society of Civil Engineers. Dr. Edwards received a B.S. in biophysics from the State University of New York and an M.S. and a Ph.D. in environmental engineering from the University of Washington.

Mark W. LeChevallier is chief scientist for innovation and technology at the American Water Corporate Center in Voorhees, NJ, which owns and operates numerous drinking water utilities throughout the United States. His research involves a wide area of issues in drinking water distribution systems, including bacterial regrowth, disinfection of biofilms, corrosion, bacterial nutrients, AOC measurement techniques, biological treatment, *Mycobacterium*, microbial recovery and identification, the impact of pressure transients on water quality, and detection, treatment, and survival of *Giardia* and *Cryptosporidium*.

He recently participated in an expert workshop on Exposure Assessment of Contamination of Distribution Systems, which resulted in several white papers that formed the basis for the current study. Dr. LeChevallier currently serves as the Chair of the AWWA Microbial/Disinfection By-Product Technical Action Workgroup and is a trustee of the AWWA Water Science and Research Division. He received his B.S. and M.S. degrees in microbiology from Oregon State University and his Ph.D. in microbiology from Montana State University.

L. D. McMullen is the CEO and general manager of the Des Moines, IA, Water Works where, among other accomplishments, he supervised one of the Upper Midwest's largest "design and build" concept water plant projects. Dr. McMullen served two terms as Chair of the National Drinking Water Advisory council of EPA and on the Science Advisory Board's Drinking Water Committee. He has served as a diplomat for water supply/wastewater issues for the American Academy of Environmental Engineers, as the Water Industry Delegation Leader to China for the Citizen Ambassador Program, and was an executive committee member of the Board of Trustees of the American Water Works Association Research Foundation. In 1994 Dr. McMullen received the President's Award of the Association of Metropolitan Water Agencies. He received his B.S. in civil engineering and an M.S. and Ph.D. in environmental engineering from the University of Iowa.

Christine L. Moe is an associate professor of infectious diseases in the Department of International Health at the Rollins School of Public Health at Emory University. Previously she was an assistant professor in the Department of Epidemiology, University of North Carolina, Chapel Hill. She received her Ph.D. in environmental sciences and engineering from the University of North Carolina and has done extensive laboratory and field research on waterborne transmission of infectious agents and diagnosis and epidemiology of enteric virus infections. She is a member of the Water Science and Technology Board and also served as a member for the NRC Committee on Evaluation of the Viability of Augmenting Potable Water Supplied with Reclaimed Water and the Committee on Watershed Management for New York City.

Eva C. Nieminski is an environmental research engineer at the Utah Department of Environmental Quality Division of Drinking Water, where she provides technical assistance to 50 water treatment plants in Utah. She is also an adjunct associate professor in the department of civil and environmental engineering at Utah State University. Her research focuses primarily on treatment of drinking water, including pathogen passage during filtration, UV disinfection, the application of surrogate measures to improve treatment plant performance, *Giardia* and *Cryptosporidium* removal via conventional treatment and direct filtration, ozone pilot studies, disinfection byproduct studies, and a

TOC study in surface water treatment plants. She serves as a trustee of the AWWA Water Quality Technology Division. In the regulatory arena, she has represented the Association of State Drinking Water Administrators on negotiated rule making for the Disinfection Byproducts and Enhanced Surface Water Treatment Rule, as well as ECOS for Stage II Disinfection By-products and Long-Term Enhanced Surface Water Treatment Rule. Dr. Nieminski received her M.S. in civil and environmental engineering from Warsaw Technical University in Poland, her M.S. in environmental health engineering from the University of Notre Dame, and her Ph.D. in civil and environmental engineering from Utah State University.

Charlotte D. Smith is president of Charlotte Smith & Associates, Inc, which provides consulting services to drinking water utilities nationwide. Before establishing Charlotte Smith & Associates, Inc., Ms. Smith worked with the New York City Department of Environmental Protection's Drinking Water Quality Division, and she was director of Water Quality for United Water Resources (formerly General Waterworks Corp.), which operated 35 drinking water utilities in 15 states. Ms. Smith's expertise with respect to distribution systems has spanned from understanding the effect of treatment plant practices and chemical and biological stability on distribution system water quality, to disinfectant residual studies, corrosion studies, nitrification control, and distribution system tracer studies. She is a member of the American Water Works Association Distribution System Water Quality Committee (immediate past chair) and Microbial/Disinfection By-Product Technical Advisory Group. Ms. Smith led the development of a Distribution System Self-Assessment Workbook for drinking water utilities. She holds a B.S. in microbiology from the University of Michigan and an M.S. in community health from the City University of New York.

David P. Spath serves as a consultant on drinking water issues to the California Department of Health Services. He has worked for the Department for more than 30 years. From 1996 until 2005 he was chief of the Department's Division of Drinking Water and Environmental Management. In that position, he was responsible for overseeing California's Public Water System Regulatory Program, its Medical Waste Regulatory Program, and the state's Nuclear Emergency Response Program. He is past chair of the National Drinking Water Advisory Council and also served on the California Recycled Water Task Force. He is a past president of the Association of State Drinking Water Administrators and served on a steering committee for EPA's environmental technology verification program related to small water systems. He was a member of the recently-concluded NRC study on Water System Security Research. Dr. Spath received his B.S. in civil engineering from Tufts University and his M.S. and Ph.D. in civil and environmental engineering from the University of Cincinnati.

Richard L. Valentine is a professor in the civil and environmental engineering department at the University of Iowa. He is also a member of the Center of Health Effects of Environmental Contamination. Dr. Valentine's current research interests are in the general areas of environmental chemistry and physical and chemical processes in natural and engineered systems, especially water and wastewater treatment process design and modeling; environmental chemistry/reaction kinetics; processes to remove trace contamination from water; and fate and transformation of hazardous chemicals. His current research related to distribution system issues includes the chemistry of disinfectants and radium and radon in drinking water; mineral dissolution processes; the use of metal oxides as adsorbents in drinking water treatment; and role of the pipe-water interface in the determination of drinking water quality. He received B.S. degrees in chemical engineering and chemistry and an M.S. in chemical engineering from the University of Michigan; and an M.S. and a Ph.D. in civil and environmental engineering from the University of California at Berkeley.

Laura J. Ehlers is a senior staff officer for the Water Science and Technology Board of the National Research Council. Since joining the NRC in 1997, she has served as study director for eleven committees, including the Committee to Review the New York City Watershed Management Strategy, the Committee on Bioavailability of Contaminants in Soils and Sediment, and the Committee on Assessment of Water Resources Research. She received her B.S. from the California Institute of Technology, majoring in biology and engineering and applied science. She earned both an M.S.E. and a Ph.D. in environmental engineering at the Johns Hopkins University. Her dissertation, entitled RP4 Plasmid Transfer among Strains of *Pseudomonas* in a Biofilm, was awarded the 1998 Parsons Engineering/Association of Environmental Engineering Professors award for best doctoral thesis.